Main table (each cell: atomic number, symbol, and Gmelin System Number as superscript):

1 H^2																	2 He^1
3 Li^{20}	4 Be^{26}											5 B^{13}	6 C^{14}	7 N^4	8 O^3	9 F^5	10 Ne^1
11 Na^{21}	12 Mg^{27}											13 Al^{35}	14 Si^{15}	15 P^{16}	16 S^9	17 Cl^6	18 Ar^1
19 K^{22}	20 Ca^{28}	21 Sc^{39}	22 Ti^{41}	23 V^{48}	24 Cr^{52}	25 Mn^{56}	26 Fe^{59}	27 Co^{58}	28 Ni^{57}	29 Cu^{60}	30 Zn^{32}	31 Ga^{36}	32 Ge^{45}	33 As^{17}	34 Se^{10}	35 Br^7	36 Kr^1
37 Rb^{24}	38 Sr^{29}	39 Y^{39}	40 Zr^{42}	41 Nb^{49}	42 Mo^{53}	43 Tc^{69}	44 Ru^{63}	45 Rh^{64}	46 Pd^{65}	47 Ag^{61}	48 Cd^{33}	49 In^{37}	50 Sn^{46}	51 Sb^{18}	52 Te^{11}	53 I^8	54 Xe^1
55 Cs^{25}	56 Ba^{30}	57** La^{39}	72 Hf^{43}	73 Ta^{50}	74 W^{54}	75 Re^{70}	76 Os^{66}	77 Ir^{67}	78 Pt^{68}	79 Au^{62}	80 Hg^{34}	81 Tl^{38}	82 Pb^{47}	83 Bi^{19}	84 Po^{12}	85 At^{8a}	86 Rn^1
87 Fr^{25a}	88 Ra^{31}	89*** Ac^{40}	104^{71}	105^{71}													

* NH$_4$^{23}

**Lanthanides 39:

58 Ce^{44}	59 Pr^{51}	60 Nd^{55}	61 Pm^{71}	62 Sm^{71}	63 Eu^{71}	64 Gd^{71}	65 Tb^{71}	66 Dy^{71}	67 Ho^{71}	68 Er^{71}	69 Tm^{71}	70 Yb^{71}	71 Lu^{71}

***Actinides:

90 Th^{44}	91 Pa^{51}	92 U^{55}	93 Np^{71}	94 Pu^{71}	95 Am^{71}	96 Cm^{71}	97 Bk^{71}	98 Cf^{71}	99 Es^{71}	100 Fm^{71}	101 Md^{71}	102 No^{71}	103 Lr^{71}

Gmelin Handbook of Inorganic Chemistry

8th Edition

Gmelin Handbook of Inorganic Chemistry

8th Edition

Gmelin Handbuch der Anorganischen Chemie

Achte, völlig neu bearbeitete Auflage

Prepared
and issued by

Gmelin-Institut für Anorganische Chemie
der Max-Planck-Gesellschaft
zur Förderung der Wissenschaften

Director: Ekkehard Fluck

Founded by

Leopold Gmelin

8th Edition

8th Edition begun under the auspices of the
Deutsche Chemische Gesellschaft by R. J. Meyer

Continued by

E.H.E. Pietsch and A. Kotowski, and by
Margot Becke-Goehring

Springer-Verlag Berlin Heidelberg GmbH 1986

Gmelin-Institut für Anorganische Chemie
der Max-Planck-Gesellschaft zur Förderung der Wissenschaften

Organometallic Compounds in the Gmelin Handbook

The following listing indicates in which volumes theses compounds are discussed or are referred to:

Ag   Silber B 5 (1975)

Au   Organogold Compounds (1980)

Bi   Bismut-Organische Verbindungen (1977)

Co   Kobalt-Organische Verbindungen 1 (1973), 2 (1973), Kobalt Erg.-Bd. A (1961), B 1 (1963), B 2 (1964)

Cr   Chrom-Organische Verbindungen (1971)

Cu   Organocopper Compounds 1 (1985), 2 (1983), 3 (1986) **present volume**

Fe   Eisen-Organische Verbindungen A 1 (1974), A 2 (1977), A 3 (1978), A 4 (1980), A 5 (1981), A 6 (1977), A 7 (1980), A 8 (1985), B 1 (partly in English; 1976), Organoiron Compounds B 2 (1978), Eisen-Organische Verbindungen B 3 (partly in English; 1979), B 4 (1978), B 5 (1978), Organoiron Compounds B 6 (1981), B 7 (1981), B 8 to B 10 (1985), B 11 (1983), B 12 (1984), Eisen-Organische Verbindungen C 1 (1979), C 2 (1979), Organoiron Compounds C 3 (1980), C 4 (1981), C 5 (1981), C 7 (1985), and Eisen B (1929-1932)

Hf   Organohafnium Compounds (1973)

Nb   Niob B 4 (1973)

Ni   Nickel-Organische Verbindungen 1 (1975), 2 (1974), Register (1975), Nickel B 3 (1966), and C 1 (1968), C 2 (1969)

Np, Pu   Transurane C (partly in English; 1972)

Pt   Platin C (1939) and D (1957)

Ru   Ruthenium Erg.-Bd. (1970)

Sb   Organoantimony Compounds 1 (1981), 2 (1981), 3 (1982), 4 (1986)

Sc, Y,   D 6 (1983)
La to Lu

Sn   Zinn-Organische Verbindungen 1 (1975), 2 (1975), 3 (1976), 4 (1976), 5 (1978), 6 (1979), Organotin Compounds 7 (1980), 8 (1981), 9 (1982), 10 (1983), 11 (1984), 12 (1985), 13 (1986)

Ta   Tantal B 2 (1971)

Ti   Titan-Organische Verbindungen 1 (1977), 2 (1980), 3 (1984), 4 and Register (1984)

U   Uranium Suppl. Vol. E 2 (1980)

V   Vanadium-Organische Verbindungen (1971), Vanadium B (1967)

Zr   Organozirconium Compounds (1973)

Gmelin Handbook
of Inorganic Chemistry

8th Edition

Cu

Organocopper Compounds

Part 3

With 18 illustrations

AUTHORS

Helmut Bauer, Max-Planck-Institut
für medizinische Forschung, Heidelberg
Jürgen Faust, Rolf Froböse, Johannes Füssel

EDITOR

Jürgen Faust, Johannes Füssel

CHIEF EDITOR

Johannes Füssel

System Number 60

Springer-Verlag Berlin Heidelberg GmbH 1986

LITERATURE CLOSING DATE: 1985

IN SOME CASES MORE RECENT DATA HAVE BEEN CONSIDERED

Library of Congress Catalog Card Number: Agr 25-1383

ISBN 978-3-662-06441-2   ISBN 978-3-662-06439-9 (eBook)
DOI 10.1007/978-3-662-06439-9

© by Springer-Verlag Berlin Heidelberg 1986
Originally published by Springer-Verlag Berlin Heidelberg New York Tokyo in 1986
Softcover reprint of the hardcover 8th edition 1986

Typesetting

Preface

The present volume finalizes the coverage of mononuclear organocopper compounds with ligands bonded by one carbon atom. As structural elucidation has not yet received much attention in organocopper chemistry and the aggregation of most products is unknown, the term "mononuclear" has been used as explained in "Organocopper Compounds" 1, pp. 3/4.

Compounds with alkyl, alkenyl, and aryl ligands have already been described in Volumes 1 (published in 1985) and 2 (published in 1983).

The present Volume 3 contains all compounds with alkynyl, carbonyl, isocyanide, and additional ligands bonded by one carbon atom. The largest part of this volume deals with copper acetylides which are widely used as reagents in organic syntheses although many of them are only poorly characterized.

For abbreviations and dimensions used throughout this volume, see p. X.

Frankfurt, September 1986   Johannes Füssel

Remarks on Abbreviations and Dimensions

Most compounds and reagents in this volume are presented in tables. Unless otherwise stated, they are arranged in the tables according to the numbers of C atoms in the empirical formula. For the sake of conciseness, some abbreviations are used and some dimensions are omitted in the tables. This necessitates the following clarification.

Geometric isomers are designated according to the IUPAC rules. Structural labels are missing when authors fail to report structural details.

Temperatures are given in °C, otherwise K stands for Kelvin. Abbreviations used with temperatures are m.p. for melting point, b.p. for boiling point, and dec. for decomposition.

Nuclear magnetic resonance is abbreviated by NMR. Chemical shifts are given as δ values in ppm with the positive sign for downfield shifts. Reference substances are $Si(CH_3)_4$ for ^1H and ^{13}C NMR, $CFCl_3$ for ^{19}F NMR, and H_3PO_4 for ^{31}P NMR.

Multiplicities of the signals are abbreviated as s, d, t, q (singlet to quartet), quint, sext, sept (quintet to septet), and m (multiplet); terms like dd (double doublet) and t's (triplets) are also used. Assignments referring to labelled structural formulas are given in the form C-4, H-3,5. Coupling constants ^nJ in Hz are given as $J(A, B)$ or as $J(1,3)$ referring to labelled structural formulas, n is the number of bonds between the coupled nuclei.

Optical spectra are labelled as IR (infrared), Raman, and UV (electronic spectrum including the visible region). IR bands and Raman lines are given in cm^{-1}, the assigned bands are usually labelled with the symbols ν for stretching vibration and δ for deformation vibration. Intensities are indicated by the common qualitative terms (vs, s, m, w, vw) or as numerical relative intensities in parentheses. The UV absorption maxima, λ_{max}, are given in nm followed by the extinction coefficient $\varepsilon(L \cdot cm^{-1} \cdot mol^{-1})$ or log ε in parentheses; sh means shoulder, br means broad. If reported, solvents or the physical state are given in parentheses immediately after the spectral symbol.

Electron spin resonance is abbreviated as ESR, hyperfine coupling constants are given as $a(X)$.

Further abbreviations:

d_c	calculated density
d_m	experimental density
$[M]^+$	molecular ion in mass spectroscopy
aq.	aqueous
conc.	concentrated
emf	electromotive force
soln.	solution
THF	tetrahydrofuran
$i\text{-}C_3H_7$	isopropyl $CH(CH_3)_2$
$s\text{-}C_4H_9$	sec-butyl $CH(CH_3)C_2H_5$
$t\text{-}C_4H_9$	tert-butyl $C(CH_3)_3$

Table of Contents

XII

Organocopper Compounds

Part 3

1.1.2 Acetylides

For acetylides of the types (RC≡C)R'CuLi and (RC≡C)R'CuMgX where R' are alkyl, alkenyl, and aryl the RC≡C group is less reactive than R'. The transfer of the group R' is an important reaction and for practical reasons these compounds are described in "Organocopper Compounds" 2, 1983, Sections 1.1.1.2.4 and 1.1.1.2.6 (pp. 174ff. and 227ff.).

General References:

A. M. Sladkov, I. R. Gol'ding, Reactions of Copper and Silver Organoacetylenides, Usp. Khim.
    48 [1979] 1625/83; Russ. Chem. Rev. **48** [1979] 868/96.
A. C. Hopkinson, Acidity, Hydrogen Bonding and Complex Formation, in: S. Patai, Chemistry
    of the Carbon–Carbon Triple Bond, Vol. 1, Wiley, Chichester 1978, pp. 75/136, 97/9,
    117/9; C.A. **89** [1978] No. 196383.
A. E. Jukes, The Organic Chemistry of Copper, Advan. Organometal. Chem. **12** [1974]
    215/322.
W. Ziegenbein, P. Cadiot, W. Chodkiewicz, in: H. G. Viehe, Chemistry of Acetylenes, Dekker,
    New York 1969, pp. 169/263, 237, 597/647.
T. F. Rutledge, Acetylenic Compounds, Reinhold, New York 1968, pp. 84/95, 207/10, 240/68.
A. M. Sladkov, L. Yu. Ukhin, Copper and Silver Acetylides in Organic Synthesis, Usp. Khim.
    37 [1968] 1750/81; Russ. Chem. Rev. **37** [1968] 748/63.

Semiconduction of RC≡CCu:

V. S. Myl'nikov, Photonics of Organometallic Semiconductors in the Form of Copper Organo-
    acetylenides, Usp. Khim. **50** [1981] 1872/91; Russ. Chem. Rev. **50** [1981] 979/90.
V. S. Myl'nikov, Electrical and Photoelectric Properties of Organic Polymeric Semiconduc-
    tors, Usp. Khim. **37** [1968] 78/103; Russ. Chem. Rev. **37** [1968] 25/38, 30.

Complex Acetylides:

G. W. Parshall, Homogeneous Catalysis: The Application and Chemistry of Catalysis by Solu-
    ble Transition Metal Complexes, Wiley, New York 1980, pp. 1/240.
B. L. Shaw, N. I. Tucker, Organotransition Metal Compounds and Related Aspects of Homo-
    geneous Catalysis, Pergamon, Elmsford, N.Y., 1975, pp. 1/214.
R. Nast, Komplexe Acetylide von Übergangsmetallen, Angew. Chem. **72** [1960] 26/31.
G. E. Coates, Copper and Silver Acetylene Complexes, Organometallic Compounds, 2nd Ed.,
    Methuen, London 1960, pp. 352/5.
R. Nast, Complex Acetylides of Transition Metals, Chem. Soc. [London] Spec. Publ. No. 13
    [1959] 103/12.
J. W. Copenhaver, M. H. Bigelow, Acetylene and Carbon Monoxide Chemistry, Reinhold,
    New York 1949, pp. 1/373.

Further:

K. Müller, Functional Group Determination of Olefin and Acetylene Unsaturation [copper
    acetylides in the analysis], Academic, London 1975, pp. 1/283.
M. Tramontini, Advances in the Chemistry of Mannich Bases [copper acetylides in Mannich
    reactions], Synthesis **1973** 703/75.

M. L. H. Green, Organometallic Compounds II: The Transition Elements [structural problems of copper acetylides], 3rd Ed., Methuen, London 1968, pp. 271/6.

G. Eglinton, W. McCrae, The Coupling of Acetylenic Compounds [copper acetylides in Grignard reactions], Advan. Org. Chem. **4** [1963] 225/328.

1.1.2.1   Compounds of the Type RC≡CCu

In the following sections, copper acetylides are formulated as RC≡CCu even though highly polymeric structures must be assumed in most cases. Oligomers have been also found with certain R groups. The degree of association and its dependence on the solvent and on the preparation method has been determined in some cases (see Nos. 55 and 98 on pp. 45 and 54).

History

Böttger prepared the first copper acetylide in 1856 by passing illuminating gas through an ammoniacal solution of CuCl [1]. In 1860 Berthelot found that the true structure of the violet-brown substance obtained is Cu_2C_2 [2, 3]. The first substituted ("organic") copper acetylide was $CH_3C≡CCu$, prepared by Berthelot in 1866 [4]. Nevertheless, the acetylides RC≡CCu obtained their present high importance in organic synthesis only in the last 30 years. By now they have come to be regarded as common reagents in organic synthesis, the use of which is in many cases more advantageous than that of the usual organometallic compounds. Their application in preparative chemistry is often based on empirical facts, and the reaction mechanism have hitherto been relatively little studied.

1.1.2.1.1   Preparation, Properties, and Selected Reactions

Preparation

The compounds listed in Table 1, pp. 17/41, can be prepared by the following methods. Ia is the standard procedure.

Method I: RC≡CH is reacted with Cu^I.

        a. An aqueous solution of $[Cu(NH_3)_n]^+$, called "Ilosvay's Reagent" [19, 20], reacts with RC≡CH directly or with solutions thereof in solvents like C_2H_5OH, CH_3OH, acetone, dimethylformamide, dioxane, etc. [121]. Two-phase solvent systems with ether [217] and toluene [164] have also been used.

The water-insoluble RC≡CCu precipitates in most cases immediately. Sometimes it is precipitated by addition of water [350]. After filtration it is usually washed successively by water, C_2H_5OH, and ether, and then thoroughly dried in vacuum at moderate temperatures (but see under explosiveness, pp. 11/2).

The purity of the acetylide depends much on the counter ion in "Ilosvay's Reagent". Often $[Cu(NH_3)_n]^+$ is produced from Cu^{II} salts and $[NH_3OH]Cl$. A small excess of $[NH_3OH]Cl$ prevents air oxidation and thus formation of oxidative byproducts. A recommended method starts with $CuSO_4$ and is said to give very pure acetylides [123]. Starting with CuCl [150] also yields pure acetylides, CuI [123, 134] gives less pure products.

The ammonia has the function of making the Cu^I soluble in aqueous media and of neutralizing the H^+ ions produced in the course of the reaction. Some RC≡CCu compounds show a considerable solubility in aqueous ammonia. In these cases, freshly prepared CuCl is dissolved in aqueous NH_4Cl solution, and the mixture is made weakly alkaline with CH_3CO_2Na [133].

References on pp. 57/67

If the substituent R contains hydrolyzable groups like $C_2H_5O_2C$ (No. 30) or CH_3O_2C (No. 93), adjusting the pH to 7.5 to 8 or adding excess $(NH_4)_2CO_3$ is necessary [163]. The preparation of $RC\equiv CCu$ compounds which contain the tetrahydropyranyloxy group (e.g., Nos. 68, 101, 139) should be carried out under ice cooling to prevent hydrolysis. If R contains double and triple bonds (Nos. 24, 25, 43), ice cooling is necessary to prevent polymerization. Preparation method Ia is often used for purification purposes. Nos. 127, 153, 154, and 167 were reprecipitated from ethereal solutions [205, 221].

Some $RC\equiv CCu$ are not precipitated by method Ia, if an excess of "Ilosvay's Reagent" is present [4].

b. An ammine complex $[Cu(NH_3)_n]^+$ is generated in an anhydrous medium. The Cu^I salt (CuI, CuCl) is suspended in a solvent like C_2H_5OH or dimethylformamide, a solution of $RC\equiv CH$ is added, and gaseous NH_3 is introduced.

c. The reaction is carried out in liquid ammonia. CuI is used because of its satisfactory solubility in NH_3.

d. The reaction is carried out in the presence of proton acceptors like $(C_2H_5)_2NH$, pyridine, dimethylformamide, hexamethylphosphoric triamide, dimethylsulfoxide, or K_2CO_3.

A recommendable method is the conversion of $RC\equiv CH$ to $RC\equiv CCu$ by the so-called "complex $CuCl-C_2H_5NH_2-HCl$". This reagent is soluble in CH_3OH or in $CH_3OH/C_2H_5NH_2$ and the usual purification problems of heterogenic preparations do not occur.

The so-called "acid preparation method" is useful for preparing $RC\equiv CCu$ compounds which are sensitive to bases (cf. Nos. 35 and 119). It can be understood as a special method Id in which H_2O acts as a proton acceptor. CuCl is suspended in H_2O and the acetylene is added. Quantitative conversion is achieved after prolonged stirring [27]. C_2H_5OH can be used as a solubilizer [60]. The products are said to be very pure.

e. The acetylene $RC\equiv CH$ is reacted with a solution of $CuOC(CH_3)_3$ in an organic solvent.

f. $RC\equiv CCu$ is prepared according to the equation $RC\equiv CH + (C_2H_5O)_3P \cdot CuCl \rightarrow RC\equiv CCu + C_2H_5Cl + HPO(OC_2H_5)_2$ with or without a solvent, with heating. Only C_2H_2 reacts at room temperature. In general, the $RC\equiv CCu$ compounds formed are soluble in excess $P(OC_2H_5)_3$. $(C_6H_5)_3P \cdot CuCl$ is also claimed to be effective for the transformation of $RC\equiv CH$ to $RC\equiv CCu$, but no experimental data are given [131].

g. With copper alkyls according to the equation $R^1Cu + RC\equiv CH \rightarrow RC\equiv CCu + R^1H$, usually in ether or in dimethylsulfoxide.

Method II: Salts like $RC\equiv CLi$ or $RC\equiv CK$ are reacted with salts of monovalent copper:

a. in liquid ammonia,

b. in an ether or in ether/alcohol mixtures.

Complexes $M[B(C\equiv CR)_4]$ (M is a metal ion or NH_4) are also able to react in high yield to form $RC\equiv CCu$ (see No. 90) [85].

               References on pp. 57/67

Method III: RC≡CH is reacted with a bivalent copper salt. The reaction product is a mixture of RC≡CCu and R(C≡C)$_2$R, which are readily separated due to their very different solubilities. RC≡CCu is said to be formed from Cu^I and RC≡CH. Cu^I is the reaction product of Cu^{II} and RC≡CH or RC≡CCu. RC≡CH can also be converted with Fehlings solution to a mixture containing RC≡CCu. This case was hypothesized to involve an unstable Cu^{II} acetylide which decomposes to the radical RC≡C. This could be the source of the R(C≡C)$_2$R also formed [106].

Method IV: Haloacetylenes and monovalent copper react according to the equation RC≡CX + 3Cu$^+$ → RC≡CCu^I + [Cu^{II}X]$^+$ + Cu^{++} (X = halogen).

Method V: Decarboxylation reactions of (substituted) Cu^I alk-2-ynoates.

The reaction RC≡CCO$_2$Cu^I → RC≡CCu + CO$_2$ corresponds to the inverse of Reaction Type 6 (reversible binding of CO$_2$, see Section 1.1.2.1.2) and is sometimes a balanced reaction. Proper conditions for the (not balanced) reaction are, for example, in dimethylformamide at 35 °C.

A similar reaction leads to the formation of RC≡CCu from (RC≡CCO$_2$)$_2$Cu^{II}. This reaction is more complex and involves redox processes. It can be accomplished with steam at 100 °C.

Method VI: Decomposition of complexes of RC≡CCu. Adducts of NH$_3$, or amines, or PR$_3$ (see Section 1.1.2.4) can be cleaved with formation of RC≡CCu. The reaction can occur simply on standing in air or in N$_2$ at room temperature. More stable adducts must be heated in vacuum (see Section 1.1.2.4, Table 23). Adducts of heavy metal salts like HgBr$_2$ (see Section 1.1.2.5) are broken down by addition of suitable reagents, in this case of aqueous KI (see Section 1.1.2.5, Table 24).

Method VII: From complexes RC≡CHCuCl of the acetylenes.

The colorless compounds RC≡CHCuCl are very sensitive to solvents and bases. Their reaction with H$_2$O, C$_2$H$_5$OH, or NH$_3$/H$_2$O causes a yellow coloration with subsequent formation of RC≡CCu.

Method VIII: Decomposition of cuprates of the [(RC≡C)$_n$Cu]$^{m-}$ type (zerovalent or monovalent copper) and of their complexes, e.g., with NH$_3$ or ethylene diamine.

The principal reactions are

[(RC≡C)$_2$Cu^I]$^-$ + CuI → 2RC≡CCu + I$^-$

[(RC≡C)$_3$Cu^I]$^{2-}$ + H$_2$O → RC≡CCu + further products

[(RC≡C)$_3$Cu^I]$^{2-}$ + 2CH$_3$SO$_2$N(CH$_3$)$_2$ → RC≡CCu + 2RC≡CN(CH$_3$)$_2$ + 2CH$_3$SO$_2^-$

[(RC≡C)$_3$Cu^I]$^{2-}$ + 2(C$_6$H$_5$)$_2$PO$_2$N(CH$_3$)$_2$ → RC≡CCu + 2RC≡CN(CH$_3$)$_2$ + 2(C$_6$H$_5$)$_2$PO$_2^-$

[(RC≡C)$_2$Cu^I]$_2$[Cu^{II}D$_n$] + H$_2$O → RC≡CCu + further products

[(RC≡C)$_3$Cu^0]$^{3-}$ + O$_2$ → RC≡CCu + [(RC≡C)$_3$Cu^I]$^{2-}$ + further products

Other preparation methods are the hydrolytic decomposition of cuprates of the type [RC≡CCu^{I}R^1]$^-$M$^+$ (R^1 = alkyl or aryl, M = Li or MgBr; see "Organocopper Compounds" 2, 1983, Sections 1.1.1.2.4 and 1.1.1.2.6 on pp. 174ff. and 227ff.), the synthesis of the acetylides of diynes according to the equation RC≡CCu + BrC≡CH → R(C≡C)$_2$Cu + HBr, and the reaction of acetylenes with copper from copper electrodes. The production of copper acetylides on the surface of copper metal, which can cause severe explosions in the manufacturing of C_2H_2, has not yet been reported for $RC\equiv CH$.

Supposedly some $RC\equiv CCu$ is formed according to method Id in the systems CuCl/ $C_2H_5NH_2$/HCl/CH_3OH [115] or $(C_2H_5)_2NH$/pyridine/CuI/Pd$[P(C_6H_5)_3]Cl_2$ [269, 284], which are used to perform catalytic reactions according to Reaction Type 2 (see Section 1.1.2.1.3).

Gaseous $RC\equiv CH$ are shaken in a sealed tube [65] or bubbled into the reaction medium. In this way acetylenes like but-1-yne can be quantitatively washed out from gas mixtures. The non-acetylenic components are not precipitated [28]. Similar separations with liquid mixtures containing $RC\equiv CH$ are described for Nos. 46 and 48.

The production of protons in the reaction of $RC\equiv CH$ and Cu^+ has been used to determine $RC\equiv CH$ quantitatively. The pyridinium ions formed according to the equation $RC\equiv CH + CuCl + py \rightarrow RC\equiv CCu + [pyH]^+ + Cl^{\ominus}$ (py = pyridine) are titrated by NaOH. The mixture is colored and inhomogenous; thus, the end point determination is done by a glass electrode. This procedure can be used for samples containing groups like CHO that interfere with the usual silver method [39, 109].

Thermochemical Data of Formation are available only for Nos. 54 ($R = n\text{-}C_4H_9$) and 90 ($R = C_6H_5$), see pp. 44 and 48.

Structural Properties

Most $RC\equiv CCu$ prepared are described as "microcrystalline" or "amorphous". X-ray examinations prove that "amorphous" $C_6H_5C\equiv CCu$ powder consists of microcrystals [155]. Two coarsely crystalline $RC\equiv CCu$ compounds (Nos. 44 and 90) were obtained for X-ray analyses by crystallization from liquid NH_3. Other $RC\equiv CCu$ are known as oils (Nos. 55, 57, 141).

In general, $RC\equiv CCu$ show a certain heterogeneity which is caused by surface oxidation and aggregation into small lumps [63]. The crystalline regions are of limited size [274].

Most copper acetylides are polymers consisting of a very high number of $RC\equiv CCu$ units. They could be better written as $(RC\equiv CCu)_n$. Very little is known concerning the value of n. At any rate, these polymers are totally insoluble in most solvents. On the other hand, some copper acetylides are soluble in ethers and CCl_4; even 1% solutions (of No. 55, $R = (CH_3)_3C$) in pentane have been obtained [218]. In these rare cases oligomeric or even monomeric structures must be assumed. Cryoscopic, ebullioscopic, and osmometric measurements have been done only for Nos. 55 and 98 ($R = cyclo\text{-}C_6H_{11}$), where the degree of association obviously depends on the preparation method, see pp. 45 and 54.

The smallness of the particles of most $RC\equiv CCu$ compounds prevented detailed X-ray diffraction studies. The only $RC\equiv CCu$ for which a complete X-ray structural determination has been performed is No. 90 ($R = C_6H_5$). This compound could be crystallized as fine needles by extraction with dry diethylamine in the absence of air. Details regarding the X-ray data are given under additional information for No. 90 on p. 51. The general structure is believed to be the same for No. 90 and other highly polymeric $RC\equiv CCu$.

Each ethynyl group lies roughly in the plane of the copper atom chain, and is "side-on"bonded to one of the copper atoms, with the bond to the phenyl group distorted away from this atom. The terminal carbon atom of the phenylethynyl group forms a bridge bond with two adjacent copper atoms so that each ethynyl group is bonded to three copper atoms, see **Fig. 1**, p. 6 [114].

                        References on pp. 57/67

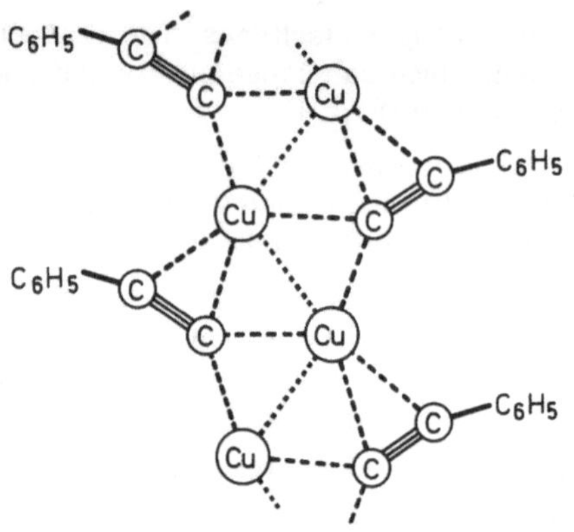

Fig. 1.   Polymeric structure of $C_6H_5C\equiv CCu$.

The polymeric structure indicated in Fig. 1 may be substituted by different R groups in the same polymeric unit. In the literature, "compounds" of this type are called "mixed acetylides" (see Section 1.1.2.2).

X-ray data indicate an average transfer of a charge 0.4 e from the copper atoms in copper acetylides [227].

Thermal Behavior

Some thermal analyses have been done with copper acetylides. Nos. 90, 98, 102, 127, 167, and 181 show one exothermic peak. With Nos. 98 ($R=$cyclo-C_6H_{11}) and 102 ($R=$n-C_6H_{13}) transformation to a different solid phase (IR spectrum unchanged) was observed in addition. No. 188 ($R=(C_6H_5)_2C(CN)CH_2$) shows some exothermic peaks. Nos. 54 ($R=$n-C_4H_9) and 90 ($R=C_6H_5$) are reported to decompose with considerable liberation of heat [261]. No. 54 first melts, than it sets to a solid, then it melts again [83]. Most $RC\equiv CCu$ compounds melt with decomposition.

Electrical Properties

There are general contradictions between the work of Mylnikov et al. [107, 120, 155] and that of Okamoto et al. [191, 220, 221]. For instance the first postulate photocurrents on surface cells which are lower in dry air than under vacuum. The latter reports higher values and explains it by electron trapping by oxygen.

Obviously both intrinsic and extrinsic photoconductivity is involved in the electrical con-duction behavior of $RC\equiv CCu$ compounds (see also conflicting results after ultraviolet pre-irradiation and long relaxation times [155]). A significant contribution by the surface states is indicated by the long wave shift of the photoconductivity spectrum compared with the UV and photo-emf spectra [126, 147].

In many cases the band gaps from absorption and photoconductivity spectra agree (intrin-sic photoconductors) [231, 278]. On the other hand, it has been stated that the sign of the photocurrent carriers in all $RC\equiv CCu$ examined is positive. After the primary act (excitation of a molecule by absorption of a photon), the exciton migrates until it collides with a structural or chemical defect and forms a pair of free photocurrent carriers. In $RC\equiv CCu$, pair formation does not require an activation energy, as shown by the temperature dependence of the photoconductivity at very high frequencies [119].

The acetylides RC≡CCu Nos. 11, 20, 22, 54, 55, 59, 60, 81, 86, 89, 90, 91, 94, 116, 117, 118, 127, 131, 140, 145, 153, 154, 158, 171, 172, 173, 174, 183, 184, and 190 have been examined for photosensitivity. All except Nos. 22, 55, 140, and 158 showed a significant photoconduction effect. Substantially no work has been done to explain these exceptions. For Nos. 22 (R = $1,2-C_2B_{10}H_{11}$), 55 (R = $(CH_3)_3C$), and 140 (R = $2-C_6H_5-1,2-C_2B_{10}H_{10}$) the absent photoeffect could result from the fact that they are not high polymers (deduced only from solubility data). But Cu_2C_2 (see Section 2.2.1.1) is a high polymer and also shows no inner photoeffect. No. 55 is octameric in benzene solution according to cryoscopical and ebullio-scopical measurements [102, 103]. However, data about the "degree of polymerization" of RC≡CCu are often inconsistent, and the values depend much on the preparation and purification methods. At any rate, the compounds No. 22, 55, 140, and 158 (R = $C_5H_5FeC_5H_4$) have a bulky substituent R; thus steric hindrance is probably present too. The borane units in Nos. 22 and 140 (all Table 1, pp. 17/41) are powerful electron acceptors and might therefore prevent the formation of the usual polymeric structure of the copper acetylides.

The shift to lower frequencies in absorption, photoconductivity, and photovoltaic spectral maxima follows the series R (in RC≡CCu compounds): $C_6H_5 <$ naphth-1-yl < naphth-2-yl < anthr-9-yl and R^1 (in $4-R^1C_6H_4C≡CCu$): $CH_3 < H < Cl < I < CH_3CO < NO_2$ [256].

The slopes in dry air and in vacuum of the plots of photocurrent I vs. light intensity L were determined using an equation of the form $I = k \cdot L^s$ [220]. For still other values no conditions are given [257, 278]. In [257], a 1060 mm Nd laser was used.

No.	R in RC≡CCu	s in dry air	s in vacuum	s (no conditions given)
20	C_2H_5S	—	—	0.55 [278]
90	C_6H_5	1.18 [220], 1.0 [221]	0.503 [220], ~0.5 [221]	0.7 [257]
91	C_6H_5O	—	—	0.65 [278]
94	C_6H_5S	—	—	1.00 [278]
116	$4-CH_3C_6H_4$	0.9 [220, 221]	0.65 [220, 221]	—
117	$C_6H_5CH_2$	—	—	1.05 [278]
153	naphth-1-yl	0.97 [220], 1.02 [221]	0.67 [220], 0.66 [221]	—
154	naphth-2-yl	1.12 [220]	0.64 [220]	—

The visible/UV absorption, photoconduction, photovoltage, and photoluminescence are obviously based upon the same band-band junctions in the RC≡CCu polymer. This follows especially from comparisons of preirradiated and non preirradiated polymeric acetylides. The high intensity of the absorption clearly proves that these bands result from characteristic electronic transitions in the polymer [125]. They are not caused by optically active impurities, as previously had been suggested in [107]. The similarity between the absorption and the photoeffect spectra clearly supports the conclusion that the first phase of photon absorption involves the formation of excitons, just as in other organic semiconductors. The competition between their dissociation and their deexcitation determines the photoelectric and lumines-cent properties. Decay of the excitons leads to the formation of free carriers, but the electrons are captured immediately by traps, which might for example be associated with structural defects in the polymer. The migration of the charge carriers (holes) is facilitated in the polymeric acetylide because the potential barriers between molecules are lowered as a result of π-complex bridges between electron-acceptor copper atoms and acetylene bonds. The luminescence of the copper acetylides probably results from deexcitation of the excitons. Preliminary irradition of RC≡CCu by ultraviolet light in air leads to a decrease in the inte-

References on pp. 57/67

grated luminescence yield and a simultaneous increase in the photoeffect. Luminescence peaks decrease rapidly and photo–emf peaks increase (e.g., in [125]).

The changes in the luminescence and photoeffect spectra which occur when the samples are irradiated by UV light are clearly caused by the disruption of weak coordination bonds in the polymer, as a result of which the ratio of polymer homologs in the material changes. When the bonds are broken, dissociation of the excitons can occur more rapidly at the defects which have thus been formed, and this leads to an increase in the photoeffect and a decrease in the luminescence yield. The low temperature experiments support this model. Water vapor reversibly suppresses the electrical conductivity and the photoconductance and increases the photovoltage [126]. The quotient of light current to dark current ("photocurrent response") is increased in vacuum [221]. The complex behavior of the conductivity of $RC{\equiv}CCu$ compounds in the dark and under illumination after freezing out water vapor, evacuation of air, and admitting of oxygen is treated by [120, 126, 155].

Photoconductivity is found at energies 5000 times lower than the threshold value. The nonlinearity exponent $s = 0.8$ excludes manyphoton processes. A relationship between forbidden band width and damage threshold is given [257]. An electron energy level scheme has been used to compare qualitatively the different $RC{\equiv}CCu$ [248, 283].

$RC{\equiv}CCu$ compounds can be made sensitive outside the range of their normal photoresponse by several cationic, anionic, and neutral organic dyes [116, 117, 154]. The sensitized photoconductivity and photo–emf spectra are similar to the absorption spectrum of the dye in solution (10^{-3} M), which suggests that the monomeric form of the dye takes part in the sensitizing act. From the view point of applying the concepts of semiconductor physics to biological polymers, some curious results have been obtained regarding the effect of nucleic acids and amino acids on the photosemiconducting properties. For instance DNA and RNA increase the photoconductivity of $RC{\equiv}CCu$ compounds by 2 to 3 orders of magnitude. Regarding further purine and pyrimidine bases see [143, 174]. The photo–emf spectra of $RC{\equiv}CCu$ compounds were found to be similar to the absorption spectra [125, 174]. A surge of the photo–emf is observed in the first moment of illumination [174]. The photo–emf is increased several-fold after sensitization of $RC{\equiv}CCu$ by purine and pyrimidine bases or nucleic acids [143, 174]. $RC{\equiv}CCu$ compounds have been found useful for preparation of electrophotographic layers. Nos. 86 ($R = 4\text{-}IC_6H_4$) and 90 ($R = C_6H_5$) gave films which charge well both positively and negatively (see pp. 45/6 and 48/50) [130, 141, 324].

Magnetic Properties

There have been no detailed investigations of the magnetic properties of $RC{\equiv}CCu$ compounds. Nos. 98 ($R = \text{cyclo-}C_6H_{11}$), 102 ($R = \text{n-}C_6H_{13}$), and 188 ($R = (C_6H_5)_2C(CN)CH_2$) are reported to be diamagnetic [238, 288]. No. 90 ($R = C_6H_5$) shows no ESR signal [297].

Color

Most $RC{\equiv}CCu$ compounds are yellow. Green or greenish shades are often reported too. Probably this is the result of some NH_3 content (cf. Nos. 2, 3, 5, 10, 64, 102). Some become clear yellow after proper purification. In the case of No. 90 ($R = C_6H_5$) it was specifically proved that the "greenish yellow" color is caused "by traces of NH_3" [111].

Substituents R like $(CH_3)_3C$, $(CH_3)_3Si$, $C_2H_5O_2C$, fur-2-yl, and especially 2- and 3-nitrophenyl have a bathochromic effect (orange to deep red acetylides). Some are reported to show varying colors (cf. No. 98 with $R = \text{cyclo-}C_6H_{11}$). Others change their color on storage. (No. 11 "darkening"; No. 5 green to brown; No. 2 khaki to brown; No. 8, "if wet in air", yellow to brown.)

References on pp. 57/67

Nuclear Magnetic Resonance Spectra

Because of the insolubility of the RC≡CCu compounds, NMR techniques are not usual means for characterization. The only RC≡CCu measured by NMR are Nos. 55 (R=(CH$_3$)$_3$C) and 140 (R=2-C$_6$H$_5$-1,2-C$_2$B$_{10}$H$_{10}$).

IR and Raman Spectra

The electronic effects discussed at p. 7 decrease the order of the C≡C bond in RC≡CCu vs. RC≡CH and lead to a decrease in its frequency. This is in most cases between 1920 and 1960 cm^{-1} and thus 270 to 300 cm^{-1} lower than in RC≡CH. At approximately 450 cm^{-1} a band is found in ring substituted copper phenylacetylides. This frequency is believed to arise from the ≡C-Cu vibration [243].

Visible/UV Spectra

A great number of visible/UV data is known for RC≡CCu compounds. In most cases a diffuse reflectance spectrum of the insoluble acetylide is reported. The usual diluent is MgO (e.g., 5×10^{-4} M RC≡CCu in MgO), but also spectra have been obtained without dilution by a nonabsorbing standard. The RC≡CCu:MgO ratio is 1:100 in **Fig. 2a** [256]. In Fig. 2b the concentration is 5×10^{-4} M [231].

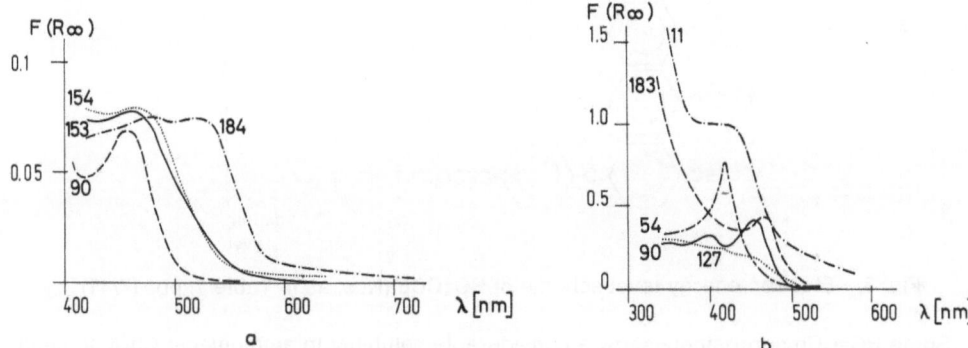

Fig. 2.   Visible/UV spectra of various RC≡CCu (Nos. as in Table 1, pp. 17/41). F(R$_\infty$)= (1-R$_\infty$)^2/2R$_\infty$, where R$_\infty$ is the reflectance of the layer relative to MgO.

Molecular Properties

Most copper acetylides are coordination polymers in which each Cu atom is π-bonded with at least 2 acetylenic groups. The electrons are drawn from the π$_u$-bonding acetylenic orbitals to the metal and from filled d-orbitals of the metal to the unoccupied π$_g$-antibonding orbitals of the acetylene. So the structure of the acetylides involves a transition term between σ- and π-bonding. The consequences of the decrease in bond order are discussed under IR spectra. The breakdown of the polymeric structure by PR$_3$, SbR$_3$, and AsR$_3$ is treated in Section 1.1.2.4. Electron energy levels have been calculated for some RC≡CCu compounds, see **Fig. 3**, p. 10, in which φ is the photoelectric work function and χ the electron affinity [248].

Solubility

In general, the RC≡CCu compounds are insoluble in H$_2$O and in nonpolar organic solvents like hexane. Most acetylides are also insoluble in common organic solvents.

References on pp. 57/67

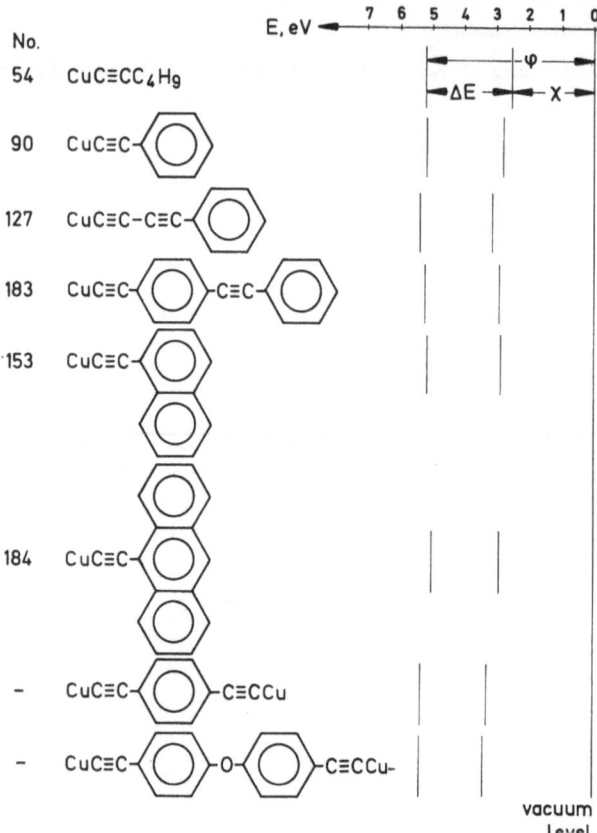

Fig. 3. Electron energy level scheme of RC≡CCu (Nos. as in Table 1, pp. 17/41).

Some RC≡CCu surprisingly show a considerable solubility in ammoniacal CuCl solution (e.g., Nos. 7 (R=CH₃), 17 (R=CH₃CH(OH)), especially if excess KCl or NH₄Cl is present (No. 7). No explanation is given. Nos. 35 (R=C₂H₅CH(OH)) and 44 (R=fur-2-yl) are soluble in aqueous NH₃. RC≡CCu with R=alkyl are more soluble in hexamethylphosphoric triamide than the aryl-substituted compounds [204]. Ethers have also been used to dissolve RC≡CCu compounds. No. 57 (R=(CH₃)₂C(OCH₃)), which contains an ether group, gives in ether a 0.1 M solution at 0 °C. No. 33 (R=n-C₃H₇) is "slightly" soluble in ether, No. 55 (R=(CH₃)₃C) can be crystallized from ether. No. 44 (R=fur-2-yl) is reported to be insoluble in ether. Tetrahydrofuran dissolves Nos. 33 ("slightly"), 40, 55, 57 ("readily"), and 74. C₂H₅OH dissolves readily some RC≡CCu compounds containing alcoholic groups like Nos. 17 and 35. In contrast, No. 44 is insoluble in that solvent. No. 45 (R=thien-2-yl) is dissolved by boiling CH₃CO₂H. The P(OR)₃ (R=C₂H₅ or CH₃) are reported to be suitable solvents for Nos. 40, 78, 90, 91, and 120. Pyridine has also been used as a solvent for RC≡CCu compounds. Nos. 16 (R=HOCH₂CH₂) and 33 (R=n-C₃H₇) give a solubility of 10⁻⁵ M, No. 8 (R=HOCH₂) of 7.5×10⁻³ M (25 °C). No. 74 is "soluble" in pyridine. Aliphatic amines were used to extract RC≡CCu from mixtures and to obtain larger crystals from a microcrystalline powder. The extraction of Nos. 55 (R=(CH₃)₃C) and 90 (R=C₆H₅) was done with (C₂H₅)₂NH or (CH₃)₂CHNH₂ in a Soxhlet extractor [102, 114]. Nos. 11 (R=CH₂=CH) and 90 are soluble in tetramethyl-ethylenediamine [280].

References on pp. 57/67

In contrast to the common solubility pattern the carbaborane-containing acetylides Nos. 22, 140, and 193 are quite soluble in organic solvents like ether, benzene, and CH_3OH. No. 193 ($R = 1,7-C_2B_{10}H_{11}$) is even soluble in CCl_4, whereas the highly chlorinated acetylide No. 9 is totally insoluble in this solvent [193, 235]. No. 193 can be recrystallized from benzene. Also a higher solubility than common $RC{\equiv}CCu$ compounds is shown by Nos. 55 ($R = (CH_3)_3C$) and 40 ($R = (CH_3)_3Si$) with a "spherical" substituent R. No. 55 is even soluble in $CHCl_3$ and in $CH_3C{\equiv}CH$ [218].

Thermal Stability, Stability in Air, and Explosiveness

The stability of $RC{\equiv}CCu$ compounds in the absence of oxygen depends much upon the group R. No. 1 ($R = H$) decomposes above $-50\,°C$ to C_2H_2 and Cu_2C_2 [91], No. 40 ($R = (CH_3)_3Si$) decomposes at about $-20\,°C$ [245], but Nos. 90 ($R = C_6H_5$) and 116 ($R = 4-CH_3C_6H_4$) are stable up to $200\,°C$ [286]. Several $RC{\equiv}CCu$ compounds do not melt when heated but decompose.

In some cases thermal decomposition forms finely divided metallic copper (Nos. 98, $R = cyclo-C_6H_{11}$; 102, $R = n-C_6H_{13}$; 188, $R = (C_6H_5)_2C(CN)CH_2$) or a copper mirror, with No. 55 ($R = (CH_3)_3C$) even at $150\,°C$. With certain $RC{\equiv}CCu$ on heating in air, an increase of weight is observed due to $CuCO_3$ formation (Nos. 98, 102). There exist acetylides stable in air at room temperature, but others are very air-sensitive like No. 28 ($R = (Z)-CH_3OCH{=}CH$). No. 30 ($R = C_2H_5O_2C$) can be air dried [188], Nos. 98 and 102 are stable in air too. Nevertheless $RC{\equiv}CCu$ compounds should, as a norm, be stored in tightly stoppered brown bottles. No. 90 ($R = C_6H_5$) has proved to be stable for years under these conditions [207]. Nos. 16 ($R = HOCH_2CH_2$), 33 ($R = n-C_3H_7$), 54 ($R = n-C_4H_9$), and 102 ($R = n-C_6H_{13}$) were thus stored "for a long time" without remarkable change [222]. No. 90 ($C_6H_5C{\equiv}CCu$) was found to be more stable than $C_6H_5C{\equiv}CAg$ and No. 54 ($n-C_4H_9C{\equiv}CCu$) [261]. If R is basically a hydrocarbon unit, the stability is decreased by substituent OH groups [183].

The information given in the literature about explosiveness seems to be based on experience with accidental explosions and with safe handling. No systematic work has been done in this field. Multiple bonds in the R group seem to enhance the tendency toward severe explosions. No. 43 ($R = CH_2{=}CHC{\equiv}C$) and No. 9 ($R = CCl_2{=}CCl$) are reported to be "highly explosive". No. 43 should only be handled when wet with ether. On the other hand, No. 74 ($R = (C_2H_5O)_2CH$) is also found to be "highly explosive". With respect to Nos. 7, 8, 74, and 117 storage under H_2O and drying before use is recommended [133]. Mixing in the dry state sometimes has been found to be dangerous. Wetting with dry acetone and the use of bone spatulas were suggested [63]. Other $RC{\equiv}CCu$ compounds show only a firework-like sparking, microexplosion behavior (e.g., No. 6, $R = NH_2CO$) or an ignition (e.g., No. 8, $R = HOCH_2$). No. 2 ($R = CF_3$) decomposes slowly on gentle heating, but gives a vigorous explosion on rapid heating [127].

A lot of copper acetylides are reported to be "nonexplosive" or the "handling was found to be quite safe" [277]. Nevertheless, the real danger cannot be estimated reliably. Curtis et al. report a severe detonation in the course of their work with a dry $RC{\equiv}CCu$ (R was not specified in the paper) [183]. In general, copper acetylides containing the carborane unit (Nos. 22, 140, 193), P or As functions (Nos. 145, 173, 174), and certain ether links (Nos. 101, 111, 118) are reported to be nonexplosive. They even show no deflagration on heating. Nos. 98 ($R = cyclo-C_6H_{11}$), 102 ($R = n-C_6H_{13}$), and 188 ($R = (C_6H_5)_2C(CN)CH_2$) are "not sensitive against shocks or percussion".

The situation is somewhat curious. No. 111 (a methylene ether) is reported to be "nonexplosive", No. 116 ($R = 4-CH_3C_6H_4$) to be "very explosive" [30], and with No. 114 ($R = 2-CH_3C_6H_4$) a 78 g synthesis batch was processed [178]. In addition, several individual

RC≡CCu compounds seem to be variable in their ease of detonation. The preparation method is an especially significant source of such variability. The RC≡CCu compounds precipitated from an acid solution (e.g., by Preparation Method Id with H_2O as a proton acceptor) and from solutions which do not contain HCHO, NH_2NH_2, or NH_2OH are more explosive than those prepared by Method Ia, for example. The reasons are not clear. Hindrance of complex formation on the surface or the change of the crystal structure have been conjectured. RC≡CCu compounds which have a metallic lustre are believed to be more explosive [156].

Chemical Reactions

The feasible reactions are listed in Table 1, pp. 17/41. They are described in Sections 1.1.2.1.2 to 1.1.2.1.25. For the simpler Reaction Types 1 to 6 (reactions with proton donors, oxidants, halogens, and CO_2 where there is no variation of groups within these reactants) the literature and the yield are given in Table 1. For some general remarks see Section 1.1.2.1.2.

The more complex reactions of Type 7 to 12 and 18 (variation of groups within the reactant) are listed in Table 1 without literature citations; for data and references see Sections 1.1.2.1.4 to 1.1.2.1.10. For Reaction Type 13 (catalysis by RC≡CCu) see Nos. 33, 40, and 90 on pp. 44 and 53/4. For the Reaction Types 14 to 17 see Sections 1.1.2.3 to 1.1.2.8, where the products of these reactions are described.

Examples of reactions which are standard procedures in organic chemistry (e.g., Reaction Type 7, see p. 13) are often hidden in the literature. Therefore, possibly not all reactions which have been reported are tabulated here. Except for Reaction Types 2, 5, and 13 thorough oxygen exclusion is of high importance. Otherwise RC≡CCu is consumed in the course of a reaction of Type 2 and the reaction products are contaminated with $R(C≡C)_2R$. Some reactions even require the removal of dissolved air in solvents like pyridine.

To understand the principle reactions of RC≡CCu, a brief analysis of the feasible reactions is given. For reactions of Type 1 to 6 see also Section 1.1.2.1.2 and for the other reaction types see Section 1.1.2.1.3.

Reaction Type 1: Reactions with proton donors:

$$RC≡CCu \xrightarrow{\text{HX}} RC≡CH, \text{ see Section 1.1.2.1.2.}$$
a: with HCl,
b: with HNO_3,
c: with CN^- ions in aqueous solution,
d: with further reagents, which are listed in Table 1 in parentheses.

Reaction Type 2: Oxidation: $RC≡CCu \xrightarrow{\text{oxidant}} R(C≡C)_2R$, see Section 1.1.2.1.2.
a: with O_2 or air,
b: with Cu^{II} compounds,
c: with $[Fe(CN)_6]^{3-}$,
d: with further reagents, which are listed in Table 1 in parentheses.

Reaction Type 3: $RC≡CCu + X_2 \rightarrow RC≡CX + CuX$, see Section 1.1.2.1.2.
a: X = I,
b: X = Br.

Reaction Type 4: $RC≡CCu + 2X_2 \rightarrow RCX=CX_2 + CuX$, see Section 1.1.2.1.2.
a: X = I,
b: X = Br.

Reaction Type 5: "Straus Reaction":

$$RC≡CCu \xrightarrow[\text{anhydrous } CH_3CO_2H]{\text{air}} RCH=CHC≡CR, \text{ see Section 1.1.2.1.2.}$$

References on pp. 57/67

Reaction Type 6 (reversal of the Preparation Method V): $RC{\equiv}CCu + CO_2 \rightarrow RC{\equiv}CCO_2Cu$, see Section 1.1.2.1.2.

Reaction Type 7: $RC{\equiv}CCu + R'X \rightarrow RC{\equiv}CR' + CuX$.

This is one of the most important reactions in organic chemistry for preparation of substituted acetylenes; thus, the examples are very numerous and must be formally classified into several groups. The hybridization type of the C atom bound to the departing X group is the primary factor in the classification scheme (in the order sp^3–sp^2–sp). The heterocycles are dealt with after the carbocycles, and finally compounds with the departing group X bound to an atom other than C are described:

a: reactions with R'X where X is bonded to a sp^3 carbon, see Section 1.1.2.1.4.1,

b: reactions with acyl halides, see Section 1.1.2.1.4.2,

c: reactions with haloalkenes, haloallenes, and allenyl esters, see Section 1.1.2.1.4.3,

d: reactions with haloalkynes, see Section 1.1.2.1.4.4,

e: reactions with C_6H_5I, see Section 1.1.2.1.4.5,

f: reactions with 2-substituted halobenzenes, see Section 1.1.2.1.4.6,

g: reactions with 3-substituted halobenzenes, see Section 1.1.2.1.4.6,

h: reactions with 4-substituted halobenzenes, see Section 1.1.2.1.4.6,

i: reactions with di- and polysubstituted halobenzenes, see Section 1.1.2.1.4.7,

j: reactions with compounds containing more than one iodobenzene moiety, see Section 1.1.2.1.4.8,

k: reactions with iodonaphthalenes, see Section 1.1.2.1.4.9,

l: reactions with other carbocycles R'X, see Section 1.1.2.1.4.10,

m: reactions with 5-membered heterocycles R'X, see Section 1.1.2.1.4.11,

n: reactions with 6-membered heterocycles R'X, see Section 1.1.2.1.4.12,

o: reactions with condensed heterocycles R'X, see Section 1.1.2.1.4.13,

p: reactions with R'X where X is bonded to S, P, Si or Sn, see Section 1.1.2.1.4.14.

Reaction Type 8: $RC{\equiv}CCu + R'X \rightarrow$ ring closure product of $RC{\equiv}CR' + CuX$

The first step corresponds to Reaction Type 7, but R'X contains a structural element $C(ZH){=}CX$ ($Z = NR'$, O or S and X = halogen) in a suitable position, which in a second step adds to the triple bond to form a new heterocyclic ring.

a: ring closure giving furans, see Section 1.1.2.1.5.1,

b: ring closure giving pyrroles or thiophenes, see Section 1.1.2.1.5.2,

c: ring closure giving other heterocycles, see Section 1.1.2.1.5.3,

d: ring closure giving carbocycles, see Section 1.1.2.1.5.4.

References on pp. 57/67

Reaction Type 9: $R'X \xrightarrow{RC \equiv CCu} R'H$ X = I or Br

The halogen of the substrate is replaced by H. Usually there is no information about the destiny of the $RC \equiv C$ unit. $C_6H_5C \equiv CCu$ is able to cause a
replacement of iodine in fluoroaromatic compounds [240], of iodine in
naphthalenes, and of bromine in phenacyl bromides [123] by H, if a proper
solvent is used. Probably the H is abstracted from this solvent, but no
investigations about its origin have been conducted, see Section 1.1.2.1.6.

Reaction Type 10: Mannich type reactions: $RC \equiv CCu + [CH_2 \text{===} NR_2^1]^+ \longrightarrow RC \equiv CCH_2NR_2^1 + Cu^+$,
see Section 1.1.2.1.7.

Reaction Type 11: Reaction with copper alkyls and copper aryls, see Section 1.1.2.1.8.

Reaction Type 12: Reaction with nitrones to form β-lactams:

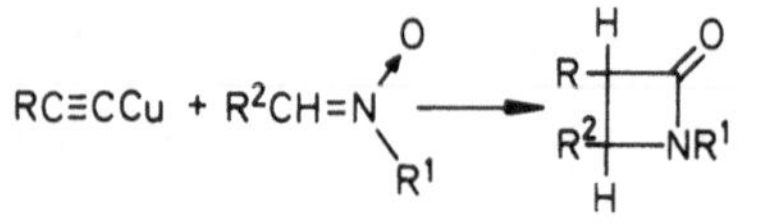

The primary product seems to be the cis-β-lactam, which is partially
rearranged to the trans-β-lactam, see Section 1.1.2.1.9.

Reaction Type 13: $RC \equiv CCu$ as a catalyst (see Nos. 33, 40, and 90 on pp. 44 and 53/4).

Reaction Type 14: Adduct formation with compounds of main group IV and V elements (see
Section 1.1.2.4) and with compounds of subgroup elements (see Section 1.1.2.5).

Reaction Type 15: Heterocuprate formation according to the equation
$RC \equiv CCu + MX \rightarrow [RC \equiv CCuX]M$ (or $RC \equiv CCu \cdot MX$; M=alkali metal or MgX;
X=F, Cl, Br, I; see Section 1.1.2.3).

Reaction Type 16: Homocuprate formation according to the equation
$RC \equiv CCu + R'M \rightarrow [RC \equiv CCuR']M$ (M=alkali metal or MgX).

a: R'=organic group other than $R^1C \equiv C$, see "Organocopper Compounds" 2, 1983, pp. 174/87 and 225/37,

b: $R' = R^1C \equiv C$, see Section 1.1.2.7.

To prepare $[RC \equiv CCuR']M$ ("Gilman's Reagent") according to Reaction
Type 16 a soluble acetylide is advantageous. This can be selected freely,
for the acetylene moiety is not of importance in the course of so-called
"Gilman Reactions". The quite soluble but very expensive No. 55 (R=
$(CH_3)_3C$) [308] and the soluble No. 57 (R=$(CH_3)_2C(OCH_3)$) [271, 310, 314]
have been suggested. But most reactions are performed with No. 33
(R=n-C_3H_7) [301, 303 to 307, 309, 311 to 313].

Reaction Type 17: Homocuprate formation according to the equation
$RC \equiv CCu + 2R'M \rightarrow [RC \equiv CCuR_2']M_2$ (M=Li or K).

a: R'=organic group other than $R^1C \equiv C$, see "Organocopper Compounds" 2, 1983, p. 241,

b: $R' = R^1C \equiv C$, see Section 1.1.2.8.

See also the remarks under Reaction Type 16.

Reaction Type 18: Reactions with other transition metal compounds, see Section 1.1.2.1.10.

References on pp. 57/67

"In situ" Reactions. There exist some reactions in which RC≡CCu seems to be an intermediate, e.g., in the so-called "Glaser coupling"

$$2RC\equiv CH + O_2 \xrightarrow{Cu^+} R(C\equiv C)_2R + H_2O$$

and in the so-called "Cadiot-Chodkiewicz coupling"

$$RC\equiv CH + BrC\equiv CR^1 \xrightarrow{Cu^+} R(C\equiv C)_2R^1 + HBr$$

The first reaction corresponds to our Reaction Type 2 and the second to our Reaction Type 7 d.

Bourgain et al. have shown in a detailed investigation that with solid RC≡CCu with either catalytic (10 g CuCl per mole RC≡CH) or stoichiometric amounts of Cu^I salt the reaction $RC\equiv CH + CH_2=CHCH_2Br \rightarrow RC\equiv CCH_2CH=CH_2$ proceeds in the same way. The yields are more strongly influenced by the solvent and by basic additives than by the general method [164]. On the other hand, the activation of the R^1Br bond by Cu^+ in the alkynyl Grignard reaction

$$RC\equiv CMgBr + R'Br \xrightarrow{Cu^+} RC\equiv CR' + MgBr_2$$

was also postulated, a simultaneous RC≡CCu formation not excluded [64].

The reaction

$$R^1(C\equiv C)_mH + Br(C\equiv C)_nR^2 \xrightarrow{Cu^+} R^1(C\equiv C)_{m+n}R^2$$

shows a maximum yield with ca. 2 mol% Cu^+. The yields are lower with pure $R^1(C\equiv C)_mCu$ instead of $R^1(C\equiv C)_mH/Cu^+$ due to the fact that the reaction

$$R^1(C\equiv C)_mCu + Br(C\equiv C)_nR^2 \rightarrow R^1(C\equiv C)_{m+n}R^2$$

produces large amounts of Cu^+, which causes side reactions with $Br(C\equiv C)_nR^2$. That means, though catalytic amounts of Cu^+ are preferable, the reaction could nevertheless have RC≡CCu as an intermediate.

The "in situ" reactions listed above are dealt with in the following sections only for acetylides already listed in Table 1, and when also the reaction is of special importance.

Remarks Concerning Table 1 on pp. 17/41

In principle Table 1 contains only characterized RC≡CCu compounds, but there are borderline cases. Literature data are often so vague that it is hardly possible to decide whether the compound was prepared or not. In many cases the only further characterization is by the reaction products of the compounds. It is often very difficult to determine from the papers just what was actually done. Further, there are often reaction conditions in which it is known that RC≡CCu must be precipitated; the article, however, contains no remark about its formation or isolation and the characterization is only in terms of consecutive reactions. In all these cases the preparation method is listed in Table 1 with an asterisk (e.g., *1a).

The compounds $ClC\equiv CCu$, $C_2H_5OCH_2C\equiv CCu$, and $(CH_3)_2NCH_2C\equiv CCu$ exist only as "LiI or NaI adducts" (see Section 1.1.2.3, Table 21), and $C_3H_7CH(OH)CH_2C\equiv CCu$ exists only as $C_3H_7CH(OH)CH_2C\equiv CCu \cdot CuCl$ (see Section 1.1.2.5, Table 24). They are therefore not included into Table 1.

All hydrates are listed in this table.

Acetylenes with a second acid proton are reported to form "copper acetylides", too. In most cases there was no decision about where the copper atom is bound. This case refers for example to acetylides with CO_2H groups (Nos. 5 and 31), but probably not to CH_2OH (e.g., No. 8). A similar problem arises with the CO_2Na group. Na and Cu here could be in interchanged locations (Nos. 4, 23).

Gmelin Handbook
Cu-Org. Comp. 3

Table 1
Compounds of the Type RC≡CCu.
The listing sequence is according to Hill's system applied to the molecular formula (C first, H second, then the other elements in alphabetical order), not to the individual R groups. Further information for compound numbers preceded by an asterisk is given at the end of the table, pp. 42/57. In situ preparations are marked by an asterisk at the method number (see "Remarks Concerning Table 1", p. 15).
For abbreviations and dimensions see p. X.

No.	R in RC≡CCu	preparation method (yield in %)	properties and remarks	Ref.	feasible reactions (see pp. 12/4)
*1	H	Ia?, IIa, V?	Method IIa from CuI and CH≡CK at −78° flocculent orange, later referred to as vermilion stable up to −50°, above −45° dec. into Cu_2C_2 and C_2H_2 / Reaction Type 17b with CH≡CK in liquid NH_3 to give [(CH≡C)₃Cu]K₂	[36, 55, 72, 91, 95, 104, 260, 317]	17b
*2	CF₃	Ia	khaki-colored, becomes brown and darkens on standing in air / dec. on heating or explosion, soluble in ether, C_2H_5OH	[65]	1b [65]
3	N≡C–	Ia (100)	olive-green; ignites on heating or rubbing see Section 1.1.2.2 for "mixed acetylide" with HOCH₂CH=CHC≡CCu	[35, 79, 90]	2c (30%) [79, 90]
4	NaO₂C (?)	IIb	alternative structure CuO₂CC≡CNa not strictly excluded / preparation from CuOH + monosodium salt of HO₂CC≡CH in H_2O at ~25° yellow / K₃[Fe(CN)₆] in H_2O yields 60% NaO₂C(C≡C)₂-CO₂Na	[13]	2c (60%) [13]

Table 1 [continued]

No.	R in RC≡CCu	preparation method (yield in %)	properties and remarks	Ref.	feasible reactions (see pp. 12/4)
5	HO_2C (?)	Ia, *Ig	alternative structures $CuO_2CC≡CH$ and even $CuO_2CC≡CCu$ not strictly excluded unstable, green, changing to brown explodes on warming; in aqueous solution dec. to Cu_2C_2, C_2H_2, and CO_2	[11 to 13, 36, 41, 189]	—
6	NH_2CO	Ia	poor elemental analysis ignition or sparking on heating in air	[35]	—
*7	CH_3	Ia (52), *Id $(HN(C_2H_5)_2/CuI)$, IIa (80), III (100), VI, VIII	yellow; soluble in large excess of reagent used in Method Ia (especially if NH_4Cl or KCl is added) IR (KBr): 82, 173, 180, 218, 318, 340, 935 and 943 (C–C), 1020 (CH_3), 1360, 1431 (CH_3), 1867, 1956 or 1957 (C≡C), 2832, 2901, and 2937 (C–H) Raman (He–Ne laser): 262 (π–Cu–C≡C), 367 and 435 (Cu–C≡), 944 (C–C), 1020 (CH_3), 1865, 1950 (C≡C)	[4, 84, 102, 103, 133, 236, 242, 269, 291, 328]	2a [51], 2a (74%) [61], 2b (48%) [61], 2b (80%) [71], 2c (50%) [61], 2c [45], 7c, 7d, 7g, 7h, 7i, 7k, 7n, 14, 15, 16b, 17b, 18
*8	$HOCH_2$	Ia (92), Id (pyridine/ $H_2O/CuCl$), *Id $(HN(C_2H_5)_2/CuI)$, *Ig	Method Ia: evaporate reaction mixture partly before filtration yellow, turning brown in wet air soluble in pyridine, sparingly in H_2O ignites on heating or contact with HNO_3	[7, 16, 69, 106, 109, 146, 189, 269, 343]	1a [7], 2a (88%) [77], 2b [49], 2c [15, 16], 3a [16, 23], 3a (50 to 80%) [121], 4a [23, 28], 7a, 7n, 8b, 8c, 10
*9	$CCl_2=CCl$	IV (100)	Method IV with or without a solubilizer (C_2H_5OH) and in presence of $(NH_4)_2CO_3$ orange-red; highly explosive in dry state storage under CCl_4, in which totally insoluble	[193, 235]	1a (26%) [235], 2c (40%) [235], 2d (Br_2, 20%) [235], 3a (82%) [235], 3b (37%) [235], [193, 235], 4b (20%) [235]

References on pp. 57/67

No.	Group	Type	Notes	Ref.	Ref.
10	C_2F_5	Ia	preparation in a sealed tube khaki-colored	[67]	1b [67]
	CH≡C see No. 200 (supplement)				
*11	$CH_2=CH$	Ia, Id, III	yellow, darkening in air IR (polyethylene; measured 100 to 400): 128, 160, 234, 285, 307, 405 (≡CCu); (KBr): 1935 (C≡C) visible/UV: 435 (2.5 eV band gap; whole spectrum given)	[31, 69, 202, 203, 228, 231, 242, 283]	7n, 14, 18
12	CH_3CO	Ia	—	[134]	7n
13	CH_3O_2C	Ia	primary product of Reaction Type 5 adds spontaneously CH_3CO_2H and $CH_3O_2CCH=CHC(O_2CCH_3)=CHCO_2CH_3$ is isolated	[92, 136]	5 [92], 7n
14	$ClCH_2CH(OH)$	not given	attempted reactions of Type 7n unsuccessful	[122]	–
15	C_2H_5	Ia	yellow IR (KBr): 1952 (C≡C) used to isolate $C_2H_5C≡CH$ from gas streams	[25, 28, 32, 102, 103, 123]	7c, 7f, 7i, 7k, 8b
16	$HO(CH_2)_2$	Ia	yellow, solubility in pyridine at 25° 10^{-5} M	[24, 150, 165]	2a (82%) [77], 2a [53], 4a [24, 28], 7n, 8a
17	$CH_3CH(OH)$	Ia	yellow, considerable solubility in ammoniacal CuCl solution	[26, 51]	2a (94%) [77], 2a [53], 2c [28], 4a [26, 28], 7a, 7d
18	C_2H_5O	not given	—	–	2c [100], 2d (H_2O_2) [100], 7a
19	CH_3OCH_2	Ia	yellow	–	2c [16], 3a [16], 7o, 7p

Table 1 [continued]

No.	R in RC≡CCu	preparation method (yield in %)	properties and remarks	Ref.	feasible reactions (see pp. 12/4)
	CH_3SCH_2 see No. 199 (supplement)				
*20	C_2H_5S	Ia	IR: 1890 (C≡C) visible/UV: 485 (2.32 eV band gap) photoconductivity maximum 530 nm (2.20 eV band gap)	[246, 278, 283]	—
21	$(CH_3)_2N$	Ia, Id	not isolated as a pure substance	[49]	2b [49]
22	$1,2-C_2B_{10}H_{11}$ (1,2-dicarba-closo-dodecarboran(12)-1-yl)	Ia (100)	yellow, does not melt when heated degree of association in benzene: 4.15 no photosensitivity in visible/UV and near IR IR (KBr): 1900, 2075 (C≡C) non explosive; soluble in organic solvents no reaction with $HgCl_2$, $HgBr_2$, $TlCl_3$	[197, 202, 203]	1d (C_2H_2) [197], 3a (68%) [197], 18
	$1,7-C_2B_{10}H_{11}$ see No. 193 (supplement)				
23	$NaO_2CC≡C(?)$	IIb	structure $CuO_2C(C≡C)_2Na$ not excluded preparation from monosodium salt of $HO_2C(C≡C)_2H + CuOH$ in aqueous medium at 80 to 90° dark red $K_3[Fe(CN)_6]$ in H_2O yields $NaO_2C(C≡C)_4CO_2Na$	[13]	2c [13]
24	$CH_3C≡C$	Ia (70 to 80)	preparation under ice cooling; Preparation Method Ia can be used to isolate $CH_3(C≡C)_2H$ from mixtures with triynes yellow	[57, 61, 150]	7d, 7n
25	$CH_3CH=CH$	Ia (70 to 80)	preparation under ice cooling dark yellow; very explosive (E)/(Z)-mixture from reaction of Type 7n	[150]	7n

No.	R	preparation	properties	Ref.	further refs.
26	CH$_2$=C(CH$_3$)	Ia	bright yellow IR (liquid): 169, 438 ($\equiv$C–Cu), 525, 550, 635, 905, 1011, 1218, 1240, 1262, 1370, 1427, 1440, 1615, 1720, 1818, 2865, 2926, 2961, 2982, 3102, 3305; similar in KBr Raman (liquid): 185, 260, 390, 525, 626, 765, 950, 1010, 1370, 1385, 1613, 2100	[186, 236, 242, 268, 355]	7i, 7o, 8a, 9, 14
27	cyclo-C$_3$H$_5$	Ia (41)	yellow	[201]	7h, 15
28	(Z)-CH$_3$OCH=CH	Ia (75)	yellow, air-sensitive slightly soluble in C$_2$H$_5$OH yields with o–bromoanilines no pyrroles, but only tarry products; for in situ reactions see [179]	[276]	7a, 7b, 7c, 7e, 7f, 7h, 7i, 8a
29	HOCH$_2$CH=CH	Ia	for "mixed acetylide" with NCC$\equiv$CCu (No. 3) and its oxidation according to Reaction Type 2c see Section 1.1.2.2	[93]	2a [54], 2a (84%) [77], 2d (FeCl$_3$/O$_2$) [56]
30	C$_2$H$_5$O$_2$C	Ia (35), *Id (dimethyl-formamide/CuI)	in Method Ia pH control at 7.5 needed or (NH$_4$)$_2$CO$_3$ must be added canary yellow or orange IR (mineral oil): 1681, 1706, 1923 (C$\equiv$C)	[12, 142, 153, 181, 188]	7i, 7j, 7n, 8b, 8c, 12
31	HO$_2$C(CH$_2$)$_2$ (?)	Ia	yellow, gelatinous	[22]	—
32	CH$_3$CO$_2$CH$_2$	Ia, *Id	oil	[92, 326]	5 (75%) [92], 7c
*33	n–C$_3$H$_7$	Ia (88), III, VIII	yellow; slightly soluble in tetrahydrofuran, ether; solubility in pyridine at 25° 10^{-5} M, insoluble in dimethylformamide IR (KBr): 1942 (C$\equiv$C)	[25, 102, 103, 111, 165, 271, 289, 316, 331]	2d (FeCl$_3$/O$_2$) [56], 7b, 7c, 7d, 7f, 7i, 7o, 7p, 8a, 8b, 8c, 13, 14, 16a
34	i–C$_3$H$_7$	Ia	yellow	[201]	7h, 7i
*35	C$_2$H$_5$CH(OH)	Ia (low yield), Id (CuCl/H$_2$O)	yellow; readily soluble in aqueous NH$_3$ and in C$_2$H$_5$OH	[27, 28]	4a [27, 28]
36	CH$_3$CH(OH)CH$_2$	*Ia	not isolated as a pure substance	[77]	2a (98%) [77], cf. [53, 78]

Table 1 [continued]

No.	R in RC≡CCu	preparation method (yield in %)	properties and remarks	Ref.	feasible reactions (see pp. 12/4)
37	$(CH_3)_2C(OH)$	Ia	also formed by anodic oxidation of Cu in alkaline $(CH_3)_2C(OH)C\equiv CH$	[50, 82]	2a [50], 7a, 7d, 7f
38	$CH_3O(CH_2)_2$	Ia	—	[28]	4a [28]
39	$(CH_3O)_2CH$	Ia (60 to 80)	in Method Ia undiluted acetal $(CH_3O)_2CHC\equiv CH$ is treated with a solution of CuCl in aqueous NH_3/C_2H_5OH (5:1); careful stirring causes solidification, the crystalline mass is successively washed with H_2O, CH_3OH, and ether yellow crystals	[150, 212]	7n
*40	$(CH_3)_3Si$	If (80 to 90), IIb, VIII	orange-red, unstable, dec. even at $-20°$ solutions in tetrahydrofuran stable for some time at 0°; soluble in $P(OC_2H_5)_3$ IR (Nujol): 855 (CH_3), 1250, 1890 (C≡C)	[131, 245, 266, 316, 325, 334]	7a, 7b, 7c, 7d, 13
41	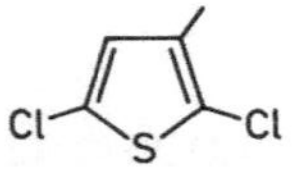 (thiophene structure)	Ia	in Preparation Method Ia the solution buffered with $(NH_4)_2CO_3$ yellow, voluminous solid	[75, 76]	2b [76], 2c [76]
42	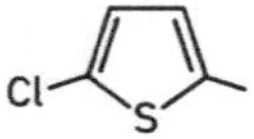 (thiophene structure)	Ia	in Preparation Method Ia the solution buffered with $(NH_4)_2CO_3$	[75]	—
43	$CH_2=CHC\equiv C$	Ia (70 to 80)	preparation under ice cooling very explosive, handle only wet with ether	[150]	7n

No.	Formula	Preparation	Properties	Ref.	
44	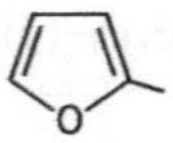 (2-methylfuran ring with O)	Ia (75)	light yellow, fine needles / stable at room temperature, explodes on contact with HNO_3 or at heating / insoluble in ether, H_2O, C_2H_5OH; readily soluble in aqueous ammonia	[40]	1c [40]
45	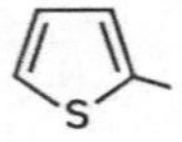 (2-methylthiophene ring with S)	Ia (56)	in Method Ia a solution buffered with $(NH_4)_2CO_3$ / yellow, voluminous solid / forms orange solution in boiling CH_3CO_2H	[75, 76, 225]	2b [76], 2c [76], 7n, 14
46	$CH_2=CHCH=CH$	Ia	yellow or orange, vacuum dried under inert gas / Preparation Method Ia used for separation of hexadienynes / forms a Cu^{2+} solution with HNO_3, used for quantitative determination	[43, 60]	1a (35%) [43], 1b [43], 1c [60]
47	$CH_3C\equiv CCH_2$	Ia	yellow	[14]	—
48	$CH_2=C=CHCH_2$	Ia	separation of $CH_2=C=CHCH_2C\equiv CH$ from $CH_2=C=CHCH=C=CH_2$ by acetylide formation	[217]	—
49	$CH_2=CHCO_2CH_2$	Ia	yellow / IR: 1930 (C≡C)	[232]	14
50	$CH_2=CHBr(CH_2)_2$	Ia	Reaction Type 2c impossible	[28]	1d ($K_3[Fe(CN)_6]$, low yield) [28]
51	$CH_2=C(CH_3)CH_2$	Ia	yellow	[81]	—
52	$CH_3CH=CHCH(OH)$	Ia	not isolated as a pure substance	[77]	2a (93%) [77]
53	$(Z)-C_2H_5OCH=CH$	Ia	—	[157, 179]	7e
*54	$n\text{-}C_4H_9$	Ia, Ic, *Id $(HN(C_2H_5)_2/CuI)$, Ie, Ig, IIa, VII, VIII	Method VII with $2CuO_3SCF_3 \cdot C_6H_6$ in CD_3NO_2/CH_2Cl_2, yellow; m.p. 140 to 150°, on further temperature rise sets to a solid, melts again at 220° / IR (KBr): 450 (≡C–Cu), 720, 1107, 1430, 1471, 1930 or 1940 (C≡C), 2937, 2962 / IR (polyethylene): 123, 158 (both Cu–Cu) / Raman: 450, 941, 1925 (C≡C), 1950	[38, 83, 189, 203, 226, 228, 230, 236, 242, 269, 316, 329, 330, 341]	7a, 7b, 7c, 7f, 7h, 7i, 7k, 7l. 7m, 7n, 7p, 8a, 8b, 8c, 12, 14, 15, 16a, 16b, 18

References on pp. 57/67

Table 1 [continued]

No.	R in RC≡CCu	preparation method (yield in %)	properties and remarks	Ref.	feasible reactions (see pp. 12/4)
	C$_2$H$_5$C(CH$_3$) see No. 201 (supplement)				
*55	(CH$_3$)$_3$C	Ia, IIb, VII, VIII	Method VII with CuO$_3$SCF$_3$·C$_6$H$_6$ in CD$_3$NO$_2$/CH$_2$Cl$_2$ yellow, orange, or red no photoconductivity in near IR or visible/UV does not react with pentafluoroiodobenzene; Reaction Type 2c yields a mixture of (CH$_3$)$_3$C(C≡C)$_2$C(CH$_3$)$_3$ and (CH$_3$)$_3$CCH$_2$CO$_2$H	[46, 185, 203, 218, 237, 294, 316, 341]	1d (NH$_3$) [218], 2a [46], 2c [34], 7e, 14, 16a
56	C$_2$H$_5$C(CH$_3$)(OH)	—	formed by anodic oxidation in alkaline C$_2$H$_5$C(CH$_3$)(OH)C≡CH	[50]	2a [50]
57	(CH$_3$)$_2$C(OCH$_3$)	IIb, III	red oil from the solution in tetrahydrofuran, solidifies on treatment with hexane readily soluble in tetrahydrofuran, soluble in ether (~0.1 M at 0°), insoluble in hexane	[200, 270, 271, 290, 338]	10, 16a, 17a
58	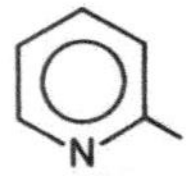	Ia (51)	orange	[123, 153]	7f, 8a
59	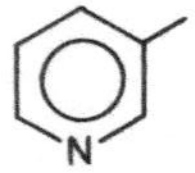	Ia	m.p. 233° (dec.) resistivity at 25°: 6.7 × 10^{11} Ω · cm photocurrent in vacuum lower than in air IR: 1950 (C≡C)	[191, 220, 221]	7g, 7h
60		Ia	m.p. 221° (dec.) resistivity at 25°: 5.2 × 10^9 Ω · cm IR: 1938 (C≡C)	[191, 220, 221]	—
61	(E)–CH$_3$CH=CHC≡C	Ia	—	[138]	7e, 7i

No.					
62	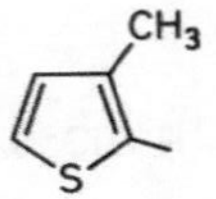 ![3-methyl-2-thienyl] CH₃ / S	Ia	yellow, voluminous	[75, 76]	2b [76], 2c [76]
63	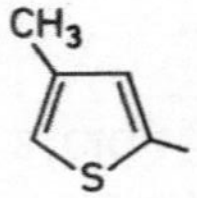 ![methylthienyl] CH₃ / S	Ia	yellow, voluminous	[75]	—
64	$CH_2=C(CH_3)CO_2CH_2$	Ia	yellow-green IR: 1930 (C≡C)	[232]	14
65	$(CH_3)_2C(CN)CH_2$ (?)	VI	preparation from $(CH_3)_2C(CN)CH_2C{\equiv}CCu \cdot {}^1/_2 NH_3$ by heating	[238]	—
	$t\text{-}C_4H_9O_2C$ see No. 208 (supplement)				
66	$CH_2=CHC(CH_3)_2$	Ia, Id (dimethylformamide/N-methylpiperidine)	yellow	[192, 265]	8b
67	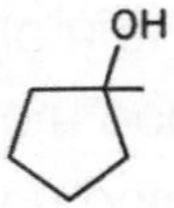 ![1-methylcyclopentanol] OH	III	—	[200]	10
68	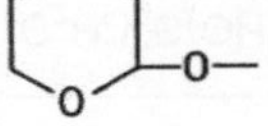 ![2-methoxytetrahydropyran] O—O	Ia	—	[161]	7e, 7k, 7n
69	$n\text{-}C_5H_{11}$	Ia, Id (hexamethylphosphoric triamide), IV (100)	yellow	[25, 42, 164, 326]	1a (56%) [42], 1c [175], 5 [92], 7a, 7b, 7c, 7e, 8c, 14, 15, 16a

Table 1 [continued]

No.	R in RC≡CCu	preparation method (yield in %)	properties and remarks	Ref.	feasible reactions (see pp. 12/4)
70	$(C_2H_5)_2C(OH)$	*Ia, III	—	[88, 200]	7a, 10
71	$n-C_3H_7C(CH_3)(OH)$	III	—	[200]	10
72	$CH_3OC(CH_3)(C_2H_5)$	III	—	[200]	10
73	$C_2H_5OC(CH_3)_2$	III	—	[200]	10
74	$(C_2H_5O)_2CH$	Ia (50 to 80), Ie, *Ig, IIb	yellow; very explosive, don't use fully dried yellow–green solutions in pyridine and tetrahydrofuran products of Reaction Type 7n usually isolated as aldehydes	[133, 183, 189, 266]	7d, 7e, 7f, 7g, 7h, 7k, 7n
75	$(C_2H_5)_2NCH_2$	Ia	—	[255]	12
76	C_6F_5	Ia (80 to 90)	bright yellow, m.p. 277° (dec.)	[250, 251]	2a [167, 251], 7e, 18
77	$4-BrC_6H_4$	Ia	solid IR (KBr): 1942 (C≡C)	[46, 102, 103, 121]	3a (50 to 80%) [121], 3b (70%) [121], 7d, 7h, 14, 18
78	$4-BrC_6H_4O$	If (80 to 90)	soluble in $P(OC_2H_5)_3$	[131]	—
79	$2-ClC_6H_4$	not given	IR (polyethylene; measured 100 to 400): 147 (Cu–Cu), 211, 247, 258, 286, 321 for reaction of Type 2b the acetylide prepared in situ	[52, 228]	2b [52]
80	$3-ClC_6H_4$	*V	IR (polyethylene; measured 100 to 400): 147 (Cu–Cu), 211, 247, 258, 286, 321 for reaction of Type 11 the acetylide prepared in situ	[52, 228, 229]	11

References on pp. 57/67

No.	compound	preparation (yield %)	properties	references
81	4-ClC$_6$H$_4$	*Ia, V	IR (polyethylene; measured 100 to 400): 147 (Cu-Cu), 211, 247, 258, 286, 321 IR (KBr): 456 ($\equiv$C-Cu), 522, 535, 790, 819, 1016, 1093, 1393, 1483, 1588, 1929, 2925, 3050, 3080 Raman: 270, 456, 532, 632, 787, 1926 visible/UV: 450 see [256] for photoconductivity and photovoltaic behavior (no numerical data)	[46, 148, 228, 229, 236, 242] 7c, 7d, 7h, 11, 14, 18
82	3-FC$_6$H$_4$	*V	Reaction Type 11 with in situ prepared acetylide	[229] 11
83	4-FC$_6$H$_4$	Ia (80 to 90)	yellow-orange; m.p. 237° (dec.)	[250, 251] 7h, 18
84	2-IC$_6$H$_4$	not given	IR (polyethylene; measured 100 to 400): $\sim$150 (Cu-Cu), 222, 245, 280, 314 heating in dimethylformamide yields besides carbocycles (see Section 1.1.2.1.5.4) an iodine-containing oligomer (2-C$_6$H$_4$C$\equiv$C)$_n$, n$\geqq$4	[182, 228] 8d
85	3-IC$_6$H$_4$	Ia (39)	IR (polyethylene; measured 100 to 400): $\sim$150 (Cu-Cu), 222, 245, 280, 314 reacts on heating in dimethylformamide like No. 84, no cyclic product isolated, the linear oligomer is (3-C$_6$H$_4$C$\equiv$C)$_n$, n$\geqq$4 ("satisfactory" yield)	[182, 228, 233] 8d
*86	4-IC$_6$H$_4$	Ia	IR (polyethylene; measured 100 to 400): $\sim$150 (Cu-Cu), 222, 245, 280, 314 diffuse reflectance maximum $\sim$450 nm reacts on heating in dimethylformamide or pyridine like No. 84 to give (4-C$_6$H$_4$C$\equiv$C)$_n$, n$\geqq$4	[121, 182, 228, 256] 2d ((CN)$_2$C=C(CN)$_2$) [170], 3a (50 to 80%) [121], 3b (70%) [121], 7a, 7c
*87	2-O$_2$NC$_6$H$_4$	Ia (99)	red; ignites on heating does not react with O$_3$, KMnO$_4$, or I$_2$ per Reaction Type 2d for "mixed acetylide" with C$_6$H$_5$C$\equiv$CCu see Section 1.1.2.2	[8, 9, 162] 2c [9], 7f, 7k, 7m

Table 1 [continued]

No.	R in RC≡CCu	preparation method (yield in %)	properties and remarks	Ref.	feasible reactions (see pp. 12/4)
88	$3-O_2NC_6H_4$	Ia	brick red	[33]	2c [9]
89	$4-O_2NC_6H_4$	Ia	formed from $4-O_2NC_6H_4N_2^+$ and Cu_2C_2 in a 14% yield m.p. 301° (dec.) resistivity $6.2 \times 10^{14}\ \Omega \cdot$ cm at 25° IR (KBr): 667, 742, 850, 1103, 1173, 1308, 1340, 1515, 1587, 1925 or 1929 or 1940 (C≡C), 3105, 3255 diffuse visible/UV reflectance plateau 410 to 480 reaction with I_2 said to give only $4-O_2NC_6H_4C≡CCl$ (Reaction Type 3a) [110, 121] or only $4-O_2NC_6H_4(C≡C)_2C_6H_4NO_2-4$ (Reaction Type 2d) [51]	[46, 102, 103, 121, 191, 220, 221, 236, 256, 302]	2a [46], 2d (I_2) [51], 3a (70%) [110, 121], 3b (70%) [121], 7d, 7h, 7m, 7p, 11, 14
*90	C_6H_5	Ia, Ib, Id, Ie, If, II, IV, V, VI, VII, VIII	preparation methods and feasible reactions show great variety and complexity; this compound has been extensively studied and many findings are reported under "further information" including comments and references, see pp. 46/54 yellow "amorphous" powder, yellow microcrystals, or golden coarse crystals	[101, 103, 207]	1a, 1b, 1d, 2a, 2b, 2c, 2d, 3a, 3b, 5, 6, 7a, 7b, 7c, 7d, 7e, 7f, 7g, 7h, 7i, 7j, 7k, 7l, 7m, 7n, 7o, 7p, 8a, 8b, 8c, 8d, 9, 10, 11, 12, 13, 14, 15, 16a, 18

References on pp. 57/67

No.	Group	Method	Properties	Ref.	Type
91	C_6H_5O	Ia, If (80 to 90)	soluble in $P(OC_2H_5)_3$; photoconduction spectrum: $\lambda_{max} = 520$ nm; gap width = 2.12 eV; IR (KBr): 1990 (C≡C); visible/UV: 485 (gap width = 2.13 eV)	[131, 246, 278, 283, 339]	7b
92	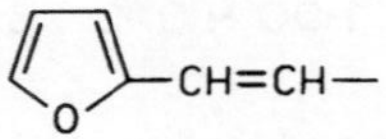 (furan)–CH=CH—	not given	—	[112]	7n
93	CH_3CO_2–(thiophene)	Ia (59)	in Method Ia pH 8 required	[212]	7n
94	C_6H_5S	VIII	photoconduction spectrum: $\lambda_{max} = 530$ nm; gap width = 2.08 eV; IR: 1900 (C≡C); visible/UV: 450 (gap width = 2.42 eV)	[246, 278, 283, 316]	—
95	$3\text{-}NH_2C_6H_4$	Ia	yellow	[10]	—
96	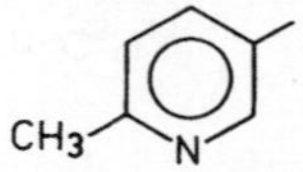	not given	Reaction Type 7n catalyzed by metallic Cu	[152]	7n
97	$CH_2=CHCH_2C\equiv CCH_2$	Ia	yellow	[81]	—
	cyclohexen-1-yl see No. 197 (supplement)				
	$(CH_3)_2C=CH(CH_2)_2$ see No. 202 (supplement)				
*98	cyclo-C_6H_{11}	Ia (20), IIa (43), IIb (70), VIII	light yellow, green-yellow, orange, or dark beige; IR (Nujol or KBr): 2164, 2176 (both C≡C); stable in air, insensitive to shock or percussion, hydrophobic, stable in dilute acids	[288, 316, 353]	14
99	(Z)-n-$C_4H_9OCH=CH$	Ia	—	[157, 179]	7d, 7e

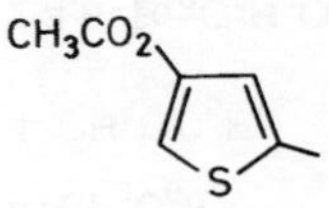

References on pp. 57/67

Table 1 [continued]

No.	R in RC≡CCu	preparation method (yield in %)	properties and remarks	Ref.	feasible reactions (see pp. 12/4)
100	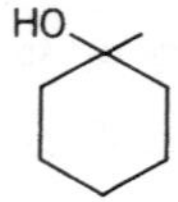 HO—C(CH₃)₂— cyclohexyl	*Ia	not isolated as a pure substance	[53]	2a [53]
101	tetrahydropyranyl —OCH₂—	Ia (50 to 80), *Id	bright yellow; handling quite safe products of Reaction Type 7 often isolated after saponification to R'C≡CCH₂OH; Reaction Type 7i yields byproducts	[133, 150, 277, 292, 326]	7c, 7d, 7e, 7f, 7g, 7h, 7i, 7j, 7k, 7m, 7n, 7o
*102	n-C_6H_{13}	Ia (87), IIa (80), VIII	yellow or green extensive spectral data on p. 55 stable in air, insensitive to shock, hydrophobic, stable in dilute acids	[288, 316]	8a, 8c, 14
103	n-$C_4H_9C(CH_3)(OH)$	III	—	[200]	10
104	$C_2H_5OC(CH_3)(C_2H_5)$	III	—	[200]	10
105	4-$CF_3C_6F_4$	Ia	—	[253]	—
106	4-$CH_3OC_6F_4$	Ia	—	[253]	—
	$2,4,6$-$Br_3C_6H_2OCH_2$ see No. 203 (supplement)				
107	$2,4,6$-$Cl_3C_6H_2OCH_2$	Ia	—	[198, 199, 339]	8a
108	$2,4$-$Cl_2C_6H_3OCH_2$	Ia	—	[198, 199]	8a
109	$2,5$-$(O_2N)_2C_6H_3OCH_2$	—	sulfur-colored, m.p. 210° (dec.), insoluble in organic solvents	[219]	3a (100%) [219]
110	C_6H_5CO	—	—	[145]	7e, 7f, 7g, 7h

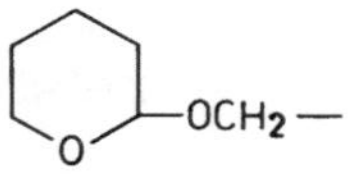

References on pp. 57/67

No.	Compound	Preparation	Properties / Spectral data	References	
111	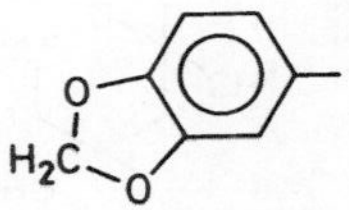 (methylenedioxytoluene structure)	Ia, VII	dark yellow, nonexplosive	[29, 247, 258]	8a
	2-BrC$_6$H$_4$OCH$_2$ see No. 194 (supplement)				
	4-BrC$_6$H$_4$OCH$_2$ see No. 195 (supplement)				
112	2-ClC$_6$H$_4$OCH$_2$	Ia	—	[198, 199]	8a
113	4-ClC$_6$H$_4$OCH$_2$	Ia	—	[198, 199]	8a
	4-O$_2$NC$_6$H$_4$OCH$_2$ see No. 207 (supplement)				
114	2-CH$_3$C$_6$H$_4$	Ia	IR (polyethylene): 133, 144, 151, 167 (all $\equiv$CCu), 207, 268, 305	[132, 228]	7f, 7j, 7k, 8d
115	3-CH$_3$C$_6$H$_4$	not given	IR (polyethylene): 133, 144, 151, 167 (all $\equiv$CCu), 207, 268, 305	[228]	11
*116	4-CH$_3$C$_6$H$_4$	Ia (80 to 90), *V	yellow; m.p. 215 to 219° (dec.) [251], 236° (dec.) [191, 220, 221] very explosive IR (KBr): 450 or 451 ($\equiv$C–Cu), 525, 806, 1503, 1932 or 1935, 3020, 3080 IR (polyethylene): 133, 144, 151, 167 (all $\equiv$CCu), 207, 268, 305 Raman (solid): 154, 286, 451, 540, 702, 721, 824, 1174, 1191, 1931 visible/UV (without dilution by a nonabsorbing standard): 430	[30, 191, 220, 221, 228, 229, 236, 242, 243, 251]	2a [46], 2c [30], 7h, 7k, 7p, 8d, 11, 12, 14, 18
117	C$_6$H$_5$CH$_2$	Ia (50 to 80)	photoconduction spectrum: λ_{max} = 445 nm, gap width = 2.65 eV IR: 572 ($\equiv$CCu) and 1946 (C$\equiv$C) UV: 415 (gap width = 2.69 eV)	[133, 242, 246, 278, 283]	7n

Table 1 [continued]

No.	R in RC≡CCu	preparation method (yield in %)	properties and remarks	Ref.	feasible reactions (see pp. 12/4)
118	4-CH₃OC₆H₄	Ia (92), VII	no ignition on heating; m.p. 260° (dec.) resistivity at 25° 2.3 × 10¹⁰ Ω · cm; photoconducting IR: 1890 (C≡C)	[29, 30, 191, 220, 221, 239]	2c [30], 7f, 7h, 7i, 7k, 7m, 8d, 11
*119	C₆H₅CH(OH)	*Ia, Id (CuCl/H₂O)	yellow	[60, 66, 77]	2a [53], 2a (88%) [77], 2b (55%) [60, 66]
120	3-CH₃C₆H₄O	If (80 to 90)	soluble in P(OC₂H₅)₃	[131]	—
121	C₆H₅OCH₂	Ia	—	[121, 339]	3a (50 to 80%) [121], 7b, 7c
*122	n-C₄H₉C≡CCH₂	Ia (79)	light yellow; m.p. 97 to 107° (dec.)	[94]	1a [94], 1d (NH₄Cl or C₂H₂) [94]
123	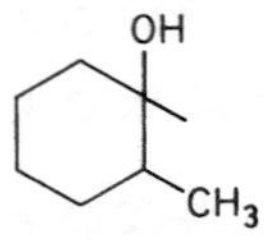 [structure: cyclohexane with OH, CH₃]	III	—	[200]	10
124	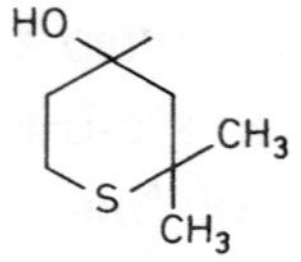 [structure: thiopyran with CH₃, CH₃, HO]	III	—	[200]	10
125	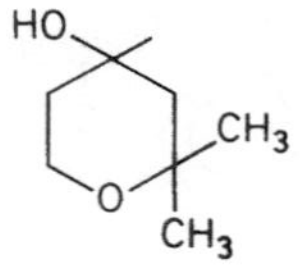 [structure: pyran with CH₃, CH₃, HO]	III	—	[200]	10

References on pp. 57/67

No.	R	method	properties	Ref.	
126	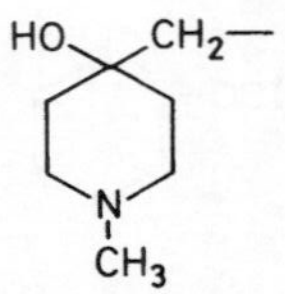 (structure: HO and CH₂— on piperidine ring with N–CH₃)	Ia	—	[255]	12
*127	$C_6H_5C\equiv C$	Ia (100)	from No. 90 and $BrC\equiv CH$ at 0° in dimethylformamide orange; m.p. 170° (dec.) IR (polyethylene; measured 100 to 400): 139, 151, 159, 168, 188 (all Cu···Cu); unidentified bands 102, 112, 284, 361 [228] IR (Nujol?): 2170 (broad, $C\equiv C$) [205, 221] IR (Nujol): 2180 ($C\equiv C$) [102, 103] IR (KBr): 1970 [203], 2180 [102, 103, 203], (both $C\equiv C$)	[59, 102, 103, 203, 205, 221, 228]	1a [221], 2a [74], 2b [58, 59, 66], 7n, 14
128	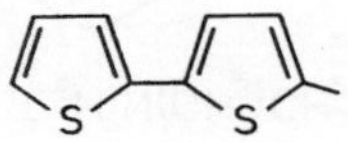 (thiophene structure)	—	—	[183, 359]	7 d
129	(E)-2-$IC_6H_4CH=CH$	Ia	(E)-structure derived from the intermolecular reaction of 3 or 4 molecules to give cyclic compounds, see Section 1.1.2.1.5.4	[172]	8 d
130	$C_6H_5CH=CH$ (E/Z mixture?)	Ia (80 to 90)	preparation from CuCl in saturated aqueous $(NH_4)_2CO_3$ orange	[97]	5 (all-trans, 40%) [92, 97]
131	4-$CH_3COC_6H_4$	not given	photoconductivity, photo–emf, and longwave visible/UV absorption maxima "between the values of 4-$IC_6H_4C\equiv CCu$ and 4-$O_2NC_6H_4C\equiv CCu$"	[256]	—
132	$C_6H_5CO_2CH_2$	Ia	—	[121]	3a (50 to 80%) [121]
133	2-$CH_3CONHC_6H_4$	Ia	yellow	[10]	2c [10]

References on pp. 57/67

Table 1 [continued]

No.	R in RC≡CCu	preparation method (yield in %)	properties and remarks	Ref.	feasible reactions (see pp. 12/4)
134	4-$C_2H_5C_6H_4$	Ia	yellow	[30]	2c [30]
135	2,4-$(CH_3)_2C_6H_3$	Ia	—	[259]	7i, 7k, 8d, 11
	4-$CH_3C_6H_4OCH_2$ see No. 204 (supplement)				
	2,4-$(CH_3O)_2C_6H_3$ see No. 196 (supplement)				
136	3,4-$(CH_3O)_2C_6H_3$	Ia	bright yellow	[258]	8a
137	4-$CH_3OC_6H_4OCH_2$	Ia	contradictory results in the reaction with C_6H_5COCl (same solvent): either Reaction Type 2a [339] or 7b [241]	[241, 339]	2a (72%) [339], 7b
138	$C_6H_5N(CH_3)CH_2$	Ia	—	[255]	12
139	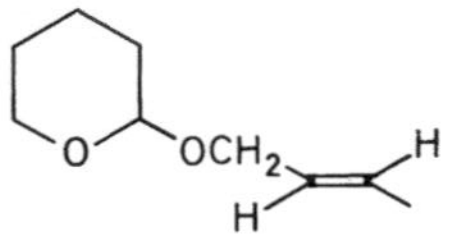	Ia (62)	for preparation CuCl is suspended in a mixture of concentrated ammonia and ice and the alkyne is added as an alcoholic solution; isolation after standing for 10 h at 4°	[211, 262]	2a (in situ) [70], 7n
140	2-C_6H_5-1,2-$C_2B_{10}H_{10}$ (2-phenyl-1,2-dicarba-closo-dodecaboran(12)-1-yl)	Ia (100)	yellow; no photosensitivity in UV/visible and near IR; NMR, compared with RC≡CH: diminution in the internal shift between signals of the ring protons (no values given) IR (KBr): 1900, 2050 (C≡C) nonexplosive, does not melt when heated; soluble in organic solvents	[197]	—
141	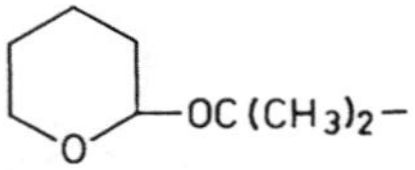 -$OC(CH_3)_2$-	Ia	sticky, rather unstable viscous oil; freeze dried before use	[272, 332, 355]	7o, 8a

References on pp. 57/67

No.	Formula / Structure	Method	Properties	Ref.	
142	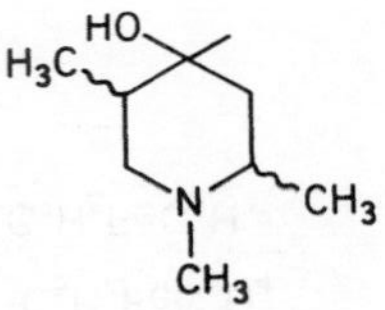	III	—	[200]	10
143	n-C$_8$H$_{17}$	Ia	—	[92]	5 [92]
144	n-C$_6$H$_{13}$C(CH$_3$)(OH)	III	—	[200]	10
*145	(n-C$_4$H$_9$)$_2$P	Id (40)	Preparation Method Id performed with CuCl in dimethylsulfoxide yellow–green; does not melt IR: 1999 (C≡C) UV: 370 (band gap = 2.74 or 2.75 eV) nonexplosive; insoluble in organic solvents	[166, 195, 283]	—
146	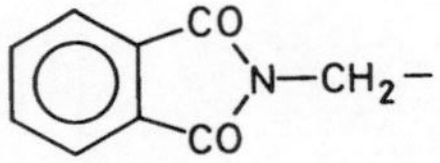	Ia	—	[255]	12
147	2,4,6-(CH$_3$)$_3$C$_6$H$_2$	Ia	IR (Nujol): 1933 vw (C≡C)	[102, 103]	7b, 7k, 8d, 11
	2,5-(CH$_3$)$_2$C$_6$H$_3$OCH$_2$ see No. 205 (supplement)				
148	C$_6$H$_5$N(C$_2$H$_5$)CH$_2$	Ia	—	[255]	12
149	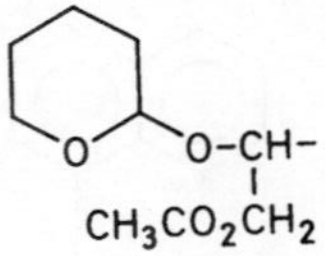	Ia (70 to 80), Id (92)	Preparation Method Id performed with CuI in dimethylsulfoxide	[135, 150]	7n
	(CH$_2$=CHCH$_2$)$_2$NC(CH$_3$)$_2$ see No. 206 (supplement)				
150	HO$_2$C(CH$_2$)$_8$(?)	Ia	alternative structure CuO$_2$C(CH$_2$)$_8$C≡CH not excluded brown soln. in H$_2$O	[86]	2c [86]

Table 1 [continued]

No.	R in RC≡CCu	preparation method (yield in %)	properties and remarks	Ref.	feasible reactions (see pp. 12/4)
151	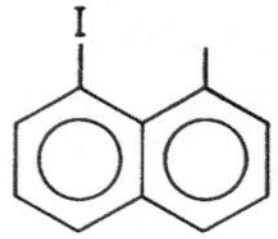 [naphthalene structure with I]	Ia	—	[159]	8d
152	[naphthalene structure with I]	Ia (?)	—	[172]	8d
*153	naphth–1–yl	Ia (52)	m.p. 250° (dec.) IR (no medium given): 1930; IR (Nujol): 1928 (both C≡C) visible/UV: 467; see [256] for spectrum in MgO (1:100) 420 to 630 nm	[191, 220, 221, 256, 263, 283]	7d, 7f, 7k, 8a, 8d
*154	naphth–2–yl	Ia	m.p. 201° (dec.) IR (no medium given): 1933; IR (Nujol): 1930 (both C≡C) visible/UV: 470; see [256] for spectrum in MgO (1:100) 420 to 600 nm	[191, 205, 220, 221, 256, 283]	7d
155	$2,5\text{-}d_2\text{-}C_5H_5FeC_5H_2D_2$	Ia (80)	—	[169]	7e
156	$BrC_5H_4FeC_5H_4$	Ia	—	[176]	2a [176]
157	$IC_5H_4FeC_5H_4$	Ia	—	[176]	2a [176]
158	$C_5H_5FeC_5H_4$	Ia	yellow, no photoconductivity and no photovoltage IR (polyethylene; measured 100 to 400): 132, 145, 154, 166 (all Cu–Cu), 240, 275, 310, 365 IR (KBr): 1935w and 2000 (both C≡C)	[129, 169, 202, 203, 228]	7e, 7h, 7k, 7l, 7n, 8a, 14

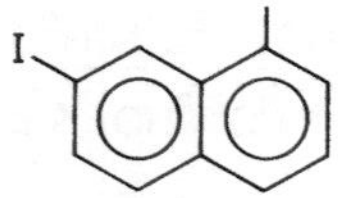

References on pp. 57/67

No.		Method	Description	Ref.	
159	$C_5H_5RuC_5H_4$	1a	—	[168]	2a [168], 7l
160	$2,4\text{-}(CH_3O)_2(3\text{-}CH_3CO_2)C_6H_2$	1a (88)	—	[264]	8a
161	$4\text{-}C_2H_5O_2CC_6H_4NHCH_2$	1a	—	[255]	12
162	$3\text{-}i\text{-}C_3H_7(4\text{-}CH_3O)C_6H_3$	1a	orange	[239]	8a
163	$CH_3O_2C(CH_2)_8$	1a	yellow-green Reaction Type 5: crude product saponified to (E)-$HO_2C(CH_2)_8CH=CHC\equiv C(CH_2)_8CO_2H$	[73, 92]	2a [73], 2c [86], 5 (75%) [92]
164	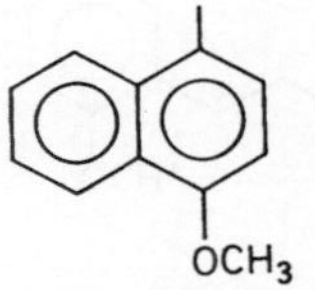 OCH₃ structure	1a (63)	—	[263]	8a
165	$n\text{-}C_9H_{19}C(CH_3)(OH)$	*1a	—	[53]	2a [53]
166	$C_6H_5(C\equiv C)_3$	*1a	—	[74]	2a [74]
*167	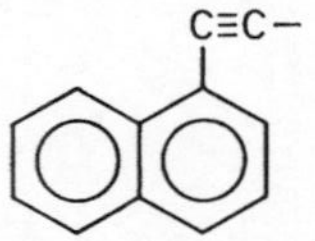 $C\equiv C-$ structure	1a	m.p. 193° (dec.) thermal analysis shows one exothermic peak at 193° IR (KBr): 2165 (C≡C)	[205, 221]	1a [205, 221], 7d
168	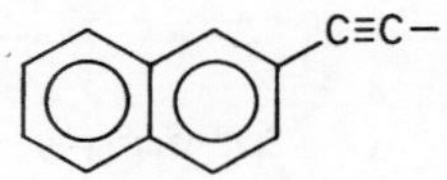 $C\equiv C-$ structure	1a	preparation like No. 167 (Method I a); also from No. 154 and BrC≡CH at 0° in dimethylformamide m.p. 183° (dec.) IR (KBr): 2280 (C≡C)	[205, 221]	1a [205, 221]
169	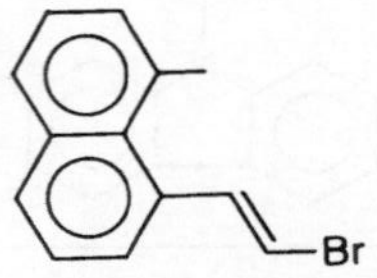 structure (Br)	1a	bright yellow	[173]	7k
170	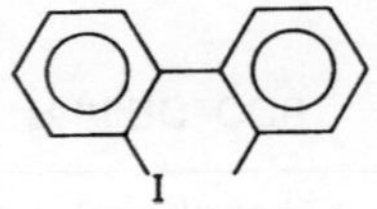 structure (I)	1a	—	[169, 209]	2a [209], 8d

References on pp. 57/67

Table 1 [continued]

No.	R in RC≡CCu	preparation method (yield in %)	properties and remarks	Ref.	feasible reactions (see pp. 12/4)
171	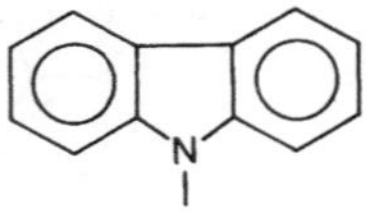	not given	m.p. 272° (dec.) resistivity at 25° 6.7 × 10^7 Ω · cm; photoconducting IR (KBr): 1985, 2010sh (both C≡C)	[191, 220, 221]	—
172	4-$C_6H_5C_6H_4$	not given	visible/UV: 468 (band gap = 2.37 eV) diagram of photoconduction kinetics given IR (KBr): 1940 (C≡C)	[246, 278, 283]	7 m
*173	$(C_6H_5)_2As$	Id (60)	Preparation Method Id with CuCl in dimethylsulfoxide; preparation by Method Ia is not possible colored, does not melt IR (no media given): 1950 or 1999 (both for C≡C) nonexplosive, insoluble in organic solvents	[166, 195, 283]	—
*174	$(C_6H_5)_2P$	not given	colored, does not melt when heated IR (KBr): 1999 (C≡C) nonexplosive, insoluble in organic solvents	[195, 283]	—
175	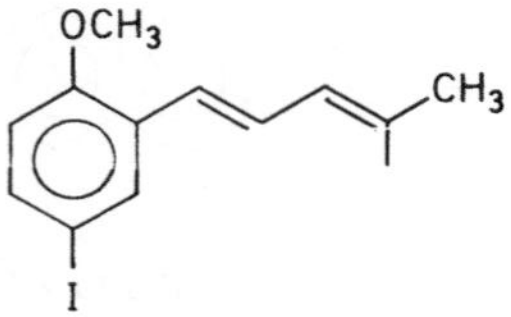	Ia (72)	yellowish orange	[223]	8 d
176	n-$C_{12}H_{25}$ (?)	*IIb	possibly only the LiI adduct is prepared and reacted yellow	[229]	11 (35%)

References on pp. 57/67

No.					
177	(fluorene, 9-CH3, 9-OH)	Ia	—	[80]	7 d
178	(C$_6$H$_5$)$_2$C(OH)	Ia	—	[80]	7 d
179	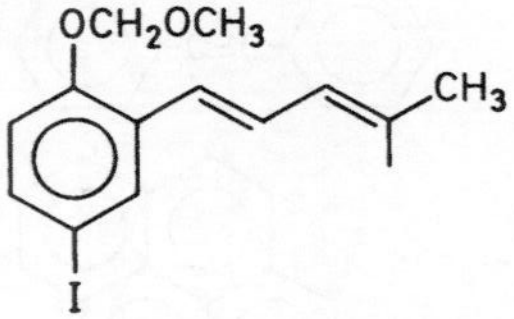 OCH$_2$OCH$_3$ (aryl with CH$_3$ side chain, I)	Ia	yellowish orange	[223]	8 d
180	4-(CH$_3$)$_2$NC$_6$H$_4$COCH$_2$N(C$_2$H$_5$)CH$_2$	Ia	—	[255]	12
181	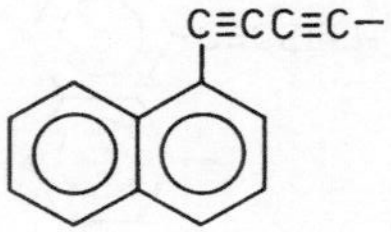 C≡CC≡C– (naphthyl)	Ia	preparation also from No. 167 and BrC≡CH at 0° in dimethylformamide m.p. 165° (dec.) thermal analysis shows one exothermic peak at 165° IR (KBr): 2180 (C≡C)	[205, 221]	1 a [205, 221]
182	4-C$_6$H$_5$C$_6$H$_4$C≡C	Ia(?)	—	[139]	7 m
183	4-C$_6$H$_5$C≡CC$_6$H$_4$	Ia(?)	IR (KBr): 1930, 2220 (both C≡C) IR (polyethylene; measured 100 to 400): ∼150 (Cu–Cu), 195, 214, 273, 293, 308, 368	[202, 203, 228]	—
184	anthr-9-yl	not given	m.p. 322° (dec.) IR (KBr): 1918 or 1920 (C≡C) IR (polyethylene; measured 100 to 400): 123, 170 (Cu–Cu), 210, 276, 304, 357 visible/UV (1 : 100 in MgO): 530	[191, 220, 221, 228, 256, 283]	—
185	4-C$_6$H$_5$C$_6$H$_4$NHCOCH$_2$	Ia	—	[215]	16 b
186	2-CH$_3$O(4-C$_6$H$_5$CH$_2$O)C$_6$H$_3$	Ia (88)	—	[264]	8 a

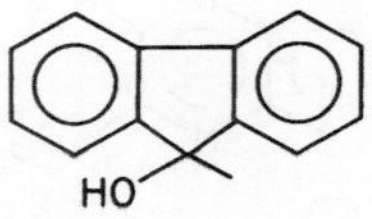

References on pp. 57/67

Table 1 [continued]

No.	R in RC≡CCu	preparation method (yield in %)	properties and remarks	Ref.	feasible reactions (see pp. 12/4)
187	2,4,6-$(C_2H_5)_3C_6H_2$	Ia	yellow, explosive reaction according to Type 2c is not feasible	[30]	—
	9-(tetrahydropyran-2-yloxy)nonyl see No. 198 (supplement)				
*188	$(C_6H_5)_2C(CN)CH_2$	Ic	dark brown, does not melt, insensitive to shock IR (solid): 272 (≡CCu), 320, 358, 375 (all ≡C-C), 476 (C≡N), 500 (C≡C), 535, 691, 752, 760, 800 (ϱCH_2), 912, 923 (both ≡C-C), 1089 (ϱCH_2), 1433 (δCH_2), 1493, 1950 (C≡C), 2238 (C≡N), 2915 ($\nu_s CH_2$), 2960 ($\nu_{as} CH_2$)	[238]	—
189	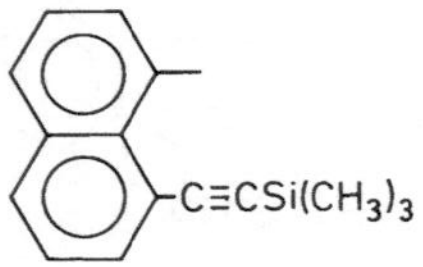	Ib	Preparation Method Ib: alkyne dissolved in C_2H_5OH, CuI suspension in dimethylformamide added, gaseous NH_3 introduced bright yellow	[173]	7k
190	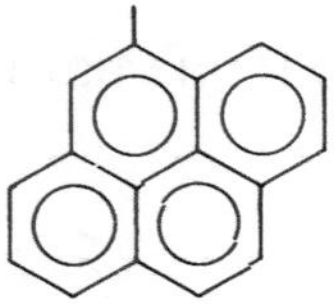	Ia(?)	m.p. 301° (dec.) resistivity at 25° 5.2 × 10⁹ Ω · cm IR (KBr): 1915 (C≡C)	[191, 220, 221]	—
191	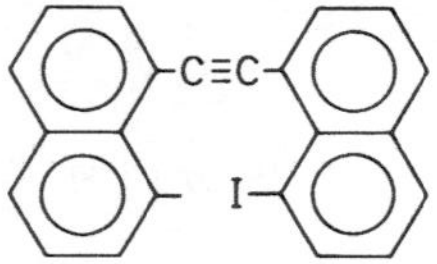	Ia(98)	Preparation Method Ia: with a CuI suspension in dimethylformamide, then aqueous ammonia added deep yellow	[173]	1a (51%) [173], 8d
192	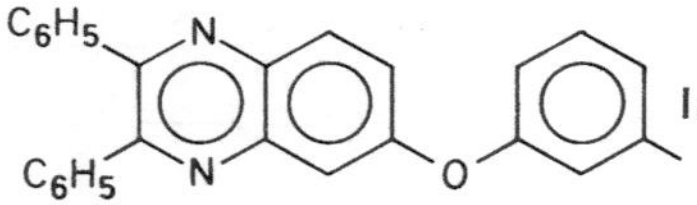	Ia	yellow	[267]	5 [267]

supplement:

No.	R / compound	Method	Remarks	Ref.	Type
193	$1,7\text{-}C_2B_{10}H_{11}$ (1,7-dicarba-closo-dodecaboran(12)-1-yl)	*Ia, VI	94% total yield from sequence: Preparation Method *Ia–Reaction Type 14–Preparation Method VI — yellow, darkening at 108 to 110° (dec. ?); does not melt on heating up to 350° nonexplosive on heating or shock soluble in ether, benzene, CH_3OH, CCl_4 IR: 2020 (C≡C), 2590 to 2640 (BH), 3080 (CH)	[295]	14
194	$2\text{-}BrC_6H_4OCH_2$	Ia	Reaction Type 7b with C_6H_5COCl not possible, only oxidation (Type 2a)	[198, 339]	2a (40%) [339], 8a
195	$4\text{-}BrC_6H_4OCH_2$	Ia	—	[198, 339]	7b, 8a
196	$2,4\text{-}(CH_3O)_2C_6H_3$	Ia (70)	bright yellow–orange	[299]	7i
197	cyclohex-1-enyl	*IIb		[301]	7d
198	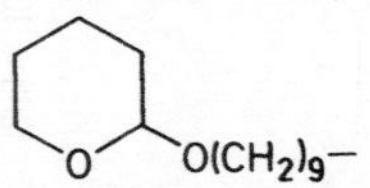 (tetrahydropyranyl)–O(CH₂)₉–	*IIb		[301]	7d
199	CH_3SCH_2	*Id	—	[326]	7c
*200	$CH{\equiv}C$	Id	—	[342]	—
201	$C_2H_5C(CH_3)$	VII	preparation with $2\,CuO_3SCF_3 \cdot C_6H_6$ in CD_3NO_2/CH_2Cl_2	[341]	—
202	$(CH_3)_2C{=}CH(CH_2)_2$	—	no details about Preparation Method I and Reaction Type 1 given	[327]	1 [327]
203	$2,4,6\text{-}Br_3C_6H_2OCH_2$	Ia	—	[339]	7b
204	$4\text{-}CH_3C_6H_4OCH_2$	Ia	—	[339]	7b
205	$2,5\text{-}(CH_3)_2C_6H_3OCH_2$	Ia	—	[339]	7b
206	$(CH_2{=}CHCH_2)_2NC(CH_3)_2$	Ia	—	[335]	8a
207	$4\text{-}O_2NC_6H_4OCH_2$	Ia	—	[351]	7b
*208	$t\text{-}C_4H_9O_2C$	Ia	—	[350]	—

*Further information:

HC≡CCu (Table 1, No. 1) is postulated as an intermediate in the formation of Cu_2C_2 from Cu^+ and C_2H_2 in aqueous media. Various equilibria were assumed, e.g., $HC≡CH+Cu^+ \rightleftharpoons HC≡CCu+H^+$ [55, 104, 260]. HC≡CCu in these solutions is even supposed to be catalytically active in the dimerization and for similar reactions of C_2H_2 [177].

Earlier reports on use of Preparation Method I a give no definitive proof of the formation of HC≡CCu [17, 55]. Its aqueous solutions are claimed to be yellow or colorless [55, 104]. High dilution conditions with a high C_2H_2/Cu^+ ratio were also used [61]. An "HC≡CCu" is reported to give on reaction with HCl at room temperature C_2H_2 and under certain conditions under cooling $2CuCl \cdot C_2H_2$; but there are no details of its preparation [17]. The equilibrium data of [17] lateron were not confirmed [48]. Conditions for preparing a "red HC≡CCu" resemble those for Cu_2C_2 synthesis; no proof of structure is given [95]. In the decomposition reaction of $HO_2CC≡CCu$ to yield Cu_2C_2, C_2H_2, and CO_2 the compound HC≡CCu could be an intermediate (see also No. 5) [36].

CF₃C≡CCu (Table 1, No. 2). On fusion with metallic sodium the fluorine is quantitatively converted to NaF [65]. $CF_3C≡CCu$ is thought to be an intermediate in the reaction of $CF_3C≡CZnCl$ and $CuCl_2$ to form $CF_3(C≡C)_2CF_3$ [127].

CH₃C≡CCu (Table 1, No. 7) is also formed by hydrolysis of $[(CH_3C≡C)_3Cu]K_2$ (see Section 1.1.2.8) and from $[(CH_3C≡C)_2Cu]K$ (see Section 1.1.2.7) and CuI in liquid ammonia [84].

Depending on the preparation and purification methods, the compound varies much in these ease of detonation. Storage under water is recommended [133]. IR and Raman band positions are listed in the table; the spectra are reproduced in **Fig. 4**.

$CH_3C≡CCu$ is not affected by pyridine, bipyridyl, o-phenanthroline, and pure $N(C_2H_5)_3$ [98]. The formation of a CO adduct from $CH_3C≡CCu$ and CO in liquid ammonia is conjectured [101]. $CH_3C≡CCu$ does not undergo the so-called "Straus Reaction" (Reaction Type 5) to give $CH_3C≡CCH=CHCH_3$. The only product was $CH_3(C≡C)_2CH_3$ [51].

HOCH₂C≡CCu (Table 1, No. 8). When Cu, in the presence of air, is immersed in an aqueous solution of $HOCH_2C≡CH$ and $CH_2=CHCH_2NH_2$ (both 0.2 M), a visible yellow film of $HOCH_2C≡CCu$ is formed within 2 min. After 10 min a thick film is obtained which is smooth

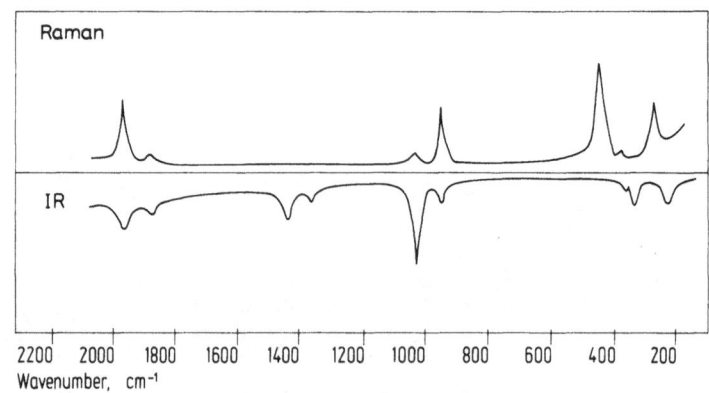

Fig. 4.   The Raman and infrared absorption spectra of $CH_3C≡CCu$ [236].

References on pp. 57/67

and fairly adherent. When cyclic voltammetry is carried out using an aqueous solution of $HOCH_2C\equiv CH$, $CH_2=CHCH_2NH_2$ and KOH (all 0.1 M), the formation of $HOCH_2C\equiv CCu$ at anodic potentials is so rapid, that it is deposited as a nonadhering yellow powder [343].

The IR spectrum in polyethylene was measured from 100 to 400 cm^{-1}. The Cu-Cu frequencies are given as 135 and 168 cm^{-1} [228]. Electrochemically grown layers of $HOCH_2C\equiv CCu$ on metallic Cu give the following Raman reflections (Ar or Kr laser): 244, 312 ($C\equiv C$-C), 401, 443 (Cu-C), 588 ($\equiv$C-C-O), 931 ($\equiv$C-C), 1920 ($C\equiv C$), all in cm^{-1} [343].

Certain reactions attack the molecule at the oxygen atom. I reacts with $HOCH_2C\equiv CCu$ to form 21% II with 5% III admixed [252, 275]. A quantitative analysis of the alkyne content in impure $HOCH_2C\equiv CH$ is possible by reaction with CuCl/pyridine (Preparation Method Id) and subsequent titration of the resulting pyridinium ions; see p. 5. The reproducibility is ±2% [109].

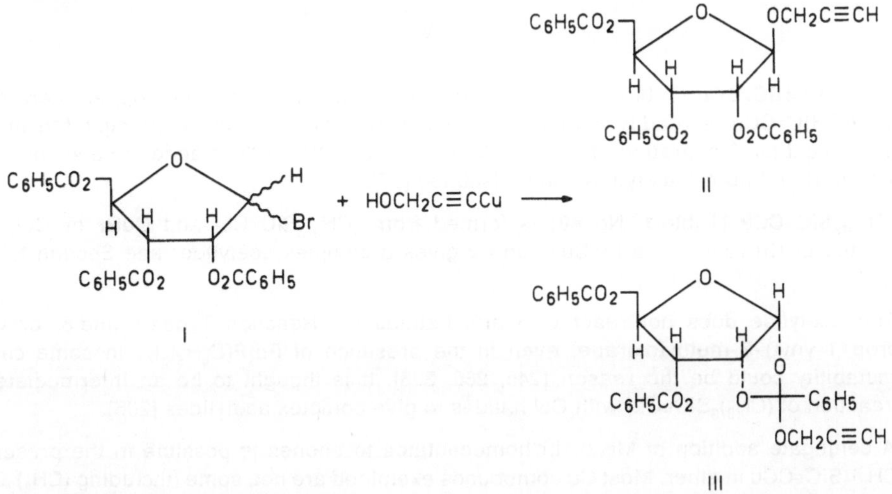

$$ I + HOCH_2C\equiv CCu \longrightarrow $$

CCl$_2$=CClC≡CCu (Table 1, No. 9). The reaction of $CCl_2=CClC\equiv CCu$ and Br_2 in CCl_4 at −10 °C yields besides CuBr 37% $CCl_2=CClC\equiv CBr$ [193, 235], 20% $CCl_2=CClCBr=CBr_2$, and 20% $(CCl_2=CClC\equiv C)_2$. The analogous reaction with I_2 in CCl_4/ether gives only 82% $CCl_2=CClC\equiv CCl$. $(CCl_2=CClC\equiv C)_2$ can also be obtained in 40% yield by oxidation with $K_3[Fe(CN)_6]$. The reaction with 3 N HCl at 50 °C yields 26% $CCl_2=CClC\equiv CH$ [235].

CH$_2$=CHC≡CCu (Table 1, No. 11). In preparation by Method Id $CH_2=CHC\equiv CCu$ precipitates over the range of 0.005 to 0.3 N H_2SO_4 [296]. Photoconductivity spectra were measured between 410 and 590 nm [202], 400 and 700 nm [203]. The long wave maximum was found at 465 nm [203, 231, 283]. These spectra give a forbidden band gap of 2.41 eV [203, 231, 283], the visible/UV data one of 2.5 eV [231, 283]; the direct transition energy is 2.53 eV [231].

The electronic spectra yield the following values for the $Cu\,2p_{3/2}$ level and (in parentheses) the C1s level (relative to Fermi levels in Cu or in "free hydrocarbons", respectively): Binding energy 934.4 eV (286.7 eV); half-width of the electron lines 2.2 eV (3.1 eV), chemical shift 2.2 eV (1.7 eV). The charge distribution is discussed [227]. For "mixed acetylides" with $C_6H_5C\equiv CCu$ see Section 1.1.2.2.

C$_2$H$_5$SC≡CCu (Table 1, No. 20). The photoconductivity spectrum was measured from 390 to 650 nm. Because of the presence of sulfur in the molecule a supplemental "intermolecular p_π-d$_\pi$ interaction" was hypothesized. A photoconductivity spectrum of $C_2H_5SC\equiv CCu$ in the

References on pp. 57/67

presence of kryptocyanine and the illuminance/electrical current characteristics were also measured [278].

n-$C_3H_7C\equiv CCu$ (Table 1, No. **33**). The reaction of 2-$IC_6H_4(CH_2)_nZOH$ to give IV is strongly catalyzed by the n-$C_3H_7C\equiv CCu$ polymer (Reaction Type 13). Monomeric Cu^I or low molecular weight aggregates of n-$C_3H_7C\equiv CCu$ in solution are not capable of rapid catalysis. The polymerically bound Cu seems to bridge the overly long distances between I and OH in 2-$IC_6H_4(CH_2)_nZOH$ to form IV. Formally this reaction can be understood as a hydrolysis of n-$C_3H_7C\equiv CCu$ to n-$C_3H_7C\equiv CH$ by HI [165].

$(Z = CH_2, CO)$

IV

$C_2H_5CH(OH)C\equiv CCu$ (Table 1, No. **35**). Like some other OH-containing copper acetylides, $C_2H_5CH(OH)C\equiv CCu$ is readily soluble in aqueous NH_3 and cannot be precipitated in the manner usual for Preparation Method Ia. When $C_2H_5CH(OH)C\equiv CH$ is added to a suspension of CuCl in H_2O, "pure" acetylide can be isolated [27].

$(CH_3)_3SiC\equiv CCu$ (Table 1, No. **40**) is formed from $(CH_3)_3SiC\equiv CLi$ and CuBr or CuI, but only if the Li:Cu ratio = 1; a Li:Cu ratio = 2 gives a complex acetylide; see Section 1.1.2.7 [325].

The acetylide does not react with aryl halides (cf. Reaction Types 7 and 8) or with 2-(prop-1-ynyl)-2-methyloxirane, even in the presence of $Pd[P(C_6H_5)_3]_4$; in some cases its instability could be the reason [245, 266, 325]. It is thought to be an intermediate in the reaction of $(CH_3)_3SiC\equiv CLi$ with Cu^I halides to give complex acetylides [206].

A conjugate addition of Mg or Li homoenolates to enones is possible in the presence of $(CH_3)_3SiC\equiv CCu$ in ether. Most Cu compounds examined are not, some (including $(CH_3)_3CC\equiv CCu$) are only slightly effective. The resulting 1,6-diones or their enolic ethers are postulated to be formed in a transmetallation reaction involving the copper acetylide, but no further copper species are mentioned [347]. Possibly the active species is the complex $(CH_3)_3C\equiv CCu \cdot S(CH_3)_2$; see Section 1.1.2.4 [358].

n-$C_4H_9C\equiv CCu$ (Table 1, No. **54**). For the reaction Cu(cryst) + 6C(graphite) + 4.5 H_2(g) → n-$C_4H_9C\equiv CCu$(cryst) thermodynamic parameters are calculated from calorimetric measurements (T = 298.15 K and p = 101.325 kPa): $\Delta H_{f298} = 73.2 \pm 3.5$ kJ · mol^{-1}, $\Delta G_{f298} = 214.6 \pm 3.7$ kJ · mol^{-1}, and $\Delta S_{f298} = -474.3 \pm 0.7$ J · mol^{-1} · K^{-1}. The specific heat C_p of the crystalline material in the range from 14 to 330 K was measured with an adiabatic vacuum calorimeter [287]; values (partly extrapolated) are:

temperature in K	C_p in J · mol^{-1} · K^{-1}	temperature in K	C_p in J · mol^{-1} · K^{-1}
5	0.1920	100	72.66
10	1.554	150	100.0
15	4.032	200	121.8
25	10.89	273.15	150.9
40	24.26	298.15	160.6
60	42.77	330	175.8

References on pp. 57/67

From electronic spectra the following values were obtained for the $Cu2p_{3/2}$ level and (in parentheses) the C1s level (relative to Fermi levels in Cu or in "free hydrocarbons", respectively): binding energy 934.3 eV (286.2 eV); half width of the electron lines 2.6 eV (3.3 eV); chemical shift 2.1 eV (1.2 eV). The charge distribution is discussed in the paper [227].

The photoconductivity spectrum (400 to 1000 nm) is known [202], $\lambda_{max}=440$ [231]. The band gap is 2.72 eV. The relative positions of the energy levels are given [248, 283]. Photoconductivity spectra are also known with sensitization by cationic dyestuffs like pinacyanol, kryptocyanine, or methylene blue or by anionic dyestuffs. Maximum sensitization (conductivity increased by 65%) has been obtained with erythrosine [283].

The visible/UV spectrum (348 to 800 nm, MgO) [231] has a maximum at 420 nm [231, 283]. The band gap is 2.73 eV, the direct transition energy 2.75 eV [231, 248], the photoelectric work function 5.25 eV [248, 257]. Relations between decomposition threshold and band gap [257] are discussed on pp. 6/8.

$n-C_4H_9C{\equiv}CCu$ is less stable than $C_6H_5C{\equiv}CCu$, but more stable than $n-C_4H_9C{\equiv}CAg$. Thermal decomposition involves a considerable liberation of heat. The products evidently result from secondary reactions of the hex-1-ynyl radical. At 200 °C the gaseous decomposition products contain 6% CH_4, 7% C_2H_6, 17% C_2H_4, traces of propane, 10% propene, 27% butane, and 33% butenes [261]. The compound can be incinerated without explosion for Cu determination [83].

The thermogram (one endo effect above 140 °C, one exo effect above 160 °C) indicates the one-stage decomposition of the aggregated oligomeric product [261]. The reaction with $2,4,6-(O_2N)_3C_6H_3$ in pyridine at room temperature yields a red solution with a visible spectrum very similar to that of the Meisenheimer complex formed from $C_6H_5C{\equiv}CCu$ (see p. 54), but after treatment with acids no solid products have been isolated [196].

$(CH_3)_3CC{\equiv}CCu$ (Table 1, No. 55). The yellow or red (or orange [185, 218]) color variations could be a consequence of aggregation-disaggregation processes, which have not been studied subsequently and about which contradictory data were published. At 80 °C [51] or on crystallization from benzene the yellow acetylide converts into a red form. Crystallization of this red form from $CHCl_3$ gives the yellow species again. The data about the dependence of color on the degree of association differ: The yellow form is said to have a higher molecular weight than the red form (cryoscopically in $CHCl_3$) [34]. The color is said to become deeper with higher association; the monomer is colorless, the dimer yellow-green, and the octamer orange-red. The benzene solution contains $[(CH_3)_3CC{\equiv}CCu]_8$ from cryoscopic and ebullioscopic measurements [102, 103]. The ether solution shows a reversible color change from red to yellow on cooling to −78 °C [237]. Evaporation of the orange pentane solution yields a liquid, which crystallizes from ether at −49 °C to give an orange precipitate [34, 218]. On storing solutions, the molecular weights fall gradually until they nearly reach the unimolecular value; some decomposition occurs to a gummy solid (finely divided copper?) [34]. Preparation according to Method Ia yields the two forms as a conglomerate, m.p. ~140 °C [135], with nonexplosive decomposition between 80 and 150 °C to form a copper mirror and a sublimate [34, 51]. The compound is soluble in ether, pentane [218], and tetrahydrofuran [271], less soluble in benzene than in ether (both forms) [34]. ^1H NMR (CCl_4): $\delta=1.37$ ppm (s) [218]. Visible/UV (heptane): points of inflection at 235 ($\varepsilon=5350$) and 316 nm (2680) with gradually diminishing absorption out to ca. 600 nm [218]. IR (KBr; C_2Cl_4): 2000 (C≡C) cm^{-1} [203, 218]. "C≡C absent" is stated in [102, 103].

$4-IC_6H_4C{\equiv}CCu$ (Table 1, No. 86) is used for electrophotographic layers (sensitivity from UV to 600 nm, enhanced after sensitization with dyes) [130]. A comparison of photoconduc-

References on pp. 57/67

tivity and photovoltage with those of other RC≡CCu has been published, but no quantitative values are given [256].

4–IC₆H₄C≡CCu reacts with $(CN)_2C=C(CN)_2$ in tetrahydrofuran at 55 °C to form polymers (probably resulting from $(CN)_2C=CCN$ radicals) and $4–IC_6H_4(C≡C)_2C_6H_4I–4$ [170].

2–O₂NC₆H₄C≡CCu (Table 1, No. **87**). On refluxing with 5–iodouracil or 2–iodo–3–hydroxy-pyridine in dry pyridine the indogenide V is formed. Both of the added compounds seem to be only catalysts. The same reaction with 2–bromopyridine yields the anhydroisatin α–anthranilide VI. With 2–IC₆H₄CO₂H a mixture of V and VI is produced. Proposals for the mechanisms of the formation of V and VI are given. The reaction with 2–IC₆H₄CONH₂ gives a mixture of V, VII (yield 6%), and VIII (yield 66%). The intermediate in the formation of VII and VIII is obviously the expected 2–O₂NC₆H₄C≡CC₆H₄CONH₂–2. The hydration of its triple bond can result in two different products. If the intermediate species 2–O₂NC₆H₄–CH₂COC₆H₄CONH₂–2 is formed, a subsequent condensation yields VIII; if 2–O₂NC₆H₄COCH₂–C₆H₄CONH₂–2 is formed, a subsequent condensation gives VII. The reaction of No. 87 with 2–phenylisatogen (IXa) in boiling pyridine yields 55% Xa. The analogous reaction with 2–(4–nitrophenyl)isatogen (IXb) gives 93% Xb. The cinnoline structure of Xa and Xb is derived from spectroscopic data [162].

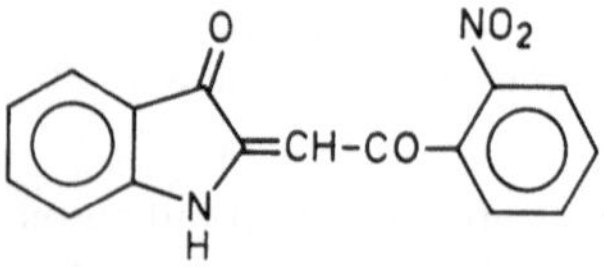

V

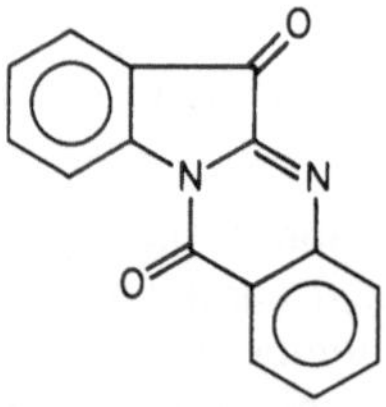

VI

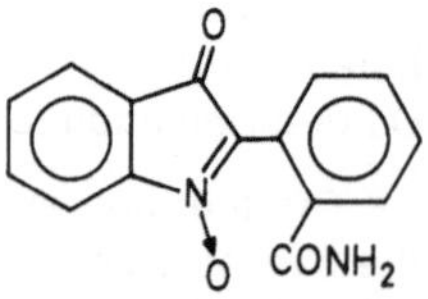

VII

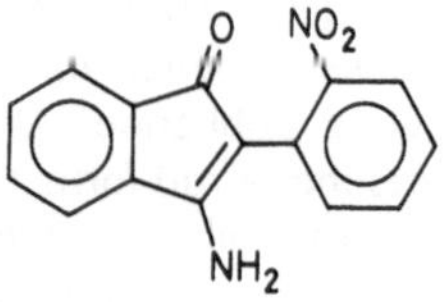

VIII

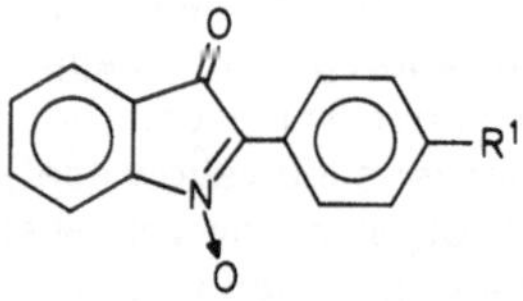

IX

a: R¹ = H

b: R¹ = NO₂

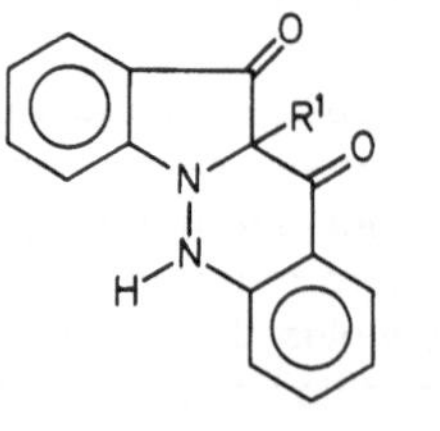

X

a: R¹ = C₆H₅

b: R¹ = 4–NO₂C₆H₄

C₆H₅C≡CCu (Table 1, No. 90). **Preparation.** Best method is Ia in C₂H₅OH/H₂O with 99% yield from Cuᴵ prepared by preceding reaction of CuSO₄·5H₂O+[NH₃OH]Cl [207], quantitative from CuI [123], 85% from CuCl [184]. The analytical use of this reaction is treated on p. 5.

For "mixed acetylides" with Cu_2C_2, $CH_2=CHC\equiv CCu$, or $2\text{-}O_2NC_6H_4Cu$, prepared according to Method Ia, see Section 1.1.2.2.

Further feasible preparation methods are Ib (in C_2H_5OH [224] or in a mixture of C_2H_5OH and dimethylformamide, yield 92% [173]) and Id (in hexamethylphosphoric triamide in the presence of a little Na_2SO_3 [164]). The H^+ ions formed cannot only be bonded by the triamide but also by K_2CO_3 and pyridine. This is a very simple preparation method, but the product is contaminated by inorganic salts (yield 30%). Nevertheless, this mixture can be advanta-geously used for certain further reactions [279]. The reaction of $C_6H_5C\equiv CH$ and CuI in $HN(C_2H_5)_2$ in the presence of $Pd[P(C_6H_5)_3]_2Cl_2$ yields a slurry of $C_6H_5C\equiv CCu$, which is without isolation further reacted [284]. In the course of the oxidative coupling reaction of $C_6H_5C\equiv CH$ to $C_6H_5(C\equiv C)_2C_6H_5$ with O_2/i-propanol/tetramethylenediamine/CuCl some $C_6H_5C\equiv CCu$ is pre-cipitated [105]. Method Ie [210, 234] and Method If [131] afford a 97 or 80 to 90% yield of $C_6H_5C\equiv CCu$.

Method II has not been used to prepare $C_6H_5C\equiv CCu$, but the similar reaction of the boron complex $NH_4[B(C\equiv CC_6H_5)_4]$ and ammoniacal Cu^I solution (H_2O/C_2H_5OH) yields within a few minutes 98% of the acetylide [85].

Method IV: $C_6H_5C\equiv CBr$ in C_2H_5OH and an ammoniacal solution of Cu^I salt react to form $C_6H_5C\equiv CCu$ in good yield [18].

Method V using $C_6H_5C\equiv CCO_2Cu^I$ goes well at 35 °C in dimethylformamide [249] and as an in situ preparation [229]. $(C_6H_5C\equiv CCO_2)_2Cu^{II}$ is converted to $C_6H_5C\equiv CCu$ by water vapor at 100 °C [21].

Method VI: $C_6H_5C\equiv CCu \cdot NH_3$ decomposes to $C_6H_5C\equiv CCu$ in a N_2 stream at room tempera-ture [84]. An analogous decomposition occurs when $C_6H_5C\equiv CCu \cdot P(C_2H_5)_3$ stands for 1 to 2 d in air [102, 103]. $2C_6H_5C\equiv CCu \cdot HgBr_2$ and KI react in H_2O/C_2H_5OH to form $C_6H_5C\equiv CCu$ [158]. $(4\text{-}CH_3C_6H_4NC)_3CuC\equiv CC_6H_5$ (see Section 1.1.4.3) decomposes to $C_6H_5C\equiv CCu$ on pro-longed washing with ether. The same product is obtained by treatment of this complex with pyridine/water (evolution of CO) or with di-p-tolylformamidine [68].

Method VII: $C_6H_5C\equiv CCu$ is isolated from the yellow decomposition product of $C_6H_5\text{-}C\equiv CHCuCl$ [29]. In contrast to other complexes of its type the paramagnetic complex XI is not very stable and a transformation occurs to $C_6H_5C\equiv CCu$ [273].

XI

Method VIII: $[(C_6H_5C\equiv C)_3Cu^0]_2Ba_3$ reacts with even traces of O_2 to form $C_6H_5C\equiv CCu$ and $[(C_6H_5C\equiv C)_3Cu^I]Ba$ [108]. Also $[(C_6H_5C\equiv C)_2Cu]Na \cdot 2NH_3$ and CuI in liquid ammonia react to form $C_6H_5C\equiv CCu \cdot NH_3$; and $[(C_6H_5C\equiv C)_2Cu^I]_2[Cu^{II}(NH_3)_4] \cdot xNH_3$ in liquid ammonia yields on evaporation $C_6H_5C\equiv CCu$ and $C_6H_5(C\equiv C)_2C_6H_5$. $[(C_6H_5C\equiv C)_2Cu][N(P(C_6H_5)_3)_2]$ reacts with $(C_6H_5)_3P \cdot AuCl$ in CH_2Cl_2 to form $C_6H_5C\equiv CCu$, $[(C_6H_5C\equiv C)_2Au][N(P(C_6H_5)_3)_2]$, and $C_6H_5C\equiv CCu \cdot P(C_6H_5)_3$. With $[C_6H_5C\equiv CAuCl][N(P(C_6H_5)_3)_2]$ only $C_6H_5C\equiv CCu$ and $[(C_6H_5C\equiv C)_2Au]\text{-}[N(P(C_6H_5)_3)_2]$ are formed. The reaction of $[(C_6H_5C\equiv C)_2Cu][N(P(C_6H_5)_3)_2]$ and $C_6H_5C\equiv CAu$ gives $C_6H_5C\equiv CCu$ and $[(C_6H_5C\equiv C)_2Au][N(P(C_6H_5)_3)_2]$ [336]. The compounds $[(C_6H_5C\equiv C)_3Cu]Li_2$

References on pp. 57/67

and $(CH_3)_2NX$ $(X=(C_6H_5)_2PO$ or $CH_3SO_3)$ give at $-20\,°C$ $C_6H_5C\equiv CN(CH_3)_2$ and $C_6H_5C\equiv CCu$ [316]. The complex $[(C_6H_5C\equiv C)_2Cu^I]_2[Cu^{II}(H_2N(CH_2)_2NH_2)_2]\cdot 2NH_3$ yields with water $C_6H_5C\equiv CCu$, on prolonged evacuation in high vacuum a mixture of $C_6H_5C\equiv CCu$ and $C_6H_5(C\equiv C)_2C_6H_5$. Water also hydrolyzes $[(C_6H_5C\equiv C)_2Cu^I]_2[Cu^{II}(H_2N(CH_2)_2NH_2)_2]\cdot 12NH_3$ to $C_6H_5C\equiv CCu$ [84]. Further methods: The hydrolysis of n-$C_4H_9(C_6H_5C\equiv C)CuLi$ (see "Organocopper Compounds" 2, 1983, pp. 178 and 185) with water yields (with gas evolution) up to 90% $C_6H_5C\equiv CCu$. The corresponding magnesium compounds $C_2H_5(C_6H_5C\equiv C)CuMgBr$ and n-$C_3H_7(C_6H_5C\equiv C)$-$CuMgBr$ react with water in tetrahydrofuran to form $C_6H_5C\equiv CCu$ plus a gaseous and a liquid byproduct, whereas with $CH_3(C_6H_5C\equiv C)CuMgCl$ and with $C_6H_5(C_6H_5C\equiv C)CuMgBr$ no $C_6H_5C\equiv CCu$ was isolated [180].

$C_6H_5C\equiv CCu$ can also be produced from copper metal by electrochemical corrosion in the presence of $C_6H_5C\equiv CH$. In dry acetone or acetonitrile containing ca. 0.05% $[N(C_2H_5)_4]ClO_4$ and 4% $C_6H_5C\equiv CH$ using a platinum cathode (20 V and up to 20 mA for 1 h) $C_6H_5C\equiv CCu$ can be collected as a powdery yellow solid and is pure after washing with acetone and petroleum ether. Based on the loss of Cu from the anode the chemical yield is quantitative and per Faraday 1.06 mol of dissolved Cu was determined [345].

Smooth, well-adhering films of $C_6H_5C\equiv CCu$ can be obtained on Cu surfaces using the electrochemical corrosion method. This is of interest for photocurrent spectroscopic investigations. There are different ways to grow adherent films. The copper potential in a $H_2O/$ CH_3OH (1:1) solution of 0.1 M KF, 0.44 M $CH_2=CHCH_2NH_2$, and 9×10^{-3} M $C_6H_5C\equiv CH$ can be cycled between -0.9 and $+0.1$ V (versus saturated calomel electrode). The thickness of the golden films of $C_6H_5C\equiv CCu$ can be varied by removing the electrode from the growth medium after a different number of cycles in each case. Open circuit corrosion of a Cu electrode using the same, but air-saturated solution, also gives well-adhering films. In this case, some control of the film thickness can be achieved by varying the immersion time. With both methods, the film thicknesses were typically in the range of 0.1 to 1.0 µm [343, 344].

A linear relationship between the net anodic charge consumed and the yield of $C_6H_5C\equiv CCu$ (coulombic yield of 50 to 65%) was observed [343]. NH_3 instead of $CH_2=CHCH_2NH_2$ leads to a rapid corrosion of Cu and to powdery, poorly adhering $C_6H_5C\equiv CCu$ [343, 344]. Addition of KOH to the solution makes oxide formation predominant [343].

Thermochemical Data of Formation. From calorimetric measurements the functions $H_T^°-H_0^°$, $S_T^°$, and $G_T^°-H_0^°$ were calculated for the range 0 to 330 K. The standard thermodynamic parameters of $C_6H_5C\equiv CCu$ are $\Delta H_{f298}^° = 296\pm2.1$ kJ/mol, $\Delta G_{f298}^° = 365\pm2.3$ kJ/mol, $\Delta S_{f298}^° = -231\pm0.7$ J·mol^{-1}·K^{-1}, defined for the reaction

$$Cu\,(cryst)+8C\,(graphite)+2.5H_2\,(gas)\ \rightarrow\ C_6H_5C\equiv CCu\,(cryst)\ [281].$$

The **molar heat capacity** at constant pressure C_p of crystalline $C_6H_5C\equiv CCu$ increases smoothly from 11 to 330 K, measured with an adiabatic vacuum calorimeter. A table of the values between 5 K (extrapolated) and 330 K is given. At 298.15 K and 1 atm $C_p^° = 154.0$ J·mol^{-1}·K^{-1} [281].

Photoelectric Properties, Conductivity (see also pp. 6/8). The dark resistivity at 25 °C is reported as $2.3\times10^{11}\,\Omega\cdot$cm (in air?) [220, 221] and as 10^9 to $10^{10}\,\Omega\cdot$cm in air with an increase up to 1.5-fold in vacuum [107].

Of all $RC\equiv CCu$ examined (see pp. 6/8), No. 90 was the acetylide with the highest photosensitivity [141]. The dependence on donor and acceptor properties of substituents introduced into the phenyl ring is treated by [221, 248]. The photosensitivity of No. 90 is generally lower than that of compounds with $2C\equiv CCu$ groups in the monomeric molecule, e.g., of

4-CuC≡CC$_6$H$_4$C≡CCu [141]. C$_6$H$_5$C≡CCu is claimed to be an intrinsic semiconductor with a positive dominant photocarrier [107, 155, 256]. On the other hand, the photocurrent response and the decay in the dark are found to be very slow (photocarriers deeply trapped in the polymer) [191].

After irradiation in O$_2$ or in air with 500 nm light C$_6$H$_5$C≡CCu shows after 2 min a resistivity stationary at 10^7 to 10^8 Ω · cm. In O$_2$ the resistivity is lower and it increases more slowly than in vacuum [221].

Contradictory results for photoconductivity spectra are obtained when measured in dry air or in vacuum [107, 202, 220, 221, 256, 278, 319]. Okamoto et al. [220, 221] give quite different curves (λ_{max} in dry air 432 and 525 nm, in vacuum 520 nm). Myl'nikov et al. [107] have found in air and in vacuum identical values (430 and 500 nm). Experimental conditions are not adequately specified for many reported maxima mentioned: 500 nm [231, 246, 283], 430 and 500 nm [107, 256, 278], and 450, 495, and 510 nm [202]; the presumption is that these were determined in air.

The high frequency photoconductivity was measured at 9.6 GHz (klystron) under visible/UV illumination (W-lamp; modulated at 150 Hz). In contrast to most organic dye stuffs the photoconductivity decreases continuously with a deactivation energy of 0.06 eV when the temperature rises from −90 to +20 °C [119].

In vacuum and in air the maxima of the photo-emf excitation are at 385 and 460 nm [107], 390 and 455 nm [155]. The photo-emf itself is in vacuum 3 to 5 times higher than in air. The explanation of this fact refers to the pumping off of traces of water [107], but this is doubtful, for dry O$_2$ also reversibly depresses the photovoltage [155]. Water vapor is reported to increase the photo-emf of C$_6$H$_5$C≡CCu (oxygen excluded?) [120]. Photovoltaic excitation spectra are also given for samples with preirradiation, e.g., constant irradiation with 365 nm light. The results are inconsistent [107, 125, 155]. The Dember effect was estimated with light chopped at 300 Hz [174].

The photodielectric effect in C$_6$H$_5$C≡CCu was studied at 10 GHz using hermetically sealed water-free samples. In contrast to most other semiconductors no illumination-induced change in the permittivity was found [321]. The photoelectric work function is reported as φ = 4.9 eV [140, 155], 5.24 eV [248, 257]. The direct transition energy of C$_6$H$_5$C≡CCu is 2.5±0.03 eV [231]. The forbidden band width was determined from the photoconductivity spectra as E$_g$ = 2.38±0.02 [187, 203, 221, 231, 246, 283] and 2.39 [257], and from the absorption spectra as E$_g$ = 2.4±0.05 [231, 248] and 2.45 eV [246, 278, 283]. The electronic spectrum of Cu-2p$_{3/2}$ in C$_6$H$_5$C≡CCu excited by MgKα radiation yields a binding energy E$_b$ = 934.2, an electronic line half-width of ΔE$_{1/2}$ = 2.3, and a chemical shift ΔE$_b$ = 2.0 eV, i.e., the binding energy is higher in the molecule than in Cu metal. The chemical shift is defined as the difference of the binding energies of the electrons between the molecule and the Fermi levels of Cu metal or of C in "hydrocarbons", respectively. The values for C-1s (no fine structure) are E$_b$ = 286.1, ΔE$_{1/2}$ = 3.0, and ΔE$_b$ = 1.1 eV vs. the hydrocarbon. The width of the Cu-2p$_{3/2}$ lines confirms the existence in the crystal of copper atoms with different effective charges [227].

The inner photoeffect of C$_6$H$_5$C≡CCu can be spectrally sensitized up to the near-infrared by absorption of cyanines [117, 154, 174, 283] or of dyes of the triphenylmethane [116, 126, 154, 283] or porphine [116, 117, 155] types. The photoconductivity and the photo-emf can be either decreased or increased [116, 117]. The resulting photo-emf often resemble the dye spectra [117, 154]. The effects of quinone, chloranile, Hg vapor [155], nitrocellulose [221], polyvinylcarbazole [130, 141, 319], RNA, DNA, alanine, adenine, thiamine [143, 174], and quinine [154] on the photoconductivity (and its relaxation time) and on the photovoltage have also been studied and discussed. No charge-transfer effects between $C_6H_5C{\equiv}CCu$ and polyvinylcarbazole have been detected [319].

Photocurrent spectra of $C_6H_5C{\equiv}CCu$ films on Cu in aqueous 0.1 M solutions of either $Na_2B_4O_7$ or LiOH in the absence of O_2 were also measured. The monochromatic photocurrent conversion efficiency at 460 nm exceeds 10^{-2}. The cathodic photocurrents are increased by the presence of O_2. Possible mechanisms for energy and charge transfer in the $C_6H_5C{\equiv}CCu$ films are discussed [344].

$C_6H_5C{\equiv}CCu$ is useful in preparation of electrophotographic layers which charge well both positively and negatively [130, 141]. The insolubility of $C_6H_5C{\equiv}CCu$ permits preparation of high quality films with polyvinylbutyral and polyvinylcarbazole (sensitivity up to 600 nm) [130, 322, 323]. The light sensitivity of 1:1 mixtures has been measured [323]. Printing carriers have been prepared with 5% $C_6H_5C{\equiv}CCu$ in poly-N-vinylcarbazole [322]. Embedding of $C_6H_5C{\equiv}CCu$ in polyethylene and in methyl methacrylate is treated by [141]. The inherent sensitivity of the copper phenylacetylide layers is to be sure only 1/5 of that of the selenium layers, but the acetylide layers work at higher temperatures. At the photosensitivity maxima the two layers are comparable [130].

Optical and Spectroscopic Properties. IR bands in KBr of more than 20% absorbance are: 266s, 366s (C-H), 515s (C-C≡C), 525s, 685s (both C-C), 745vs, 915m, 1026m, 1070m (all C-H), 1440s, 1481s, 1594m (all C-C), 1930 (C≡C), all in cm^{-1} [216]. A very similar spectrum in KBr, reported later, does not show the band at 1440 cm^{-1} [236]. In KBr [102, 103, 203] the C≡C vibration was found at 1933 cm^{-1}, in Nujol at 1930 [171] and 1933 cm^{-1} [221]. Several further authors [158, 221, 246, 256, 278] found also 1933 cm^{-1} for C≡C, others gave 1926 [243] and 1953 cm^{-1} [191] (no experimental conditions given). Certain solvents cause a breakdown of the polymeric structure of $C_6H_5C{\equiv}CCu$, therefore the wavenumber is enhanced: from 1923 in KBr to 2075 in $P(OC_2H_5)_3$ and 2083 cm^{-1} in hexamethylphosphoric triamide [164]. The C-Cu vibration (in the solid state?) is found to belong to the absorption at 450 [243] or 422 cm^{-1} [242]. Broad asymmetric bands (140, 153, and 187 cm^{-1} in polyethylene) were classified as Cu-Cu vibrations [228]. Spectra measured in C_2H_5OH contain bands which belong to $C_6H_5(C{\equiv}C)_2C_6H_5$ [125]. The shift of approximately 180 cm^{-1} of the C≡C frequency compared with that of $C_6H_5C{\equiv}CH$ is related to the polymeric structure [242].

The Raman spectrum (He-Ne laser) of solid $C_6H_5C{\equiv}CCu$ shows bands at 200m, 308m, 422vs, 515m (C-C≡C), 529vs (C-C), 749m (C-H), 999m (ring breathing), 1172w and 1192m (both C-H), 1594m (C-C), 1926s (C≡C), 3060vw (C-H), all in cm^{-1} [216]. Very similar spectra are given in [236, 274]. The bands at 308 and at 750 to 780 cm^{-1} are believed to result from π-Cu-C≡C vibrations [236]. Electrochemically grown layers of $C_6H_5C{\equiv}CCu$ on metallic Cu give the following Raman reflections (Ar or Kr laser): 148 (π-Cu-C≡C vibration), 201 (C≡C-Cu), 310 (C≡C-C), 423 (Cu-C), 530 (ring-C-C), 747 (C-H), 782 (ring-C-C-C), 1928 (C≡C), all in cm^{-1} [343].

Visible/UV spectra: The absorption maxima in mineral oil are at 400 and 477 nm [274]. Maxima at 395 [231], 455 [125, 231], 465 [246, 278, 283], and 500 nm [203] were also found. The diffuse reflectance spectrum of pure $C_6H_5C{\equiv}CCu$ is reported to have a broad maximum between 420 and 470 nm [256]. Spectra in MgO were recorded with acetylide concentrations of 5×10^{-4} (maxima at 390 and 455 nm) [125, 231] and 1% (maximum at 465 nm) [256]. $C_6H_5C{\equiv}CCu$ photoluminesces on UV excitation. The positions of the sharp maxima do not depend on the wavelength of the exciting light between 313 and 416 nm. The luminescence spectra in air and in vacuum show differences. At 77 K the maxima in air are at 568 and 630 nm, in vacuum at 610 and 680 nm [231]. At 296 K 516 and 583 nm have been found in air [125, 231].

References on pp. 57/67

$C_6H_5C \equiv CCu$ is characterized by X-ray photoelectron spectroscopy (XPES) and by Auger spectroscopy. The core level photoelectron spectra ($Cu\,2\,p_{3/2} = 932.5$ eV), the $Cu(L_3VV)$ Auger line (a kinetic energy of 915.9 eV corresponds to that line) and the Auger parameters are discussed in terms of both the different coordination and the polarizability of the ligands [346].

Crystallographic Properties. With CuKα radiation the reflections of highest intensity are at 15.0, 9.8, and 5.1 Å. The chain diameter is approximately 18 Å [274]. An X-ray analysis was carried out using a coarsely crystalline specimen [101]. Most of the crystals exhibit twinning on the (100) plane. The structure is monoclinic, with $a = 30.89$, $b = 3.89$, $c = 20.36$ Å, $\beta = 109.8°$, $Z = 16$, space group $C2/c - C_{2h}^6$ (No. 15). Reflections with k+l odd are diffuse with streaking in the direction of c*. The copper atoms are arranged in infinite zig–zag chains parallel to the b axis, and these are related in pairs by the twofold axes. Cu $\cdots$ Cu distances along the chains are 2.42 and 2.47 Å, and between the chains 3.06 and 3.20 Å. The bonding between the copper atoms and the ethynyl groups is shown in **Fig. 5**. $C_6H_5C \equiv CCu$ consists of a rectangular tube of copper atoms with the $C_6H_5C \equiv C$ groups on the outside. Each $C_6H_5C \equiv C$ seems to be σ-bonded to one copper atom, symmetrically π-bonded to another, and asymmetrically π-bonded to a third copper atom. Each ethynyl group lies roughly in the plane of the copper atom chain, and is "side-on" bonded to one of the copper atoms, with the bond to the phenyl group distorted away from this atom. The terminal carbon atom of the phenylethynyl group forms a bridge bond with two adjacent copper atoms, so that each ethynyl group is bonded to three copper atoms. Neglecting the very short Cu–Cu distances, the coordination around each copper atom is roughly trigonal [114, 151].

The two different types of $C \equiv C$ bonds were not confirmed by IR and laser Raman data [216].

Chemical Behavior. After careful washing and drying, $C_6H_5C \equiv CCu$ can be stored in tightly stoppered bottles for a long time [222]. Stored under N_2 at room temperature in a brown bottle it is stable for years [207]. $C_6H_5C \equiv CCu$ is stable to air up to 200 °C [191, 286] but ignites on further heating [6]. The melting point (with decomposition) is 229 °C [191, 221]. $C_6H_5C \equiv CCu$ is more stable than $n\text{-}C_4H_9C \equiv CCu$ [261]. An exothermic peak was found at 171 °C [221]. One endothermic effect at 205 °C indicates decomposition in the solid state. Above 220 °C a considerable liberation of heat has been observed. The decomposition yields a copper mirror and mainly $C_6H_5(C \equiv C)_2C_6H_5$ plus a polymer $(C_8H_5)_n$ [261]. UV irradiation

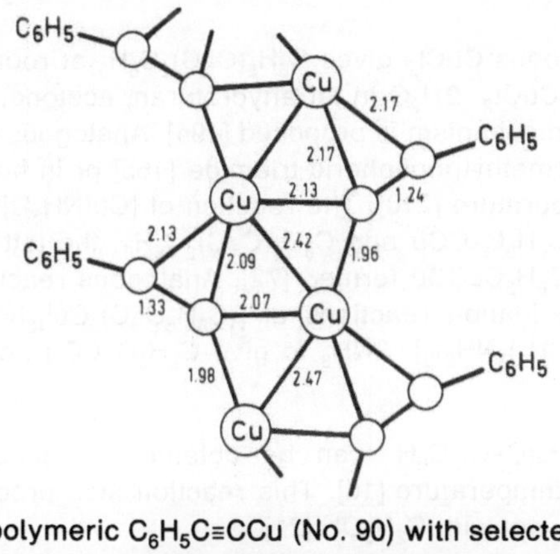

Fig. 5. Structure of polymeric $C_6H_5C \equiv CCu$ (No. 90) with selected bond lengths (in Å).

References on pp. 57/67

of $C_6H_5C\equiv CCu$ powder [155] or solutions in C_2H_5OH [125] is reported to form $C_6H_5(C\equiv C)_2C_6H_5$, but in both cases no specifications about oxygen exclusion are given.

$C_6H_5C\equiv CCu$ is insoluble in water and benzene [274]. The solubility in pyridine at 25 °C is 7.5×10^{-3} M [165]. $C_6H_5C\equiv CCu$ can be extracted by $HN(C_2H_5)_2$ [114] or $(CH_3)_3CNH_2$ [102] and is also slightly soluble in $(CH_3)_2CHOH$ [105]. It is soluble in $P(OCH_3)_3$ [234], $P(OC_2H_5)_3$ [131, 144, 164], $(CH_3)_2N(CH_2)_2N(CH_3)_2$ [280], and in liquid ammonia as $C_6H_5C\equiv CCu \cdot NH_3$ (see Section 1.1.2.4) [101]. It is readily soluble in $P(N(CH_3)_2)_3$ [164], a few percent can be dissolved in N-vinylcarbazole (for thermal polymerization to electrophotographic layers, see p. 50) [130]. Known reaction media for $C_6H_5C\equiv CCu$ also include CH_3CN [165], dimethylform-amide [165], and N-methylpyrrolidone [87, 137]. $C_6H_5NO_2$ and 1-methylnaphthalene are used as diluents [137].

The following Reaction Types 1 to 13 are explained on pp. 12/4.

Reaction Type 1 is carried out with diluted HCl or H_2SO_4 [6] and with gaseous HCl in absolute ether [21]. Concentrated HCl or HNO_3 (no exact concentrations given) cause defla-gration [6]. The reaction with o-bromothiophenol to give thianthrene and $C_6H_5C\equiv CH$ can also be understood as solvolysis by HBr [123]. $C_6H_5C\equiv CCu$ and R^1OH (R^1 = aryl) give in pyridine a dark violet solution of supposedly R^1OCu and $C_6H_5C\equiv CH$ [337].

The reaction $C_6H_5C\equiv CCu + Fe^{3+} + H^+ \rightarrow C_6H_5C\equiv CH + Cu^{2+} + Fe^{2+}$ is used for quantitative determination of $C_6H_5C\equiv CCu$. No oxidation according to Reaction Type 2 was observed [39]. A quantitative reaction with isopropylbenzene to form $C_6H_5C\equiv CH$ is described (no further information) [89].

Reaction Type 2a: According to photo-emf measurements (see p. 49), dry $C_6H_5C\equiv CCu$ is not oxidized by O_2 at room temperature [120]. Under all other conditions it reacts readily to give $C_6H_5(C\equiv C)_2C_6H_5$. Even prolonged washing in air forms considerable amounts of the butadiyne [128]. In some cases it is not clear whether insufficient oxygen exclusion or a reagent (polyhalogenophenols, iodoketones) causes the formation of $C_6H_5(C\equiv C)_2C_6H_5$ [123, 165]. Regarding the effects of UV irradiation in air see [155]. The preparation of $C_6H_5(C\equiv C)_2C_6H_5$ by oxidation with air is accomplished by shaking $C_6H_5C\equiv CCu$ with aqueous [5, 51] or ethanolic [6] ammonia at room temperature. Heating in water in the presence of air gives a mixture of $C_6H_5(C\equiv C)_2C_6H_5$ with its CuCl complex [44]. The diphenyl butadiyne is preferably prepared from a mixture of $C_6H_5C\equiv CH$, CuCl, tetramethylethylenediamine, and isopropanol by oxidation at 28 °C with air. Some $C_6H_5C\equiv CCu$ is deposited and dissolves in the course of the reaction. The rate of this exothermic and almost quantitative reaction has been determined [105].

Reaction Type 2b: Aqueous $CuCl_2$ gives $C_6H_5(C\equiv C)_2C_6H_5$ at room temperature [62] and at 100 °C (yield 88%) [37]. $CuCl_2 \cdot 2H_2O$ in tetrahydrofuran, acetone, or dioxane (yield 45%) gives the same product. A mechanism is proposed [194]. Analogous reactions are described with cupric acetate in hexamethylphosphoric triamide [165] or in hot pyridine [99] and with $Cu(OC_4H_9-t)_2$ at room temperature [210]. The reaction of $[Cu(NH_3)_4][SCN]_2$ with $C_6H_5C\equiv CNa$ in liquid ammonia yields $C_6H_5C\equiv CCu$ and $C_6H_5(C\equiv C)_2C_6H_5$, the latter obviously from Cu^{2+} oxidation of a part of the $C_6H_5C\equiv CCu$ formed [72]. Analogous reactions with other counter ions and intramolecular oxidation reactions of $[(C_6H_5C\equiv C)_2Cu^I]_2[Cu^{II}(NH_3)_4] \cdot xNH_3$ and of $[(C_6H_5C\equiv C)_2Cu^I]_2[Cu^{II}(H_2N(CH_2)_2NH_2)_2] \cdot 2NH_3$ to give $C_6H_5C\equiv CCu$ and $C_6H_5(C\equiv C)_2C_6H_5$ have also been published [84].

Reaction Type 2c: $C_6H_5(C\equiv C)_2C_6H_5$ can be obtained in good yields with aqueous $K_3[Fe(CN)_6]$/KOH at room temperature [10]. This reaction also proceeds without KOH; the copper acetylide should be wet with C_2H_5OH [21].

References on pp. 57/67

Reaction Type 2d is performed with $SiCl_4$, $(C_6H_5)_2SiCl_2$, or $SnCl_4$ in tetrahydrofuran (a mechanism is proposed) [208] and with $SOCl_2$ [204]. Mixing with $TlCl_3 \cdot 4H_2O$ causes ignition. On the other hand, with $TlCl_3 \cdot 4H_2O$ in C_2H_5OH or tetrahydrofuran up to 75% $C_6H_5(C\equiv C)_2C_6H_5$ is formed via a white, insoluble substance of unknown structure [158]. $KMnO_4$ [21] and Fe^{3+} [39] are not suitable.

For reactions with halogens X_2 or interhalogen compounds via $C_6H_5C\equiv CX$ to form $C_6H_5(C\equiv C)_2C_6H_5$ see Reaction Types 3 and 4. $C_6H_5C\equiv CCu$ reacts with acetyl bromide or $(CH_3)_2POCl$ in pyridine to give $C_6H_5(C\equiv C)_2C_6H_5$. A mechanism is proposed [208]. $C_6H_5C\equiv CCu$ and p-$(NC)_2C=C_6H_4=C(CN)_2$ yield in refluxing CH_3CN the radical anion $[p-(NC)_2C=C_6H_4=C(CN)_2]^-$ and $C_6H_5(C\equiv C)_2C_6H_5$, the latter obviously formed from $2C_6H_5C\equiv C$ radicals [170]. N-bromosuccinimide [165], BrCN [254], and $BrC\equiv CH$ oxidize it to $C_6H_5(C\equiv C)_2C_6H_5$. Iodine vapor or a solution of iodine in CH_2Cl_2 or n-hexane converts solid $C_6H_5C\equiv CCu$ into a mixture of the butadiyne and CuI [285]. Earlier an incorrect explanation and product analysis of this reaction was given [286]. Quinone and chloranil vapors have some unexpected effects on the photoemf and photoconductivity of dry $C_6H_5C\equiv CCu$; the visible/UV spectra show the formation of $C_6H_5(C\equiv C)_2C_6H_5$ [120]. The reaction with C_2H_5OH is claimed to yield $C_6H_5(C\equiv C)_2C_6H_5$, but this result is possibly due to incomplete air exclusion [125].

Reaction Type 3: I_2 in ether yields $C_6H_5C\equiv CI$ [110]. Bromination of $C_6H_5C\equiv CCu$ (25% excess) in ether at 2 to 5°C forms $C_6H_5C\equiv CBr$, $C_6H_5(C\equiv C)_2C_6H_5$, and CuBr [118]. With no excess of the copper compound $CuBr_2$ is formed [285]. The butadiyne should result from the reaction of $C_6H_5C\equiv CBr$ with excessive $C_6H_5C\equiv CCu$.

No formation of $C_6H_5CX=CX_2$ (X = Br or I) according to Reaction Type 4 has been observed.

Reaction Type 5 [21, 51, 123]: The adduct $C_6H_5C\equiv CCu \cdot CH_3CO_2Cu$ (see Section 1.1.2.5) is supposed to be an intermediate, for it gives $C_6H_5C\equiv CCH=CHC_6H_5$ under the same conditions [21, 51].

Reaction Type 6: The insertion of CO_2 at 80°C in tetrahydrofuran/pyridine gives lower yields than with certain P-complexes of $C_6H_5C\equiv CCu$ (see Section 1.1.2.4) [234]. An analogous reaction of $C_6H_5C\equiv CCu$ and CS_2 to give $C_6H_5C\equiv CCS_2Cu$ is not feasible [320].

Reaction Type 7: For heat-resistant polymers from $C_6H_5C\equiv CCu$ and diiododiphenyldicarbonic acid dichlorides [333] see Section 1.1.2.1.4.2.

Reaction Type 10: The reaction of $C_6H_5C\equiv CCu$ with 1-amino-benzimidazolium iodide is treated in Section 1.1.2.1.7 on p. 136 because of the Mannich cation type structure of the reacting heterocyclic salt [354].

Reaction Type 13: $C_6H_5C\equiv CCu$ catalyzes air oxidation of Fe^{2+}, of hydroquinone, and of the leuco forms of dyes. The mechanism of the catalytic action of $C_6H_5C\equiv CCu$ is probably associated with its strongly developed surface states (cf. [120]) [124]. $C_6H_5C\equiv CCu$ catalyzes the cyclization of alkynylpyrazol carbonic acids containing the structural unit $HO_2CC=CC\equiv C$ to give pyranopyrazoles. A benzo[c]furan derivative has been obtained from 2-ethynyl benzoic acid [318]. The cyclization of $2-IC_6H_4(CH_2)_nOH$ to give XII (n = 2 or 3), formally a hydrolysis of the acetylide by HI, is strongly catalyzed only by high molecular weight aggregates of $C_6H_5C\equiv CCu$. This effect is related to bridging the excessively long distance between

XII

References on pp. 57/67

I and OH in 2-$IC_6H_4(CH_2)_nOH$ on the surface of the polymeric structure. Consequently, mono-meric Cu^I or low molecular weight aggregates of $C_6H_5C\equiv CCu$ in solution are not capable of rapid catalysis [165].

Equimolar $C_6H_5C\equiv CCu$ proved to be highly effective in preparation of XIII from XIV and n-C_4H_9Li by suppressing reduction to XV. The suppressing mechanism obviously involves the complexation of a Li carbenoid by Cu^I to achieve a decrease in the basicity of the carbanion [244].

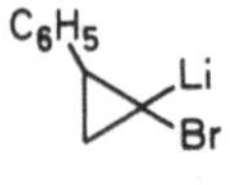

XIII	XIV	XV

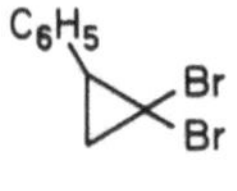

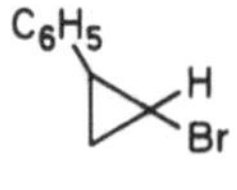

Other reactions: The reaction of $C_6H_5C\equiv CCu$ and 2-$IC_6H_4CONH_2$ in boiling pyridine gives 74% XVI. Obviously the expected primary product $C_6H_5C\equiv CC_6H_4CONH_2$-2 is hydrated to $C_6H_5COCH_2C_6H_4CONH_2$-2, which undergoes a reaction of the Knoevenagel type. The course of the reaction with XVII depends on the solvent used. In boiling pyridine the indolone XVIII (yield 62%) is formed, whereas in boiling xylene a 34% yield of the quinolone XIX has been isolated [162]. The reaction of $C_6H_5C\equiv CCu$ and 2,4,6-$(O_2N)_3C_6H_3$ at room tempera-ture in pyridine yields first the red complex XX of the Meisenheimer type. After treatment with acids 5% 2,4,6-$(O_2N)_3C_6H_2C\equiv CC_6H_5$ and 6% XXI can be isolated. If the red Meisenheimer complex is exposed to air and daylight, a red precipitate of unknown structure is observed. Treatment with acids yields the isatogen XXII [196]. The reaction of $C_6H_5C\equiv CCu$ with $C_6H_5C\equiv CAu$ and $[(n-C_4H_9)_4N][(C_6H_5C\equiv C)_2Au]$ gives $[(n-C_4H_9)_4N][(C_6H_5C\equiv C)_6Cu_2Au_3]$ [349].

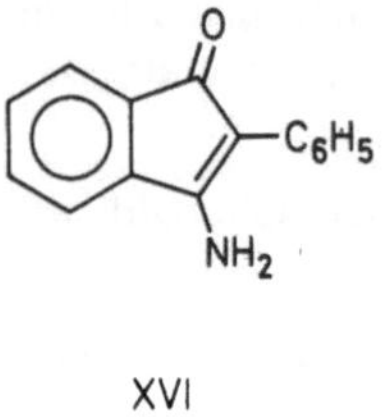

XVI	XVII	XVIII	XIX

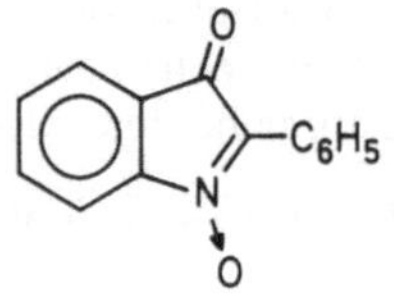

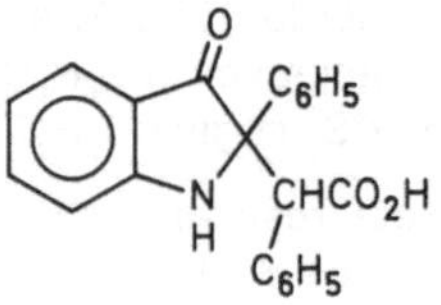

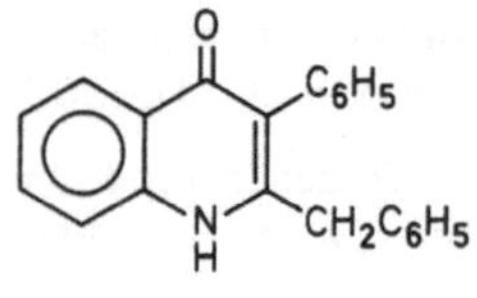

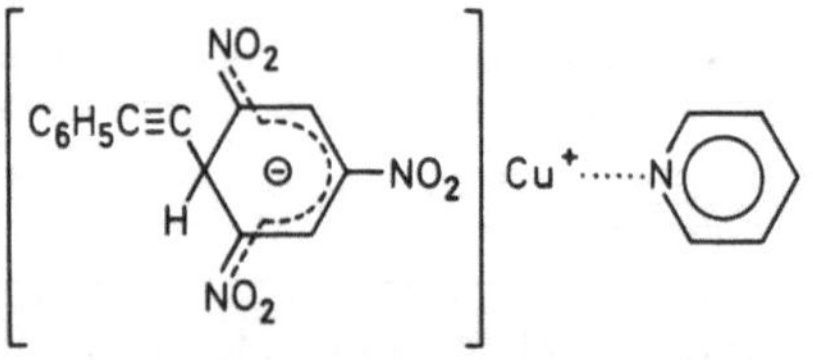

XX

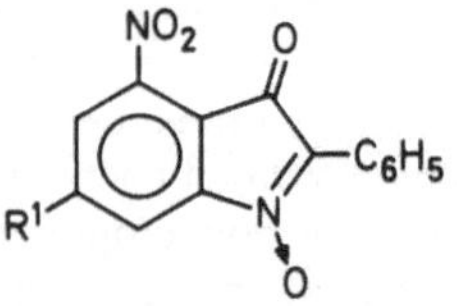

XXI : R^1 = NO_2

XXII : R^1 = OH

cyclo-$C_6H_{11}C\equiv CCu$ (Table 1, No. **98**). The degree of association n has been estimated osmometrically in $CHCl_3$. It depends on the preparation method. The product obtained by Method Ia could not be measured because of its insolubility; n is believed to be very high. Preparation Method II produces oligomers (n=4.5 to 8.5). With Method IIa n depends on the Cu^I concentration in the reaction batch (0.02 M CuI gives n=8.5 and 0.04 M gives n=5.4). In Method IIb cyclo-$C_6H_{11}C\equiv CNa$ is suspended in ether and stirred with a suspension of

CuCl in ammonia. The separation of the acetylide is possible after addition of CH_3OH. The high C≡C frequency (2176 cm^{-1}) is said to result from the absence of d_π-p_π^* intermolecular interaction. Differential thermal analysis shows an endothermic peak below 125 °C (transformation to a different solid phase, IR spectrum unchanged). Thermogravimetric analysis in air shows $CuCO_3$ formation at 200 °C. Up to 550 °C an appreciable amount of metallic copper is formed [288].

n-C_6H_{13}C≡CCu (Table 1, No. **102**). Differential thermal analysis shows an endothermic peak below 125 °C and a transformation to a different solid phase (IR spectrum unchanged). At 300 °C an appreciable amount of metallic copper is formed. The thermogravimetric analysis in air shows $CuCO_3$ formation beginning at 200 °C [288].

IR (Nujol or KBr): 1922 (C≡C) [288]. IR (KBr): 717, 1429, 1460, 1472 (all s), 1930 (m, C≡C) [236]. Raman (solid): 450 (vs, ≡C–Cu) [236, 242], 945(vs), 1875 (C≡C), 1929 (vs, C≡C), 1955 (C≡C), all wave numbers in cm^{-1}. The relative intensities of the last three bands are 0.6:10:1.2 [236].

4-$CH_3C_6H_4$C≡CCu (Table 1, No. **116**). The resistivity at 25 °C is $1.6 \times 10^{11} \, \Omega \cdot$ cm [191, 220]. Dark- and photoconductivities in polyethylene films [221] and in nitrocellulose were also measured. The electrical resistivity of mixtures of the acetylide and nitrocellulose has a minimum at approximately 20% nitrocellulose [190, 221]. The photocurrent decays very slowly. The effect of oxygen on photoconduction [191, 220] and the dependence of the photocurrent on the light intensity [220, 221] were measured in vacuum and in dry air. For the formation of a mixed Au–Cu complex upon reaction with $[(C_6H_5C≡C)_3Au_2]^-$ see Section 1.1.2.5 [348]. In contrast to C_6H_5C≡CCu, No. 116 does not react with $[n-C_4H_9)_4N][(C_6H_5C≡C)_2Au]$ [349].

C_6H_5CH(OH)C≡CCu (Table 1, No. **119**) is soluble in aqueous NH_3 and cannot be precipitated in the usual manner with Ilosvay's reagent by Preparation Method Ia. Method Id: C_6H_5CH(OH)C≡CH is dissolved in C_2H_5OH and mixed with a suspension of CuCl in H_2O. The pH value is maintained at 7. After 3 h the conversion is complete [60, 66].

n-C_4H_9C≡CCH$_2$C≡CCu (Table 1, No. **122**). Because of the solubility conditions no quantitative conversion is possible by Method Ia. The mixture of acetylide and alkyne obtained by stirring n-C_4H_9C≡CCH$_2$C≡CH and CuCl in $H_2O/C_2H_5OH/NH_3$ is washed successively with H_2O, C_2H_5OH, and ether and then extracted with ether until no more yellow n-C_4H_9C≡CCH$_2$C≡CH dissolves [94].

C_6H_5(C≡C)$_2$Cu (Table 1, No. **127**). The photoconductivity spectrum [203] has the longwave maximum at 500 nm [203, 231, 283]. The band gap is 2.24±0.05 (from absorption spectra) [231, 283] or 2.22±0.02 eV (from photoconductivity spectra) [203, 231, 248]. The direct transition energy is 2.48±0.03 eV and the electron affinity 3.17 or 3.19 eV, respectively [231, 248]. The vibrational structure of the photoconductivity spectrum (λ_{max}=450, 480, and 500 nm) and of the absorption spectrum (λ_{max}=415 and 455 nm [231] or 465 nm [283]), an electron energy scheme, and the photoelectric work function (5.41 eV) have been reported [248]. The differential thermal analysis shows one exothermic peak at 240 °C [221].

(n-C_4H_9)$_2$PC≡CCu. (Table 1, No. **145**). The conductance of photoresistors made from this compound was at 50 to 180 V in the range of 10^{-12} to $10^{-13} \, \Omega^{-1}$ cm^{-1}(?) [166]. The photoconductivity spectrum maxima are at 440 nm [283], 470 nm [166], 390, 440 nm [195]. The photocurrent carriers are positively charged [195]. The gap width from the photoconductivity spectrum is 2.78 eV [195, 283]. A kinetic photoconductivity curve is given [166].

1-$C_{10}H_7$C≡CCu (Table 1, No. **153**). The resistivity at 25 °C is $2.1 \times 10^{11} \, \Omega \cdot$ cm. Dark and light currents are determined in polyethylene films and in nitrocellulose. The resistivity

References on pp. 57/67

shows a minimum at ca. 20 wt% nitrocellulose [190, 221]. The spectral dependence of the photocurrent has been measured in vacuum ($\lambda_{max} = 546$ nm) and in dry air ($\lambda_{max} = 553$ nm, more intense) [221]. The kinetics of the photoconductivity in air and in vacuum are given [220, 221]. The photoconductivity spectrum [256] shows a maximum at 507 nm [256, 283], the band gap is 2.22 eV. From the absorption spectrum a value of 2.30 eV has been obtained [248, 283]. The photovoltaic spectrum has a maximum at 465 nm [256].

The photoelectric work function is 5.22, and the electron affinity is 2.92 (absorption) or 3.00 eV (photoemission), respectively. An electron energy level scheme is given [248].

2-$C_{10}H_7C\equiv CCu$ (Table 1, No. **154**). The resistivity at 25 °C is $6.0 \times 10^{10}\ \Omega \cdot$ cm [191, 220, 221]. The photocurrent varies linearly with the light intensity in vacuum or dry air [220]. The photoconductivity spectrum [256] shows a maximum at 512 nm [256, 283], and the band gap is 2.22 eV. The value from the absorption spectrum is 2.30 eV [283]. The photo-emf spectrum has a maximum at 470 nm [256].

1-$C_{10}H_7(C\equiv C)_2Cu$ (Table 1, No. **167**; see XXIII). Following Preparation Method I a an ethereal solution of impure XXIV is shaken with CuCl in aqueous ammonia. This procedure is very useful for purification of XXIV. The compound is also prepared from No. 153 and $BrC\equiv CH$ at 0 °C in dimethylformamide [205, 221].

$(C\equiv C)_2Cu$ $(C\equiv C)_2H$

XXIII   XXIV

$(C_6H_5)_2AsC\equiv CCu$ (Table 1, No. **173**). Visible/UV absorption maxima have been found at 570 nm in MgO [283], and at 490 nm in MgO/BaSO$_4$ (concentrations between 0.1 and 10%) [195]; the band gap is 1.88 eV [195, 283]. Broad overlapping maxima of the photoconductivity spectrum are reported at 440, 500, and 620 nm. The low photosensitivity tail extends to 1.2 µm, which may explain the presence of a quasicontinuous distribution of vacancies in the forbidden band. $(C_6H_5)_2AsC\equiv CCu$ has an appreciably greater growth and decay time for photoconductivity than other copper acetylides [195]. The gap width from photoconductivity measurements is 1.75 eV [195, 283]. Photoresistors with a conductivity of 10^{-12} to $10^{-13}\ \Omega^{-1}$ cm^{-1}(?) at 50 to 180 V were made from the acetylide [166].

$(C_6H_5)_2PC\equiv CCu$ (Table 1, No. **174**). Two photoconductivity maxima at 375 and 430 nm [195] as well as one maximum at 450 nm [283] are reported; the band gap is 2.78 ± 0.02 eV [195], but cf. [283]. The visible/UV maximum has been found at 370 nm [283]; the gap width (from UV) is 2.74 eV [195, 283]. For these measurements an optimum concentration in MgO and BaSO$_4$ of 1:100 was determined [195].

4-$C_6H_5C\equiv CC_6H_4C\equiv CCu$ (Table 1, No. **183**). The joint analysis of the photoconductivity and the stretching frequencies of the ethynyl groups has shown that coordination bonds are created by the ethynyl group nearest to the copper atom [202].

The photoconductivity spectrum [202, 203] shows maxima at 440, 490, and 510 nm [231]; the band gap from photoconductivity is 2.26 eV [231, 248, 283]. The photoelectric work function is 5.24, and the electron affinity is 2.94 eV (from UV absorption) or 2.98 eV (from photoconductivity). An electron energy level scheme is given [248].

UV spectrum in MgO: $\lambda_{max} = 465$ nm [231] or 475 nm, the band gap (from UV) is 2.3 eV [283].

$C_{14}H_9C\equiv CCu$ (Table 1, No. 184). The resistivity at 25 °C is $5.0 \times 10^9 \ \Omega \cdot$ cm [191, 220, 221]. The photoconductivity spectrum [256] shows a maximum at 550 nm [283]; the band gap is 2.14 eV [248, 257, 283]. The photo-emf spectrum has a broad plateau between 400 and 520 nm and the maximum value is at $\sim$510 nm [256]. The photoelectric work function is 5.03 eV; an electron energy level scheme is given [248].

$(C_6H_5)_2C(CN)CH_2C\equiv CCu$ (Table 1, No. 188) is stable in air and insoluble in H_2O and in organic solvents. It is not sensitive to acids and decomposes to a dark matter at 215 to 230 °C without melting. At 500 °C in vacuum elemental copper is formed. The differential thermal analysis shows exothermic peaks at 200 °C (Cu–C cleavage) and at 240 and 260 °C (both C–C cleavage) [238].

$H(C\equiv C)_2Cu$ (Table 1, No. 200). $H(C\equiv C)_2H$ is bubbled into a solution of CuCl in $CH_3CON(CH_3)_2$. The electrical conductivity is in the range of insulators but upon doping with I_2 it is increased by 10 orders of magnitude to a metallic conductor level. The temperature dependence of the conductivity is discussed in detail. After doping, the I_2 is removed under reduced pressure resulting in a conductivity decrease by 1 to 2 orders of magnitude [342]. The doping can also be accomplished with different electron acceptors and/or donors [356, 357]. The actual semiconductor can be prepared by sealing the polymer in an ampule and then heating [357].

Fourier transform IR spectroscopy gives very weak, diffuse peaks between 1900 and 2200 cm^{-1}.

According to X-ray powder diffraction measurements the black acetylide is nearly amorphous [342]. It is insoluble in organic solvents, even in strong acids and bases and not fusible. The acetylide decomposes explosively by heat or mechanical shock, therefore elemental analyses are not possible. It is gradually degraded in air at room temperature. According to IR spectroscopy compounds containing aromatic rings and quinone carbonyl are formed [342].

$t\text{-}C_4H_9O_2CC\equiv CCu$ (Table 1, No. 208) is prepared, according to Method Ia, in C_2H_5OH/H_2O (1:3) and precipitated with H_2O. The reaction with XXV in hexamethylphosphoric triamide, dimethylformamide, or CH_3CN at room temperature gives 10% XXVI, but no trace of the expected product from the iodine substitution. At 0 °C with O_2 bubbling, the yield is enhanced to 40%; under N_2 no acetylide is consumed. The reaction does not involve radicals, but no meachanism is known. It can also be performed with in situ produced acetylide. 4-Acetoxy-, 4-phenylsulfonyl-, or 4-allylazetidin-2-ones do not react with $t\text{-}C_4H_9O_2CC\equiv CCu$ [350].

The IR spectrum in Nujol shows maxima at 1674 (C=O) and 1925 (C≡C) cm^{-1} [350].

References:

[1] R. Böttger (Jahresber. Physik. Ver. Frankfurt **1856/57** 37/44; Dinglers Polytech. J. **152** [1859] 22/9).

[2] M. P. Berthelot (Compt. Rend. **50** [1860] 805/8).

[3] M. P. Berthelot (Ann. Chim. Phys. [3] **67** [1863] 52/77).
[4] M. P. Berthelot (Ann. Chim. Phys. [4] **9** [1866] 421/5).
[5] C. Glaser (Ber. Deut. Chem. Ges. **2** [1869] 422/4).
[6] C. Glaser (Liebigs Ann. Chem. **154** [1870] 137/71).
[7] L. Henry (Ber. Deut. Chem. Ges. **5** [1872] 569/72).
[8] A. Baeyer (Ber. Deut. Chem. Ges. **13** [1880] 2254/63).
[9] A. Baeyer (Ber. Deut. Chem. Ges. **15** [1882] 50/6).
[10] A. Baeyer, L. Landsberg (Ber. Deut. Chem. Ges. **15** [1882] 57/61).

[11] E. Bandrowsky (Ber. Deut. Chem. Ges. **15** [1882] 2698/704).
[12] A. Baeyer (Ber. Deut. Chem. Ges. **18** [1885] 674/81).
[13] A. Baeyer (Ber. Deut. Chem. Ges. **18** [1885] 2269/81).
[14] G. Griner (Ann. Chim. Phys. [6] **26** [1892] 305/94).
[15] R. Lespieau (Compt. Rend. **123** [1896] 1295/6).
[16] R. Lespieau (Ann. Chim. Phys. [7] **11** [1897] 232/88).
[17] R. Chavastelon (Compt. Rend. **126** [1898] 1810/2).
[18] J. V. Nef (Liebigs Ann. Chem. **308** [1899] 264/328, 316).
[19] L. Ilosvay von Nagyilosva (Ber. Deut. Chem. Ges. **32** [1899] 2697/9).
[20] L. Ilosvay von Nagyilosva (Z. Anal. Chem. **40** [1901] 123/4).

[21] F. Straus (Liebigs Ann. Chem. **342** [1905] 190/265).
[22] W. H. Perkin, J. L. Simonsen (J. Chem. Soc. **91** [1907] 816/40).
[23] R. Lespieau (Bull. Soc. Chim. France [4] **3** [1908] 638/40).
[24] R. Lespieau, M. Pariselle (Compt. Rend. **146** [1908] 1035/7).
[25] R. M. Schierl (Ztg. Calciumcarbid **10** [1909] 93/5).
[26] R. Lespieau (Compt. Rend. **150** [1910] 113/4).
[27] R. Lespieau (Compt. Rend. **152** [1911] 879/81).
[28] R. Lespieau (Ann. Chim. Phys. [8] **27** [1912] 137/89).
[29] W. Manchot (Liebigs Ann. Chem. **387** [1912] 257/93).
[30] F. Kunckell (Ber. Deut. Pharm. Ges. **23** [1913] 188/227).

[31] R. Willstätter, T. Wirth (Ber. Deut. Chem. Ges. **46** [1913] 535/8).
[32] M. Picon (Compt. Rend. **158** [1914] 1184/7).
[33] S. Reich (Arch. Sci. Phys. Nat. [4] **45** [1918] 259/76).
[34] A. E. Favorskii, L. Morev (Zh. Russ. Fiz. Khim. Obshchestva **50** [1920] 571/88; C.A. **1924** 2496).
[35] C. Moureu, J. C. Bongrand (Ann. Chim. [Paris] [9] **14** [1920] 47/58).
[36] F. Straus, W. Voss (Ber. Deut. Chem. Ges. **59** [1926] 1681/91).
[37] F. Straus (Ber. Deut. Chem. Ges. **59** [1926] 1664/81).
[38] H. van Risseghem (Bull. Soc. Chim. Belges **35** [1926] 328/64).
[39] F. Hein, A. Meyer (Z. Anal. Chem. **72** [1927] 30/1).
[40] C. Moureu, C. Dufraisse, J. R. Johnson (Ann. Chim. [Paris] [10] **7** [1927] 14/42, 34).

[41] F. Straus, W. Heyn, E. Schwemer (Ber. Deut. Chem. Ges. **63** [1930] 1086/92).
[42] R. Truchet (Ann. Chim. [Paris] [10] **16** [1931] 309/419, 346).
[43] A. L. Klebanskii, U. A. Dranitsyna, I. M. Dobromil'skaya (Dokl. Akad. Nauk SSSR **1935** II 229/32; C **1936** I 2528).
[44] Yu. S. Zal'kind, B. M. Fundyler (Zh. Obshch. Khim. **6** [1936] 530/3).
[45] L. Pauling, H. D. Springall, K. J. Palmer (J. Am. Chem. Soc. **61** [1939] 927/37).
[46] Yu. S. Zal'kind, B. M. Fundyler (Zh. Obshch. Khim. **9** [1939] 1725/8).
[47] A. S. Carter, Du Pont Co. (U.S. 2228752 [1939/41]; C.A. **1941** 2908).
[48] E. R. Gilliland, H. L. Bliss, C. E. Kip (J. Am. Chem. Soc. **63** [1941] 2088/90).

[49] D. Kästner, Badische Anilin- & Soda-Fabrik A.-G. (Ger. 871004 [1942/53]; C.A. **1956** 8712).

[50] A. I. Lebedeva (Dokl. Akad. Nauk SSSR **42** [1944] 71/5; Compt. Rend. Acad. Sci. USSR **42** [1944] 70/3; C.A. **1945** 4287).

[51] P. Piganiol (Acétylène Homologues et Dérivés, Masson, Paris 1945, p. 263).

[52] G. E. Coates (J. Chem. Soc. **1946** 838/9).

[53] K. Bowden, I. Heilbron, E. R. H. Jones, K. H. Sargent (J. Chem. Soc. **1947** 1579/83).

[54] I. Heilbron, E. R. H. Jones, F. Sondheimer (J. Chem. Soc. **1947** 1586/90).

[55] O. A. Chaltykyan (Zh. Obshch. Khim. **18** [1948] 1626/38).

[56] T. Bruun, C. Haug, N. A. Sörensen (Acta Chem. Scand. **4** [1950] 850/5).

[57] W. Hunsmann (Chem. Ber. **83** [1950] 213/7).

[58] M. Nakagawa (Kagaku No Ryoiki [J. Japan. Chem.] **4** [1950] 564/5; C.A. **1951** 7081).

[59] M. Nakagawa (Proc. Japan Acad. **26** No. 10 [1950] 38/42; C.A. **1951** 7081).

[60] M. Nakagawa (Proc. Japan Acad. **26** No. 10 [1950] 43/7; C.A. **1951** 8486).

[61] H. H. Schlubach, V. Wolf (Liebigs Ann. Chem. **568** [1950] 141/59).

[62] F. Bohlmann (Chem. Ber. **84** [1951] 545/56).

[63] W. C. Easterbrook, J. W. B. Erskine (J. Appl. Chem. [London] **1** [1951] Suppl. No. 1, pp. 53/61).

[64] W. J. Gensler, G. R. Thomas (J. Am. Chem. Soc. **73** [1951] 4601/4).

[65] R. N. Haszeldine (J. Chem. Soc. **1951** 588/91).

[66] M. Nakagawa (Nippon Kagaku Zasshi **72** [1951] 561/6; C.A. **1951** 6602).

[67] R. N. Haszeldine, K. Leedham (J. Chem. Soc. **1952** 3483/90).

[68] F. Klages, K. Mönkemeyer (Chem. Ber. **85** [1952] 109/22).

[69] D. Kästner, Badische Anilin- & Soda-Fabrik A.-G. (Ger. 870845 [1952/53]; C.A. **1956** 4200).

[70] R. Ahmad, B. C. L. Weedon (J. Chem. Soc. **1953** 3286/94).

[71] G. M. Mkryan, N. A. Papazyan (Dokl. Akad. Nauk Arm. SSR **16** [1953] 17/26).

[72] R. Nast (Z. Naturforsch. **8b** [1953] 381/3).

[73] J. P. Riley (J. Chem. Soc. **1953** 2193/8).

[74] J. B. Armitage, N. Entwistle, E. R. H. Jones, M. C. Whiting (J. Chem. Soc. **1954** 147/54).

[75] A. Vaitiekunas, F. F. Nord (J. Org. Chem. **19** [1954] 902/6).

[76] A. Vaitiekunas, F. F. Nord (J. Am. Chem. Soc. **76** [1954] 2733/6).

[77] J. B. Armitage, C. L. Cook, N. Entwistle, E. R. H. Jones, M. C. Whiting (J. Chem. Soc. **1952** 1998/2005).

[78] F. Bohlmann, H. G. Viehe, H. J. Mannhardt (Chem. Ber. **88** [1955] 361/70).

[79] F. J. Brockman (Can. J. Chem. **33** [1955] 507/10).

[80] W. Chodkiewicz, P. Cadiot (Compt. Rend. **241** [1955] 1055/7).

[81] P. Kurtz, Bayer A.-G. (Brit. 775723 [1955/57]; C.A. **1957** 16511).

[82] J. Colonge, R. Falcotet (Compt. Rend. **242** [1956] 1484/6).

[83] J. W. Gensler, A. P. Mahadevan (J. Org. Chem. **21** [1956] 180/2).

[84] R. Nast, W. Pfab (Chem. Ber. **89** [1956] 415/21).

[85] V. A. Sazonova, N. Ya. Kronrod (Zh. Obshch. Khim. **26** [1956] 1876/81; J. Gen. Chem. [USSR] **26** [1956] 2093/9).

[86] A. Chicoisne, G. Dupont, R. Dulou (Bull. Soc. Chim. France **1957** 1232/7).

[87] W. Chodkiewicz (Ann. Chim. [Paris] [13] **2** [1957] 819/69).

[88] J. Colonge, R. Falcotet (Bull. Soc. Chim. France **1957** 1166/9).

[89] A. L. Klebanskii, I. V. Grachev, O. M. Kuznetsova (Zh. Obshch. Khim. **27** [1957] 2977/83; J. Gen. Chem. [USSR] **27** [1957] 3008/13).

[90] A. J. Saggiomo (J. Org. Chem. **22** [1957] 1171/5).

[91] R. Nast, W. Pfab (Z. Anorg. Allgem. Chem. **292** [1957] 287/92).

[92] M. Akhtar, B. C. L. Weedon (Proc. Chem. Soc. **1958** 303).

[93] P. J. Ashworth, E. R. H. Jones (J. Chem. Soc. **1958** 950/4).

[94] W. J. Gensler, J. Casella Jr. (J. Am. Chem. Soc. **80** [1958] 1376/80).

[95] F. Wessely, E. Zbiral, E. Lahrmann (Chem. Ber. **92** [1959] 2141/51).

[96] L. P. Ellinger, British Oxygen Co., Ltd. (Brit. 784638 [1957]; C.A. **1958** 4068).

[97] M. Akhtar, T. A. Richards, B. C. L. Weedon (J. Chem. Soc. **1959** 933/40).

[98] D. Blake, G. Calvin, G. E. Coates (Proc. Chem. Soc. **1959** 396/7).

[99] G. Eglinton, A. R. Galbraith (J. Chem. Soc. **1959** 889/96).

[100] J. F. Arens (Advan. Org. Chem. **2** [1960] 117/212).

[101] R. Nast, C. Schultze (Z. Anorg. Allgem. Chem. **307** [1960] 15/21).

[102] G. E. Coates, C. Parkin (J. Inorg. Nucl. Chem. **22** [1961] 59/67).

[103] G. E. Coates, C. Parkin (in: S. Kirschner, Advances in the Chemistry of the Coordination Compounds, Plenum, New York 1961, pp. 173/9; Proc. 6th Intern. Conf. Coord. Chem., Detroit 1961, pp. 173/9).

[104] O. M. Temkin, R. M. Flid, E. D. German, T. A. Onishchenko (Kinetika Kataliz **2** [1961] 205/13; Kinet. Catal. [USSR] **2** [1961] 190/8).

[105] A. S. Hay (J. Org. Chem. **27** [1962] 3320/1).

[106] A. A. Clifford, W. A. Waters (J. Chem. Soc. **1963** 3056/62).

[107] V. S. Myl'nikov, A. N. Terenin (Dokl. Akad. Nauk SSSR **153** [1963] 1381/4; C.A. **60** [1964] 11480).

[108] R. Nast, P.-G. Kirst, G. Beck, J. Gremm (Chem. Ber. **96** [1963] 3302/5).

[109] S. Siggia (Quantitative Organic Analysis Via Functional Groups, 3rd Ed., Wiley, New York 1963, pp. 395/6).

[110] A. M. Sladkov, L. Y. Ukhin, V. V. Korshak (Izv. Akad. Nauk SSSR Ser. Khim. **1963** 2213/5; Bull. Acad. Sci. USSR Div. Chem. Sci. **1963** 2043/5).

[111] R. D. Stevens, C. E. Castro (J. Org. Chem. **28** [1963] 3313/5).

[112] R. E. Atkinson, R. F. Curtis, G. T. Phillips (Chem. Ind. [London] **1964** 2101/2).

[113] F. Bohlmann, H. Schönowsky, E. Inhoffen, G. Gran (Chem. Ber. **97** [1964] 794/800).

[114] P. W. R. Corfield, H. M. M. Shearer (Abstr. Am. Cryst. Assoc. Meeting, Bozeman, Mont., 1964, pp. 96/7).

[115] M. Gaudemar (Compt. Rend. **258** [1964] 4803/4).

[116] V. S. Myl'nikov, A. N. Terenin (Mol. Phys. **8** [1964] 387/99).

[117] V. S. Myl'nikov, A. N. Terenin (Dokl. Akad. Nauk SSSR **155** [1964] 1167/70; Dokl. Chem. Proc. Acad. Sci. USSR **154/159** [1964] 397/9).

[118] A. M. Sladkov, L. Yu. Ukhin (Izv. Akad. Nauk SSSR Ser. Khim. **1964** 392/3; Bull. Acad. Sci. USSR Div. Chem. Sci. **1964** 370).

[119] L. N. Ionov, I. A. Akimov, A. N. Terenin (Dokl. Akad. Nauk SSSR **169** [1966] 550/3; C.A. **65** [1966] 11508).

[120] V. S. Myl'nikov (Dokl. Akad. Nauk SSSR **164** [1965] 622/5; C.A. **63** [1965] 17258).

[121] A. M. Sladkov, L. Yu. Ukhin, G. N. Gorshkova, M. A. Chubarova, A. G. Makhsumov, V. I. Kasatochkin (Zh. Org. Khim. **1** [1965] 415/21; J. Org. Chem. [USSR] **1** [1965] 406/11).

[122] R. E. Atkinson, R. F. Curtis, G. T. Phillips (J. Chem. Soc. C **1966** 1101/3).

[123] C. E. Castro, E. J. Gaughan, D. C. Owsley (J. Org. Chem. **31** [1966] 4071/8).

[124] V. S. Myl'nikov, Kh. L. Arvan (Zh. Fiz. Khim. **40** [1966] 2653/4; Russ. J. Phys. Chem. **40** [1966] 1426/7).

[125] G. I. Lashkov, V. S. Myl'nikov (Opt. Spektrosk. **20** [1966] 86/91; Opt. Spectrosc. [USSR] **20** [1966] 44/7; C.A. **65** [1966] 6534).

[126] V. S. Myl'nikov (Zh. Fiz. Khim. **40** [1966] 979/84; Russ. J. Phys. Chem. **40** [1966] 527/30).

[127] W. P. Norris, W. G. Finnegan (J. Org. Chem. **31** [1966] 3292/5).

[128] M. D. Rausch, A. Siegel, L. P. Klemann (J. Org. Chem. **31** [1966] 2703/4).

[129] M. Rosenblum, N. Brawn, J. Papenmeier, M. Applebaum (J. Organometal. Chem. **6** [1966] 173/80).

[130] I. Sidaravichius, F. A. Lavina, G. Rybalko, A. M. Sladkov, V. S. Myl'nikov, Yu. P. Kudryavzev, L. Ukhin (Opt. Mekhan. Prom. **33** No. 5 [1966] 27/30; Soviet J. Opt. Technol. **33** [1966] 228/31).

[131] M. F. Shostakovskii, L. A. Polyakova, L. A. Vasil'eva, A. I. Polyakov (Zh. Org. Khim. **2** [1966] 1899; J. Org. Chem. [USSR] **2** [1966] 1865).

[132] H. Staab, F. Graf (Tetrahedron Letters **1966** 751/7).

[133] R. E. Atkinson, R. F. Curtis, J. A. Taylor (J. Chem. Soc. C **1967** 578/82).

[134] R. E. Atkinson, R. F. Curtis, G. T. Phillips (J. Chem. Soc. C **1967** 2011/5).

[135] F. Bohlmann, C. Zdero, W. Gordon (Chem. Ber. **100** [1967] 1193/9).

[136] F. Bohlmann, P.-H. Bonnet, H. Hofmeister (Chem. Ber. **100** [1967] 1200/5).

[137] K. Gump, S. W. Moje, C. E. Castro (J. Am. Chem. Soc. **89** [1967] 6770/1).

[138] E. R. H. Jones, S. Safe, V. Thaller (J. Chem. Soc. C **1967** 1038/44).

[139] I. L. Kotlyarevskii, V. N. Andrievskii, M. S. Shvartsberg (Khim. Geterotsikl. Soedin. **3** [1967] 308/9; Chem. Heterocycl. Compounds [USSR] **3** [1967] 236/7).

[140] V. M. Bentsa, F. L. Vilesov, Yu. A. Aleksandrov (Pribory Tekhn. Eksperim. **1967** No. 6, pp. 147/51; Inst. Exptl. Tech. [USSR] **1967** 1402/6).

[141] F. A. Levina, G. Rybalko, J. Sidaravicius (Reprogr. Ber. 2nd Intern. Kongr., Cologne, FRG, 1967 [1969], pp. 196/9).

[142] A. M. Malte, C. E. Castro (J. Am. Chem. Soc. **89** [1967] 6770).

[143] V. S. Myl'nikov (Dokl. Akad. Nauk SSSR **175** [1967] 726/9; C.A. **68** [1968] No. 54760).

[144] I. M. Dolgopol'skii, A. L. Klebanskii, Z. F. Dobler (Zh. Prikl. Khim. **33** [1960] 209/12).

[145] M. S. Shvartsberg, A. N. Kozhevnikova, I. L. Kotlyarevskii (Izv. Akad. Nauk SSSR Ser. Khim. **1967** 466/7; Bull. Acad. Sci. USSR Div. Chem. Sci. **1967** 460/1).

[146] A. M. Sladkov (Diss. INEOS Akad. Nauk SSSR, Moscow 1967).

[147] V. S. Myl'nikov (Usp. Khim. **37** [1968] 78/103; Russ. Chem. Rev. **37** [1968] 25/38, 30).

[148] H. Veschambre, G. Dauphin, A. Kergomard (Bull. Soc. Chim. France **1967** 2846/54).

[149] P. Cadiot, W. Chodkiewicz (in: H. G. Viehe, Chemistry of Acetylenes, Dekker, New York 1969, pp. 597/647).

[150] F. Bohlmann, P. Blaszkiewicz, E. Bresinsky (Chem. Ber. **101** [1968] 4163/9).

[151] M. L. H. Green (Organometallic Compounds II: The Transition Elements 3rd Ed., Methuen, London 1968, pp. 271/6).

[152] S. P. Korshunov, R. I. Katkevich, O. T. Shashlova, L. I. Vereshchagin (Zh. Org. Khim. **4** [1968] 676/80; J. Org. Chem. [USSR] **4** [1968] 659/62).

[153] S. A. Mladenović, C. E. Castro (J. Heterocycl. Chem. **5** [1968] 227/30).

[154] V. S. Myl'nikov (Zh. Fiz. Khim. **42** [1968] 2168/73; Russ. J. Phys. Chem. **42** [1968] 1150/3).

[155] V. S. Myl'nikov, A. N. Terenin (J. Polym. Sci. Polym. Symp. No. 16 [1967] 3655/65; C.A. **70** [1969] No. 38243).

[156] T. F. Rutledge (Acetylenic Compounds, Reinhold, New York 1968, pp. 84/95, 207/10, 240/68).

[157] M. F. Shostakovskii, A. N. Volkov, A. N. Khudyankova, Yu. M. Skvortsov, I. I. Danda (Khim. Atsetilena Tr. 3rd Vses. Konf., Dushanbe 1968 [1972], pp. 94/7; C.A. **79** [1973] No. 18257).

[158] A. M. Sladkov, L. Yu. Ukhin, Zh. I. Orlova (Izv. Akad. Nauk SSSR Ser. Khim. **1968** 2586/90; Bull. Acad. Sci. USSR Div. Chem. Sci. **1968** 2446/9).

[159] H. A. Staab, A. Nissen, J. Ipaktschi (Angew. Chem. **80** [1968] 241; Angew. Chem. Intern. Ed. Engl. **7** [1968] 226).

[160] H. A. Staab, H. Mack, E. Wehinger (Tetrahedron Letters **1968** 1465/9).

[161] R. E. Atkinson, R. F. Curtis, D. M. Jones, J. A. Taylor (J. Chem. Soc. C **1969** 2173/6).
[162] C. Bond, M. Hooper (J. Chem. Soc. C **1969** 2453/60).
[163] B. Bossenbroek, D. C. Sanders, H. M. Curry, H. Shechter (J. Am. Chem. Soc. **91** [1969] 371/9).
[164] M. Bourgain, J. F. Normant (Bull. Soc. Chim. France **1969** 2477/81).
[165] C. E. Castro, R. Havlin, U. K. Honvad, A. Malte, S. Moje (J. Am. Chem. Soc. **91** [1969] 6464/70).
[166] I. R. Gol'ding, A. M. Sladkov, V. S. Myl'nikov (Izv. Akad. Nauk SSSR Ser. Khim. **1969** 2062; Bull. Acad. Sci. USSR Div. Chem. Sci. **1969** 1918/9).
[167] M. D. Rausch, A. Siegel, L. P. Klemann (J. Org. Chem. **34** [1969] 468/70).
[168] M. D. Rausch, A. Siegel (J. Org. Chem. **34** [1969] 1974/6).
[169] M. D. Rausch, A. Siegel (J. Organometal. Chem. **17** [1969] 117/25).
[170] L. Yu. Ukhin, A. M. Sladkov, Zh. I. Orlova (Izv. Akad. Nauk SSSR Ser. Khim. **1969** 705/6; Bull. Acad. Sci. USSR Div. Chem. Sci. **1969** 637/8).
[171] A. Camus, N. Marsich (J. Organometal. Chem. **21** [1970] 249/58).
[172] K. Endo, Y. Sakata, S. Misumi (Tetrahedron Letters **1970** 2557/60).
[173] R. H. Mitchell, F. Sondheimer (Tetrahedron **26** [1970] 2141/50).
[174] V. S. Myl'nikov (J. Polym. Sci. Polym. Symp. No. 30 [1970] 673/81).
[175] J. F. Normant, M. Bourgain, A.-M. Rone (Compt. Rend. C **270** [1970] 354/7).
[176] M. Rosenblum, N. M. Brawn, D. Ciappenelli, J. Tancrede (J. Organometal. Chem. **24** [1970] 469/77).
[177] D. V. Sokol'skii, Ya. A. Dorfman, S. S. Segizbaeva (Zh. Prikl. Khim. **43** [1970] 502/7; J. Appl. Chem. [USSR] **43** [1970] 511/5).
[178] H. A. Staab, F. Graf (Chem. Ber. **103** [1970] 1107/17).
[179] A. N. Volkov, Yu. M. Skvortsov, I. I. Danda, M. F. Shostakovskii (Zh. Org. Khim. **6** [1970] 897/902; J. Org. Chem. [USSR] **6** [1970] 903/7).
[180] L. I. Zakharkin, L. P. Sorokina (Zh. Org. Khim. **6** [1970] 2470/3; J. Org. Chem. [USSR] **6** [1970] 2482/4).

[181] H. E. Zimmerman, J. R. Dodd (J. Am. Chem. Soc. **92** [1970] 6507/14).
[182] R. D. Stephens, W. C. Kray, Shell Oil Co. (U.S. 3749700 [1970/73]; C.A. **79** [1973] No. 146990).
[183] R. F. Curtis, J. A. Taylor (J. Chem. Soc. C **1971** 186/8).
[184] J. J. Eisch, C. K. Hordis (J. Am. Chem. Soc. **93** [1971] 2974/80).
[185] J. J. Eisch, M. W. Foxton (J. Org. Chem. **36** [1971] 3520/6).
[186] H. Franke (Diss. Berlin T.U. 1971).
[187] C. C. Leznoff, R. J. Hayward (Can. J. Chem. **49** [1971] 3596/601).
[188] M. S. Newman, M. W. Logue (J. Org. Chem. **36** [1971] 1398/401).
[189] J. F. Normant, M. Bourgain (Tetrahedron Letters **1971** 2583/6).
[190] Y. Okamoto, M. Mincer, A. Golubovic, N. Dimond (Nature Phys. Sci. **229** [1971] 157/8).
[191] Y. Okamoto, S. K. Kundu, A. Golubovic, N. Dimond (Am. Chem. Soc. Div. Org. Coatings Plastics Chem. Papers **31** [1971] 312/6; C.A. **78** [1973] No. 160226).
[192] H. Plieninger, J. Sirowej (Chem. Ber. **104** [1971] 2027/9).
[193] A. Roedig, V. Kimmel, W. Lippert (Tetrahedron Letters **1971** 1219/20).
[194] A. M. Sladkov, I. R. Gol'ding (Dokl. Akad. Nauk SSSR **200** [1971] 132/3; Dokl. Chem. Proc. Acad. Sci. USSR **196/201** [1971] 756/7).
[195] V. S. Myl'nikov, A. N. Dune (Opt. Spektrosk. **31** [1971] 405/9; Opt. Spectrosc. [USSR] **31** [1971] 217/9).
[196] O. Wennerström (Acta Chem. Scand. **25** [1971] 789/94).

[197] L. I. Zakharkin, V. N. Kalinin, I. R. Gol'ding, A. M. Sladkov, A. V. Grebennikov (Zh. Obshch. Khim. **41** [1971] 823/8; J. Gen. Chem. [USSR] **41** [1971] 830/3).

[198] A. G. Makhsumov, O. V. Afanas'eva, Sh. U. Abdullaev, U. A. Abidov (U.S.S.R. 389090 [1971/73]; C.A. **79** [1973] No. 136978).

[199] O. V. Afanas'eva, A. G. Makshumov, Sh. U. Abdullaev (Dokl. 4th Vses. Konf. Khim. Atsetilena, Alma–Ata 1972, Vol. 1, pp. 244/7; C.A. **79** [1973] No. 31756).

[200] I. N. Azerbaev, I. A. Poblavskaya, G. N. Kondaurov (Dokl. 4th Vses. Konf. Khim. Atsetilena, Alma–Ata 1972, Vol. 1, pp. 353/8; C.A. **79** [1973] No. 77998).

[201] C. E. Hudson, N. L. Bauld (J. Am. Chem. Soc. **94** [1972] 1158/63).

[202] V. S. Myl'nikov, A. N. Dun'e, I. R. Gol'ding, A. M. Sladkov (Tezisy Dokl. 5th Vses. Conf. Org. Metal. Soedin., Moscow 1972, Vol. 1, pp. 468/9; Summaries 5th Intern. Conf. Organometal. Compounds, Moscow 1972, pp. 470/1, Ref. No. 180).

[203] V. S. Myl'nikov, A. N. Dun'e, I. R. Gol'ding, A. M. Sladkov (Zh. Obshch. Khim. **42** [1972] 2543/6; J. Gen. Chem. [USSR] **42** [1972] 2532/5).

[204] J. F. Normant (Synthesis **1972** 63/80, 70ff.).

[205] Y. Okamoto, K. L. Chellappa, S. K. Kundu (J. Org. Chem. **37** [1972] 3185/7).

[206] R. Oliver, D. R. M. Walton (Tetrahedron Letters **1972** 5209/12).

[207] D. C. Owsley, C. E. Castro (Org. Syn. **52** [1972] 128/31).

[208] A. M. Sladkov, I. R. Gol'ding (Dokl. 4th Vses. Konf. Khim. Atsetilena, Alma–Ata 1972, Vol. 1, pp. 59/64; C.A. **79** [1973] No. 78253).

[209] H. A. Staab, E. Wehinger, W. Thorwart (Chem. Ber. **105** [1972] 2290/309).

[210] T. Tsuda, T. Hashimoto, T. Saegusa (J. Am. Chem. Soc. **94** [1972] 658/9).

[211] F. Bohlmann, P.-D. Hopf (Chem. Ber. **106** [1973] 3621/5).

[212] F. Bohlmann, W. Skuballa (Chem. Ber. **106** [1973] 497/504).

[213] M. Bourgain, J. F. Normant (Bull. Soc. Chim. France **1973** 1777/9).

[214] M. Bourgain, J. F. Normant (Bull. Soc. Chim. France **1973** 2137/42).

[215] E. J. Corey, G. W. I. Fleet, M. Kato (Tetrahedron Letters **1973** 3963/6).

[216] I. A. Garbusova, V. T. Alexanjan, L. A. Leites, I. R. Gol'ding, A. M. Sladkov (J. Organometal. Chem. **54** [1973] 341/4).

[217] H. Hopf, F. T. Lenich (Chem. Ber. **106** [1973] 3461/2).

[218] H. O. House, M. Umen (J. Org. Chem. **38** [1973] 3893/901).

[219] A. G. Makhsumov, I. R. Askarov, A. M. Sladkov (Vopr. Farm. Farmakol. [Tashkent] **1973** 316/8; C.A. **84** [1976] No. 121354).

[220] Y. Okamoto, S. K. Kundu (J. Phys. Chem. **77** [1973] 2677/80).

[221] Y. Okamoto (AD-760772 [1973] 1/84; C.A. **80** [1974] No. 37537).

[222] M. Stefanovič, L. Krstič, S. Mladenovič (Glasnik Hem. Drustva Beograd **38** [1973] 463/8; C.A. **82** [1975] No. 43645).

[223] A. Tozuka, T. Otsubo, Y. Sakata, S. Misumi (Mem. Inst. Sci. Ind. Res. Osaka Univ. **30** [1973] 83/9; C.A. **79** [1973] No. 66092).

[224] O. M. Abu Salah, M. I. Bruce (J. Chem. Soc. Dalton Trans. **1974** 2302/4).

[225] F. Bohlmann, J. Kocur (Chem. Ber. **107** [1974] 2115/9).

[226] P. L. Coe, N. E. Milner (J. Organometal. Chem. **70** [1974] 147/52).

[227] E. P. Denisov, K. K. Smirnov, V. S. Myl'nikov (Fiz. Tverd. Tela [Leningrad] **16** [1974] 1782/4; Soviet Phys.-Solid State **16** [1974] 1157/8).

[228] L. V. Konovalov, V. S. Myl'nikov (Zh. Strukt. Khim. **15** [1974] 709/11; J. Struct. Chem. [USSR] **15** [1974] 612/3).

[229] R. Levene, J. Y. Becker, J. Klein (J. Organometal. Chem. **67** [1974] 467/71).

[230] J. P. Marino, D. M. Floyd (J. Am. Chem. Soc. **96** [1974] 7138/40).

[231] V. S. Myl'nikov (Zh. Strukt. Khim. **15** [1974] 244/9; J. Struct. Chem. [USSR] **15** [1974] 224/8).

[232] S. Sh. Rashidova, V. A. Barabanov, D. D. Il'yasova (Uzb. Khim. Zh. **18** [1974] 63/6; C.A. **82** [1975] No. 17 434).

[233] H. A. Staab, K. Neunhoeffer (Synthesis **1974** 424).

[234] T. Tsuda, K. Ueda, T. Saegusa (J. Chem. Soc. Chem. Commun. **1974** 380/1).

[235] A. Roedig, V. Kimmel, W. Lippert, B. Heinrich (Liebigs Ann. Chem. **755** [1972] 106/21).

[236] V. T. Aleksanyan, I. A. Garbuzova, I. R. Gol'ding, A. M. Sladkov (Spectrochim. Acta A **31** [1975] 517/24).

[237] D. E. Bergbreiter, G. M. Whitesides (J. Org. Chem. **40** [1975] 779/82).

[238] M. C. Barral, V. Moreno, A. Santos (Anales Quim. **71** [1975] 770/4; C.A. **84** [1976] No. 180 367).

[239] M. T. Cox, J. J. Holohan (Tetrahedron **31** [1975] 633/5).

[240] R. S. Dickson, L. J. Michel (Australian J. Chem. **28** [1975] 1943/55).

[241] A. Kh. Filippova, G. S. Lyashenko, N. S. Vyazankin (Tezisy Dokl. 5th Vses. Konf. Khim. Atsetilena, Tiflis 1975, p. 104; C.A. **88** [1978] No. 190 287).

[242] I. A. Garbusova, V. T. Aleksanyan, I. R. Gol'ding, A. M. Sladkov (Tezisy Dokl. 5th Vses. Konf. Khim. Atsetilena, Tiflis 1975, pp. 457/8; C.A. **88** [1978] No. 189 456).

[243] I. R. Gol'ding, I. A. Garbusova, A. M. Sladkov, V. T. Aleksanyan (Tezisy Dokl. 5th Vses. Konf. Khim. Atsetilena, Tiflis 1975, p. 455; C.A. **88** [1978] No. 189 798).

[244] K. Kitatani, T. Hiyama, H. Nozaki (J. Am. Chem. Soc. **97** [1975] 949/51).

[245] M. W. Logue, G. L. Moore (J. Org. Chem. **40** [1975] 131/2).

[246] V. S. Myl'nikov, A. A. Kharchenko, I. R. Gol'ding, A. M. Sladkov (Tezisy Dokl. 5th Vses. Konf. Khim. Atsetilena, Tiflis 1975, p. 456; C.A. **88** [1978] No. 189 799).

[247] F. G. Schreiber, R. Stevenson (Chem. Letters **1975** 1257/8).

[248] D. A. Sukhov, V. S. Myl'nikov (Fiz. Tverd. Tela [Leningrad] **17** [1975] 923/5; Soviet Phys.–Solid State **17** [1975] 589/90).

[249] T. Tsuda, Y. Chujo, T. Saegusa (J. Chem. Soc. Chem. Commun. **1975** 963/4).

[250] O. M. Abu Salah, M. I. Bruce (Australian J. Chem. **29** [1975] 73/7).

[251] O. M. Abu Salah, M. I. Bruce (Australian J. Chem. **29** [1975] 531/41).

[252] K. Arakawa, T. Miyasaka, N. Hamamichi (4th Symp. Chem. Nucleic Acids. Res., Kyoto 1976, Vol. 2, pp. 1/4; C.A. **86** [1977] No. 140 379).

[253] I. Barrow, A. E. Pedler (Tetrahedron **32** [1976] 1829/34).

[254] P.–L. Compagnon, B. Grosjean (Synthesis **1976** 448/9).

[255] L. K. Ding, W. J. Irwin (J. Chem. Soc. Perkin Trans. I **1976** 2382/6).

[256] V. S. Myl'nikov, I. R. Gol'ding, A. M. Sladkov (Izv. Akad. Nauk SSSR Ser. Khim. **1976** 330/5; Bull. Acad. Sci. USSR Div. Chem. Sci. **1976** 311/5).

[257] V. S. Myl'nikov, A. A. Kharchenko, M. M. Sobolev (Kvantovaya Elektronika [Moscow] **3** [1976] 288/92; Soviet J. Quant. Electron. **6** [1976] 151/3).

[258] F. G. Schreiber, R. Stevenson (J. Chem. Soc. Perkin Trans. I **1976** 1514/8).

[259] S. Staicu, I. G. Dinulescu, F. Chiraleu, M. Avram (J. Organometal. Chem. **117** [1976] 385/94).

[260] G. S. Natarajan, K. A. Venkatachalam (J. Appl. Chem. Biotechnol. **22** [1972] 1019/25; C.A. **77** [1972] No. 169 523).

[261] N. N. Travkin, I. R. Gol'ding, T. A. Sladkova, B. G. Gribov, A. M. Sladkov, N. F. Konovalov (Zh. Obshch. Khim. **46** [1976] 1088/91; J. Gen. Chem. [USSR] **46** [1976] 1081/3).

[262] F. Bohlmann, C. Hühn (Chem. Ber. **110** [1977] 1183/5).

[263] M. Brenner, C. Brush (Tetrahedron Letters **1977** 419/22).

[264] R. P. Duffley, R. Stevenson (J. Chem. Soc. Perkin Trans. I **1977** 802/4).

[265] S. Inoue, N. Takamatsu, Y. Kishi (Yakugaku Zasshi [J. Pharm. Soc. Japan] **97** [1977] 564/8; C.A. **87** [1977] No. 184744).

[266] J. S. Kiely, P. Boudjouk, L. L. Nelson (J. Org. Chem. **42** [1977] 2626/9).

[267] R. F. Kovar, G. F. L. Ehlers, F. E. Arnold (J. Polym. Sci. Polym. Chem. Ed. **15** [1977] 1081/95).

[268] F. G. Schreiber, R. Stevenson (J. Chem. Soc. Perkin Trans. I **1977** 90/2).

[269] K. Sonogashira, T. Yatake, Y. Tohda, S. Takahashi, N. Hagihara (J. Chem. Soc. Chem. Commun. **1977** 291/2).

[270] E. J. Corey, M. G. Bock, A. P. Kozikowski, A. V. Rama Rao, D. Floyd, B. H. Lipshutz (Tetrahedron Letters **1978** 1051/4).

[271] E. J. Corey, D. Floyd, B. H. Lipshutz (J. Org. Chem. **43** [1978] 3418/20).

[272] R. P. Duffley, R. Stevenson (Syn. Commun. **1978** 175/80).

[273] G. A. Razuvaev, V. K. Cherkasov, G. A. Abakumov (J. Organometal. Chem. **160** [1978] 361/71).

[274] I. A. Garbuzova, I. R. Gol'ding, A. M. Sladkov, Ya. V. Genin, D. Ya. Tsvankin, V. T. Aleksanyan (Izv. Akad. Nauk SSSR Ser. Khim. **1978** 1328/31; Bull. Acad. Sci. USSR Div. Chem. Sci. **1978** 1154/6).

[275] N. Hamamichi, T. Miyasaka, K. Arakawa (Chem. Pharm. Bull. **26** [1978] 898/907).

[276] G. A. Kraus, K. Frazier (Tetrahedron Letters **1978** 3195/8).

[277] M. M. Kwatra, D. Z. Simon, R. L. Salvador, P. D. Cooper (J. Med. Chem. **21** [1978] 253/7).

[278] V. S. Myl'nikov, A. A. Kharchenko, I. R. Gol'ding, A. M. Sladkov (Izv. Akad. Nauk SSSR Ser. Khim. **1978** 822/4; Bull. Acad. Sci. USSR Div. Chem. Sci. **1978** 708/11).

[279] M. S. Shvartsberg, A. A. Moroz, A. N. Kozhevnikova (Izv. Akad. Nauk SSSR Ser. Khim. **1978** 875/9; Bull. Acad. Sci. USSR Div. Chem. Sci. **1978** 756/60).

[280] K. Sonogashira, Y. Fujikura, T. Yatake, N. Toyoshima, S. Takahashi, N. Hagihara (J. Organometal. Chem. **145** [1978] 101/8).

[281] B. V. Lebedev, T. Bykova, E. G. Kiparisova, N. N. Mukhina, A. M. Sladkov, I. R. Gol'ding (Izv. Akad. Nauk SSSR Ser. Khim. **1979** 1880/2; Bull. Acad. Sci. USSR Div. Chem. Sci. **1979** 1743/5).

[282] D. B. Ledlie, G. Miller (J. Org. Chem. **44** [1979] 1006/7).

[283] V. S. Myl'nikov (Zh. Nauchn. Prikl. Fotogr. Kinematogr. **24** [1979] 25/8; C.A. **91** [1979] No. 66221).

[284] A. N. Sinyakov, M. S. Shvartsberg (Izv. Akad. Nauk SSSR Ser. Khim. **1979** 1126/8; Bull. Acad. Sci. USSR Div. Chem. Sci. **1979** 1053/4).

[285] J. S. Miller, S. T. Matsuo (J. Polym. Sci. Polym. Letters **18** [1980] 809/10).

[286] Y. Okamoto, M. C. Wang (J. Polym. Sci. Polym. Letters **18** [1980] 249/51).

[287] B. V. Lebedev, T. A. Bykova, A. M. Sladkov, E. G. Kiparisova, I. R. Gol'ding (Zh. Obshch. Khim. **51** [1981] 724/7; J. Gen. Chem. [USSR] **51** [1981] 590/2).

[288] E. C. Royer, M. C. Barral, V. Moreno, A. Santos (J. Inorg. Nucl. Chem. **43** [1981] 705/9).

[289] C. R. Johnson, D. S. Dhanoa (J. Chem. Soc. Chem. Commun. **1982** 358/9).

[290] B. H. Lipshutz, R. S. Wilhelm, D. M. Floyd (J. Am. Chem. Soc. **103** [1981] 7672/4).

[291] G. Struve, S. Seltzer (J. Org. Chem. **47** [1982] 2109/13).

[292] R. H. Bradbury, T. L. Gilchrist, C. W. Rees (J. Chem. Soc. Perkin Trans. I **1981** 3234/8).

[293] L. Scattebøl (Acta Chem. Scand. **13** [1959] 198).

[294] E. T. Bogoradovskii, V. S. Zavgorodskii, K. S. Mingaleva, A. A. Petrov (Zh. Obshch. Khim. **44** [1974] 142/5; J. Gen. Chem. [USSR] **44** [1974] 139/41).

[295] L. I. Zakharkin, A. I. Kovredov, V. A. Ol'shevskaya (Izv. Akad. Nauk SSSR Ser. Khim. **1982** 673/5; Bull. Acad. Sci. USSR Div. Chem. Sci. **1982** 599/602).

[296] N. G. Karapetyan, A. S. Tarkhanyan, A. N. Lyubimova (Izv. Akad. Nauk Arm. SSR Ser. Khim. Nauk **18** [1965] 360/5; C.A. **63** [1965] 17883).

[297] V. V. Korshak, A. M. Sladkov, Yu. Kudryavtsev (Vysokomol. Soedin. **2** [1960] 1824/7; C.A. **1961** 26522).

[298] N. A. Bumagin, I. O. Kalinovskii, A. B. Ponomarev, I. P. Beletskaya (Dokl. Akad. Nauk SSSR **265** [1982] 1138/43; Dokl. Chem. Proc. Acad. Sci. USSR **262/267** [1982] 262/6).

[299] R. T. Scannel, R. Stevenson (J. Heterocycl. Chem. **17** [1980] 1727/8).

[300] N. A. Bumagin, I. O. Kalinovskii, I. P. Beletskaya (Izv. Akad. Nauk SSSR Ser. Khim. **1981** 2836; Bull. Acad. Sci. USSR Div. Chem. Sci. **1981** 2366).

[301] E. J. Corey, R. H. Wollenberg (J. Am. Chem. Soc. **96** [1974] 5581/3).

[302] D. B. Sokol'skii, L. N. Nikolenko (Dokl. Akad. Nauk SSSR **82** [1952] 923/5; C.A. **1953** 2723).

[303] E. J. Corey, R. H. Wollenberg (J. Org. Chem. **40** [1975] 2265/6).

[304] E. J. Corey, R. H. Wollenberg (Tetrahedron Letters **1976** 4705/8).

[305] B. F. Middleton, M. J. Weiss, C. V. Grudzinskas, S.-M. L. Chen, American Cyanamid Co. (Brit. 2006186 [1977]; Ger. 2837530 [1977/79]; C.A. **91** [1979] No. 38996).

[306] H. C. Kluender, Miles Labs. Inc. (U.S. 4065493 [1977]; U.S. 4100356 [1977]; U.S. 4100357 [1977]; U.S. 4127727 [1977/78]; C.A. **88** [1978] No. 190216).

[307] M. B. Floyd, M. J. Weiss, C. V. Grudzinskas, S.-M. L. Chen, American Cyanamid Co. (Brit. 2009173 [1978/79]; C.A. **92** [1980] No. 214968).

[308] M. F. Semmelhack, A. Yamashita, J. C. Tomesch, K. Hirotsu (J. Am. Chem. Soc. **100** [1978] 5565/7).

[309] W. D. Woessner, Miles Labs. Inc. (Brit. 2009148 [1978]; Ger. 2850465 [1978/79]; C.A. **91** [1979] No. 91249).

[310] E. J. Corey, P. B. Hopkins, S. Kim, S. Yoo, K. P. Nambiar, J. R. Falck (J. Am. Chem. Soc. **101** [1979] 7131/4).

[311] G. H. Posner, M. J. Chapdelaine, C. M. Lentz (J. Org. Chem. **44** [1979] 3661/5).

[312] J. Davies, S. M. Roberts, D. P. Reynolds, R. F. Newton (J. Chem. Soc. Perkin Trans. I **1981** 1317/20).

[313] A. J. Dixon, R. J. Taylor (J. Chem. Soc. Perkin Trans. I **1981** 1407/10).

[314] E. Piers, J. M. Chong, H. E. Morton (Tetrahedron Letters **22** [1981] 4905/8).

[315] J. A. Miller, G. Zweifel (Synthesis **1983** 128/30).

[316] G. Boche, M. Bernheim, M. Nießner (Angew. Chem. Suppl. **1982** 34/8; Angew. Chem. **95** [1983] 48).

[317] O. N. Temkin, R. M. Flid (Kataliticheskie Prevrashcheniya Atsetilenovykh Soedinenii v Rastvorakh Kompleksov Metallov [Catalytic Conversion of Acetylene Compounds in Solutions of Metal Complexes], Nauka, Moscow 1968, pp. 1/213).

[318] M. S. Shvartsberg, S. F. Vasilevskii, T. V. Anisimova, V. A. Gerasimov (Izv. Akad. Nauk SSSR Ser. Khim. **1981** 1342/8; Bull. Acad. Sci. USSR Div. Chem. Sci. **1981** 1071/6).

[319] N. A. Agal'tsova, V. S. Myl'nikov, A. M. Sladkov (Izv. Akad. Nauk SSSR Ser. Khim. **1972** 203/4; Bull. Acad. Sci. USSR Div. Chem. Sci. **1972** 199/200).

[320] A. Camus, N. Marsich, G. Nardin (J. Organometal. Chem. **188** [1980] 389/99).

[321] L. N. Ionov, I. A. Akimov (Fiz. Tekhn. Poluprov. **5** [1971] 2017/20; Soviet Phys.-Semicond. **5** [1971] 1756/8).

[322] I. I. Zhilevich, D.-I. B. Sidaravichyus (U.S.S.R. 176620 [1965]).

[323] F. A. Levina, V. S. Myl'nikov, G. I. Rybalko, D.-I. B. Sidaravichyus, A. M. Sladkov, A. N. Terenin (U.S.S.R. 169395 [1965]; C.A. **64** [1966] 16889).

[324] F. A. Levina, G. I. Rybalko, D.-I. B. Sidaravichyus, A. M. Sladkov (U.S.S.R. 165970 [1965]).

[325] H. Kleijn, J. Meijer, G. C. Overbeek, P. Vermeer (Rec. J. Roy. Neth. Chem. Soc. **101** [1982] 97/101).
[326] V. Ratovelomanana, G. Linstrumelle (Tetrahedron Letters **22** [1981] 315/8).
[327] J. F. Normant, G. Cahiez, C. Chuit, J. Villieras (J. Organometal. Chem. **77** [1974] 281/7).
[328] B. Župančić, B. Jenko, LEK Tovarna Farmacevtskih Kemičnih Izdelkov (Austrian 351511 [1979]; C.A. **91** [1979] 175017).
[329] R. A. Amos, J. A. Katzenellenbogen (J. Org. Chem. **42** [1977] 2537/45).
[330] C. G. Gordon-Gray, C. G. Whiteley (J. Chem. Soc. Perkin Trans. I **1977** 2040/6).

[331] R. J. Batten, J. D. Coyle, R. J. K. Taylor (Synthesis **1980** 910/1).
[332] G. E. Schneiders, R. Stevenson (Syn. Commun. **10** [1980] 699/705).
[333] S. Venkateso, U.S. Dept. of the Air Force (U.S. Appl. 219396 [1981]; C.A. **96** [1982] No. 7629).
[334] K. Ruitenberg, H. Kleijn, H. Westmijze, J. Meijer, P. Vermeer (Rec. Trav. Chim. **101** [1982] 405/9).
[335] G. E. Schneiders, R. Stevenson (J. Chem. Res. S **1982** 182).
[336] O. M. Abu-Salah, A. R. Al-Ohaly (J. Organometal. Chem. **255** [1983] C39/C40).
[337] A. Afzali, H. Firouzabadi, A. Khalafi-nejad (Syn. Commun. **13** [1983] 335/9).
[338] R. Baudouy, J. Sartoretti, F. Choplin (Tetrahedron **39** [1983] 3293/305).
[339] A. Kh. Filippova, G. S. Lyashenko, I. D. Kalikhman, V. Yu. Vitkovskii, G. G. Naumova, N. S. Vyazankin (Izv. Akad. Nauk SSSR Ser. Khim. **1983** 950/3; Bull. Acad. Sci. USSR Div. Chem. Sci. **1983** 861/3).
[340] A. T. Hutton, P. G. Pringle, B. L. Shaw (Organometallics **2** [1983] 1889/91).

[341] J. G. Hefner, P. M. Zizelman, L. D. Durfee, G. S. Lewandos (J. Organometal. Chem. **260** [1984] 369/80).
[342] H. Matsuda, H. Nakanishi, M. Kato (J. Polym. Sci. Polym. Letters Ed. **22** [1984] 107/11).
[343] L. M. Abrantes, M. Fleischmann, I. R. Hill, L. M. Peter, M. Mengoli, G. Zotti (J. Electroanal. Chem. **164** [1984] 177/87).
[344] L. M. Abrantes, L. M. Castillo, M. Fleischmann, I. R. Hill, L. M. Peter, G. Mengoli, G. Zotti (J. Electroanal. Chem. **177** [1984] 129/37).
[345] R. Kumar, D. G. Tuck (J. Organometal. Chem. **281** [1985] C47/C48).
[346] C. Battistoni, G. Mattogno, E. Paparazzo, L. Naldini (Inorg. Chim. Acta **102** [1985] 1/3).
[347] J. Enda, I. Kuwajima (J. Am. Chem. Soc. **107** [1985] 5495/501).
[348] O. M. Abu-Salah (J. Organometal. Chem. **270** [1984] C26/C28).
[349] O. M. Abu-Salah, A.-R. A. Al-Ohaly, C. B. Knobler (J. Chem. Soc. Chem. Commun. **1985** 1502/3).
[350] A. Balsamo, B. Macchia, F. Macchia, A. Rossello, P. Domiano (Tetrahedron Letters **26** [1985] 4141/4).

[351] L. V. Tumasheva, A. Kh. Filippova, G. S. Lyashenko, N. S. Vyazankin (Izv. Akad. Nauk SSSR Ser. Khim. **1984** 2797/8; Bull. Acad. Sci. USSR Div. Chem. Sci. **1984** 2563/4).
[352] R. T. Scannel, R. Stevenson (J. Chem. Soc. Perkin Trans. I **1983** 2927/31).
[353] E. C. Royer, M. C. Barral (Inorg. Chim. Acta **90** [1984] L47/L49).
[354] L. Yu. Ukhin, V. N. Komissarov, Zh. I. Orlova, N. A. Dolgopolova (Zh. Obshch. Khim. **54** [1984] 1676/8; J. Gen. Chem. [USSR] **54** [1984] 1493/4).
[355] J. L. Gaston, R. J. Greer, M. F. Grundon (J. Chem. Res. S **1985** 135).
[356] Agency of Industrial Sciences and Technology (Japan. Kokai Tokkyo Koho 59-202206 [84-202206] [1983/84]; C.A. **102** [1985] No. 158993).
[357] Agency of Industrial Sciences and Technology (Japan. Kokai Tokkyo Koho 59-202207 [84-202207] [1983/84]; C.A. **102** [1985] No. 158994).
[358] J. Enda, T. Matsutani, I. Kuwajima (Tetrahedron Letters **25** [1984] 5307/10).
[359] R. F. Curtis, J. A. Taylor (Tetrahedron Letters **1968** 2919/20).

1.1.2.1.2   General Remarks about Reactions of RC≡CCu with Proton Donors, Oxidants, Halogens, and CO_2

In each of these reactions the products vary only according to R in RC≡CCu. Therefore, they are included in detail in Table 1 (see pp. 17/41) with yields and references as Reaction Types 1 to 6. On the following pages, only some general remarks about these reactions are given. However, reactions differing from this simple pattern (containing a variety of further groups like R^1, R^2 etc.) need more information than can be included in Table 1 and they are therefore discussed in Sections 1.1.2.1.3 to 1.1.2.1.10.

Reactions with Proton Donors

Reaction Type 1 in Table 1 is the formation of RC≡CH from RC≡CCu. The most common reagents are HCl (Type 1a), HNO_3 (Type 1b), and CN^-/H_2O with formation of $[Cu(CN)_3]^{2-}$ (Type 1c). Further reagents like NH_3, Fe^{3+}/H_2O, H_2SO_4, C_2H_2, $K_3[Fe(CN)_6]/H_2O$, and $2-BrC_6H_4SH$ are summarized under Reaction Type 1d.

The sequence RC≡CH → RC≡CCu → RC≡CH has especially been used for purification purposes [5, 50]. Sensitive RC≡CH are often treated in aqueous two-phase systems (ether [42], benzene [26, 39], CCl_4 [44]) to avoid side reaction by rapid extraction of RC≡CH into the organic phase. The contact of the RC≡CH product with the aggressive aqueous solution can also be shortened by continuously distilling the RC≡CH with water vapor [15]. RC≡CCu can also be converted to RC≡CH with HCl in absolute ether [5]. A very mild method is stirring RC≡CCu with saturated, slightly acidified aqueous NH_4Cl and ether [26]. $C_4H_3OC≡$ CCu (C_4H_3O = fur-2-yl) could only be transformed to $C_4H_3OC≡CH$ by aqueous NaCN; the ring may be opened upon addition of aqueous HCl [13].

Based upon the dark violet color generated during the reaction and the reactivity of the solutions of $C_6H_5C≡CCu$ and some phenols in pyridine, a partial hydrolysis $C_6H_5C≡CCu +$ ArOH → $ArOCu + C_6H_5C≡CH$ has been postulated [51].

Oxidation

The reaction of RC≡CCu with oxygen (Reaction Type 2a in Table 1), with Cu^{II} compounds (Type 2b), $K_3[Fe(CN)_6]$ (Type 2c), or other oxidants (Type 2d) is a useful method to prepare diynes $R(C≡C)_2R$. For the preparation of $R^1(C≡C)_2R^2$ see under "Mixed Acetylides", Section 1.1.2.2.

Oxidation by air or oxygen (Reaction Type 2a) is frequently an undesired acetylide-consuming side reaction. To avoid it, a thorough O_2 exclusion is necessary. Air dissolved in solvents like pyridine must be removed. Nevertheless, almost all reaction products of RC≡CCu seem to be accompanied by at least traces of $R(C≡C)_2R$ and with long-time reactions in boiling solvents the amounts are said to be high. Sometimes it is doubtful whether the formation of $R(C≡C)_2R$ results from inadequate handling of the reaction mixture or whether a special reaction type leads only to the diyne [52]. In some cases there has been no conclusion as to whether $R(C≡C)_2R$ is formed from O_2 or from other oxidants [25, 34, 35, 40, 46]. Air oxidation also causes contamination of RC≡CCu by $R(C≡C)_2R$ etc. on storage and drying, but air-stable RC≡CCu seem to exist too [33, 41]. The air oxidation is enhanced by UV irradiation [36]. Reaction Type 2a as a preparative method for $R(C≡C)_2R$ is usually carried out in alcoholic aqueous solutions or suspensions of RC≡CCu and NH_3 with aeration [2, 16 to 18, 28]. The oxygen demand is much higher than calculated. Obviously NH_3 is oxidized too [18].

References on pp. 70/1

The so-called "Glaser Coupling" is the corresponding in situ reaction. Oxidation of RC≡CH in the presence of Cu^I compounds, normally in weakly acidic or neutral solutions, gives R(C≡C)$_2$R. Discovered in 1869 by Glaser, it is a simple and safe method to prepare R(C≡C)$_2$R. There exist contradictory opinions as to whether the catalytic cycle begins with the RC≡CCu formation or not. Probably the answer to this question depends on the nature of R and of the reaction conditions. In some cases a certain amount of RC≡CCu is precipitated and can be filtered off [20, 30]. In other cases, RC≡CCu could not be detected in the reaction batch by polarography [37]. From the reaction rate law, it follows that RC≡CCu cannot be the only intermediate [32]. Terminal acetylenes like 1-ethynylcyclohexanol are oxidized with up to 93% yields by bubbling O$_2$ through an acetone solution in the presence of a Cu^I complex [49].

Oxidation by Cu^{II} compounds (Reaction Type 2b). Not only salts like Cu[O$_2$CCH$_3$]$_2$ can be used, but also Cu^{II} complexes [19], and organic Cu^{II} compounds like Cu[OC(CH$_3$)$_3$]$_2$ [43]. In most cases the RC≡CCu is formed in situ during the oxidation reaction from RC≡CH with Cu^{II}. This reaction is called "Eglinton Coupling". In contrast to the "Glaser Coupling", RC≡CCu as an intermediate seems to be doubtless. As found in [28], with R = CR$_2^I$OH, the Cu^I formed during the reaction is not sufficient for RC≡CCu formation, and additional copper salt must be fed to the reaction mixture. The reaction velocity depends on the rate of the RC≡CCu formation and on the basicity of the solution (amines, mostly pyridine). The oxidation involves one-electron transfer [31]. An unstable dimeric Cu^{II} acetylide is obviously formed from RC≡CCu, which decomposes to give R(C≡C)$_2$R and completes the catalytic cycle by regeneration of Cu$^+$. This is oxidized to Cu^{II} again. The reaction of CuCl with O$_2$ in pyridine has been shown to form CuCl$_2$(pyridine)$_2$ and a soluble Cu^{II} compound. For the mechanism of the "Eglinton Coupling" see [28, 32]. The precipitation of RC≡CCu during the reactions has been observed [22], but is not desirable, for it seems to reduce the reaction rate [23].

The oxidative preparation of R(C≡C)$_2$R from RC≡CCu with K$_3$Fe(CN)$_6$ (Reaction Type 2c) has been widely used. In part, this reaction is performed in strongly alkaline (KOH) aqueous solution [10, 11, 21].

A great number of further oxidation reactions giving R(C≡C)$_2$R is summarized under Reaction Type 2d in Table 1. Oxidizing agents like concentrated HNO$_3$ are too aggressive to give R(C≡C)$_2$R and cause further oxidation or even explosions [3, 13 to 15].

Halogenation

The halogenation with retention of the C≡C bond (Reaction Type 3) can be achieved with X$_2$ (X = Br or I) in organic solvents like CH$_2$Cl$_2$ or n-hexane; RC≡CX is the product. Cl$_2$ is not suitable; it leads to decomposition of the initial alkyne. In the reaction of RC≡CCu with Br$_2$ the copper can be oxidized to Cu^{II} and CuBr$_2$ is isolated [48]. Detonations and inflammations have been observed not only with Cl$_2$, but sometimes also at contacting of RC≡CCu with Br$_2$ vapors diluted by air admixture and on mixing with elemental I$_2$ [1]. KI$_3$ is effective in the preparation of IC≡CR in aqueous solutions. It is added until a consistent brown color of the solution is maintained with stirring [4, 6]. Interhalogens can also be reacted [12, 47]. 4-O$_2$NC$_6$H$_4$C≡CCu and I$_2$ do not give 4-O$_2$NC$_6$H$_4$C≡CI, but only 4-O$_2$NC$_6$H$_4$(C≡C)$_2$C$_6$H$_4$NO$_2$-4 [17]. Halogenation of Reaction Type 3 generally occurs along with oxidation and halogen addition (Reaction Types 2d and 4).

Halogenation followed by halogen addition (Reaction Type 4) gives RCX=CX$_2$. There exists no detailed investigation about the scope and the limitation of Type 3 and 4 reactions. KI$_3$ is a useful reagent too [6 to 8]; I$_2$ can be used in warm water [9].

References on pp. 70/1

"Straus Reaction"

$RC\equiv CCu$ gives $RCH=CHC\equiv CR$ in anhydrous boiling CH_3CO_2H with admission of air; afterwards one molecule CH_3CO_2H may add to the $C\equiv C$ bond (Reaction Type 5). For example, $CH_3O_2CC\equiv CCu$ does not give the expected $CH_3O_2CCH=CHC\equiv CCO_2CH_3$, but the acetoxy muconate $CH_3O_2CCH=CHC(O_2CCH_3)=CHCO_2CH_3$ [24]. A complex $C_6H_5C\equiv CCu \cdot CH_3CO_2Cu$ (see Section 1.1.2.5) has been isolated and is supposed to be an intermediate of the "Straus Reaction" [17]. This reaction can also be carried out with $RC\equiv CCu$ formed in situ from $RC\equiv CH$ and $CuCl$ [29].

The geometry of the product $RCH=CHC\equiv CR$ has not been studied in detail. There have been reported isolated (E)-compounds [24, 29]; an (E)/(Z)-mixture is presumed [27]. From acetylides $CuC\equiv CZC\equiv CCu$ the resulting cyclic enynes do not permit an (E)-structure. Therefore only (Z)-"Straus products" are observed. If substitution reactions with $RC\equiv CCu$ are carried out in CH_3CO_2H as a solvent, oxygen must be excluded. Otherwise, the "Straus Reaction" product can be formed, cf. the reaction of $C_6H_5C\equiv CCu$ and 4,6-diiodoresorcinol [34].

Insertion of CO_2

$RC\equiv CCu$ and CO_2 give $RC\equiv CCO_2Cu$ (Reaction Type 6). This is the reverse of the Preparation Method V (see p. 4) [45].

References:

[1] R. Böttger (Jahresber. Physik. Ver. Frankfurt **1856/57** 37/4; Dinglers Polytech. J. **152** [1859] 22/9).
[2] C. Glaser (Ber. Deut. Chem.Ges. **2** [1869] 422/4).
[3] C. Glaser (Liebigs Ann. Chem. **154** [1870] 137/71).
[4] R. Lespieau (Ann. Chim. [Paris] [7] **11** [1897] 232/88).
[5] F. Straus (Liebigs Ann. Chem. **342** [1905] 190/265).
[6] R. Lespieau (Bull. Soc. Chim. France [4] **3** [1908] 638/40).
[7] R. Lespieau, M. Pariselle (Compt. Rend. **146** [1908] 1035/7).
[8] M. Lespieau (Compt. Rend. **150** [1910] 113/4).
[9] M. Lespieau (Ann. Chim. [Paris] [8] **27** [1912] 137/89).
[10] F. Kunckell (Ber. Deut. Pharm. Ges. **23** [1913] 188/227).

[11] S. Reich (Arch. Sci. Phys. Nat. [4] **45** [1918] 259/76).
[12] L. B. Howell, W. A. Noyes (J. Am. Chem. Soc. **42** [1920] 991/1010).
[13] C. Moureu, C. Dufraisse, J. R. Johnson (Ann. Chim. [Paris] [10] **7** [1927] 14/42, 34).
[14] A. B. Garcia (9th Congr. Intern. Quim. Pura Aplicada, Madrid 1934, Vol. 6, pp. 310/20; C. **1936** II 2411).
[15] A. L. Klebanskii, U. A. Dranitsyna, I. M. Dobromil'skaya (Dokl. Akad. Nauk SSSR **1935** II 229/32; C. **1936** I 2528).
[16] Yu. S. Zal'kind, B. M. Fundyler (Zh. Obshch. Khim. **6** [1936] 530/3).
[17] P. Piganiol (Acétylène, Homologues et Dérivés, Massan, Paris 1945, p. 263).
[18] H. H. Schlubach, V. Wolf (Liebigs Ann. Chem. **568** [1950] 141/59).
[19] R. Nast (Z. Naturforsch. **8b** [1953] 381/3).
[20] J. P. Riley (J. Chem. Soc. **1953** 2193/8).

[21] A. Vaitiekunas, F. F. Nord (J. Am. Chem. Soc. **76** [1954] 2733/6).
[22] J. H. Baxendale, D. T. Westcott (private communication from [23]).
[23] G. Eglinton, A. R. Galbraith (Chem. Ind. [London] **1956** 737/8).
[24] M. Akhtar, C. L. Weedon (Proc. Chem. Soc. **1958** 303).

[25] Nippon Kagaku Kai Hen, "Jikken Kagaku Koza, Kikagobutsu no Gosei to Seisei", Ganzen **9** [1958] 391; Communications of the Chemical Society of Japan, "Experimental Chemical Correspondence, Synthesis and Purification of Compounds", Complete Ed. **9** [1958] 391.

[26] W. J. Gensler, J. Casella Jr. (J. Am. Chem. Soc. **80** [1958] 1376/80).

[27] M. Akhtar, T. A. Richards, B. C. L. Weedon (J. Chem. Soc. **1959** 933/40).

[28] G. Eglinton, A. R. Galbraith (J. Chem. Soc. **1959** 889/96).

[29] L. Scatteböl (Acta Chem. Scand. **13** [1959] 198).

[30] A. S. Hay (J. Org. Chem. **27** [1962] 3320/1).

[31] A. A. Clifford, W. A. Waters (J. Chem. Soc. **1963** 3056/62).

[32] F. Bohlmann, H. Schönowsky, E. Inhoffen, G. Gran (Chem. Ber. **97** [1964] 794/800).

[33] V. S. Myl'nikov (Dokl. Akad. Nauk SSSR **164** [1965] 622/5; Dokl. Phys. Chem. Proc. Acad. Sci. USSR **160/165** [1965] 693/5; C.A. **63** [1965] 17 258).

[34] C. E. Castro, E. J. Gaughan, D. C. Owsley (J. Org. Chem. **31** [1966] 4071/8).

[35] F. W. Hill (Metall **22** [1968] 135/40).

[36] V. S. Myl'nikov, A. N. Terenin (J. Polym. Sci. Polym. Symp. No. 16 [1965/68] 3655/65; C.A. **70** [1969] No. 38 243).

[37] T. F. Rutledge (Acetylenic Compounds, Reinhold, New York – Amsterdam – London 1968, pp. 84/95, 207/10, 240/68).

[38] M. D. Rausch, A. Siegel, L. P. Klemann (J. Org. Chem. **34** [1969] 468/70).

[39] R. H. Mitchell, F. Sondheimer (Tetrahedron **26** [1970] 2141/50).

[40] M. Rosenblum, N. M. Brawn, D. Ciappenelli, J. Tancrede (J. Organometal. Chem. **24** [1970] 469/77).

[41] M. S. Newman, M. W. Logue (J. Org. Chem. **36** [1971] 1398/401).

[42] Y. Okamoto, K. L. Chellappa, S. K. Kundu (J. Org. Chem. **37** [1972] 3185/7).

[43] T. Tsuda, T. Hashimoto, T. Saegusa (J. Am. Chem. Soc. **94** [1972] 658/9).

[44] Y. Okamoto (AD-760772 [1973] 1/84; C.A. **80** [1974] No. 37 537).

[45] T. Tsuda, K. Ueda, T. Saegusa (J. Chem. Soc. Chem. Commun. **1974** 380/1).

[46] O. M. Abu Salah, M. I. Bruce (Australian J. Chem. **29** [1975] 531/41).

[47] P.-L. Compagnon, B. Grosjean (Synthesis **1976** 448/9).

[48] J. S. Miller, S. T. Matsuo (J. Polym. Sci. Polym. Letters Ed. **128** [1980] 809/10).

[49] K. Bowden, I. Heilbron, E. R. H. Jones, K. H. Sargent (J. Chem. Soc. **1947** 1579/83).

[50] J. F. Normant, G. Cahiez, C. Chuit, J. Villieras (J. Organometal. Chem. **77** [1974] 281/7).

[51] A. Afzali, H. Firouzabadi, A. Khalafinejad (Syn. Commun. **13** [1983] 335/9).

[52] A. K. Filippova, G. S. Lyashenko, I. D. Kalikhman, V. Y. Vitkovskii, G. G. Naumova, N. S. Vyazankin (Izv. Akad. Nauk SSSR Ser. Khim. **1983** 950/3; Bull. Acad. Sci. USSR Div. Chem. Sci. **32** [1983] 861/3).

1.1.2.1.3   General Remarks about Reactions of RC≡CCu with R'X to Form RC≡CR' and Its Ring Closure Products

Within the least twenty years, the reaction $RC≡CCu + XR' \rightarrow RC≡CR' + CuX$ (X = I, Br etc.) has become the most important means in organic chemistry to introduce alkynyl groups into various molecules. The number of known examples is very large, and the group R' shows a wide variation. Different reaction paths including subsequent reactions of RC≡CR' were observed. The reaction is therefore described under different "Reaction Types" (7a to 7p with formation of RC≡CR' and 8a to 8d with formation of ring closure products of RC≡CR'):

Reaction Type 7: RC≡CCu + R'X → RC≡CR' + CuX

  a: reactions with R'X where X is bonded to an sp³ carbon, see Section 1.1.2.1.4.1,

  b: reactions with acyl halides, see Section 1.1.2.1.4.2,

  c: reactions with haloalkenes, haloallenes, and allenyl esters, see Section 1.1.2.1.4.3,

  d: reactions with haloalkynes, see Section 1.1.2.1.4.4,

  e: reactions with C_6H_5I, see Section 1.1.2.1.4.5,

  f: reactions with 2-substituted halobenzenes, see Section 1.1.2.1.4.6,

  g: reactions with 3-substituted halobenzenes, see Section 1.1.2.1.4.6,

  h: reactions with 4-substituted halobenzenes, see Section 1.1.2.1.4.6,

  i: reactions with di- and polysubstituted halobenzenes, see Section 1.1.2.1.4.7,

  j: reactions with compounds containing more than one iodobenzene moiety, see Section 1.1.2.1.4.8,

  k: reactions with iodonaphthalenes, see Section 1.1.2.1.4.9,

  l: reactions with other carbocycles R'X, see Section 1.1.2.1.4.10,

  m: reactions with 5-membered heterocycles R'X, see Section 1.1.2.1.4.11,

  n: reactions with 6-membered heterocycles R'X, see Section 1.1.2.1.4.12,

  o: reactions with condensed heterocycles R'X, see Section 1.1.2.1.4.13,

  p: reactions with R'X where X is bonded to S, P, Si, or Sn, see Section 1.1.2.1.4.14.

Reaction Type 8: RC≡CCu + R'X → ring closure product of RC≡CR' + CuX

  a: ring closure giving furans, see Section 1.1.2.1.5.1,

  b: ring closure giving pyrroles or thiophenes, see Section 1.1.2.1.5.2,

  c: ring closure giving other heterocycles, see Section 1.1.2.1.5.3,

  d: ring closure giving carbocycles, see Section 1.1.2.1.5.4.

The molecules R'X, into which the alkynyl group is introduced, must have a suitable leaving group X. In most cases X is iodine. Bromine is less reactive or unreactive [8]; only bromoalkynes (R'X is $R^1C≡CBr$) are used on a broader scale. Chlorine is not a useful leaving group in this reaction, except at highly activated sites like in $2,4,6-(O_2N)_3C_3H_2Cl$ and in R^1COCl. Further leaving groups are benzothiazol-2-yloxy and CN. For polyhalo compounds, one or more halogen atoms can be substituted by RC≡C and often side reactions are observed.

The leaving group X can be bonded to an sp³-, sp²-, or sp-hybridized carbon or to another atom like Si, P, S, Sn, Pt, Hg, Ru, Rh, Re, or Ir. With sp³-hybridized carbon the reaction is restricted to certain structures like the allyl-, benzyl-, or phenacyl-type. CH_3I, for instance, can only be reacted with RC≡CCu in the presence of a special catalytic system.

References on p. 75

Bromo-ribofuranosides are understood as heterocycles (Reaction Type 7m), not as alkyl bromides.

The most common R'X within these Reaction Types 7 and 8 are halobenzenes and 2-iodothiophenes. But the range is very wide and even haloporphines were reacted.

In most cases the reaction is accomplished by refluxing the mixture of the reactants for 5 to 120 h in solvents like pyridine or dimethylformamide. Far milder conditions than usual are sufficient if the reaction is catalyzed by Pd compounds. This reaction has been studied in detail with $C_6H_5C\equiv CCu$ and C_6H_5I and is therefore described more in detail in Section 1.1.2.1.4.5. It should be noted that there are also a few substrates that permit reaction times in the range of minutes without catalysis. Besides pyridine and dimethylformamide, many other solvents were used, including even aqueous systems [9]. Auxiliary bases are normally not necessary except for Reaction Type 7b and sometimes 7a. If highly insoluble products are formed, the $RC\equiv CCu$ is better added in more than one portion to the reaction mixture [7]. $RC\equiv CCu$ is not introduced into the reaction as a pure substance in all cases. To avoid oxidative side reactions it is often better to use less than the equimolar amount of Cu^I. Often a few mol% of a Cu^I salt is sufficient for the reaction of $RC\equiv CH$ with R'X. The reason for this is that Cu^+, set free during the substitution reaction, can react to form additional $RC\equiv CCu$. Finally, $RC\equiv CH$ is totally converted to the acetylide. In other cases there is no exact proof for the formation of $RC\equiv CCu$ as an intermediate but it is presumed. For instance, $RC\equiv CH$, activated copper powder, K_2CO_3, and R'X (X=I or Br) in pyridine or dimethylformamide give in a 70 to 80% yield the same $RC\equiv CR'$ as is formed from $RC\equiv CCu$ and R'X [4 to 6].

Cu^I-catalyzed Grignard reactions of the type $RC\equiv CMgBr + XR' \rightarrow RC\equiv CR' + MgBrX$ (X= halogen) have only been entered into the tables exceptionally, though they were believed to be formally reactions of Type 7. It is an open question whether the reaction of $RC\equiv CMgBr$ and R'X in the presence of copper salts involves $RC\equiv CCu$ as an intermediate [1 to 3]. In some cases at the end of the coupling processes yellow precipitates of $RC\equiv CCu$ have been observed [2]. But the Cu^I-catalyzed Grignard reactions seem to be more complex. The anions of the added Cu^I salts also play a role [2], and the activation of the C–halogen bond in XR', without $RC\equiv CCu$ formation, has been formulated [1].

In many cases, reactions of $RC\equiv CCu$ and R'X do not yield $RC\equiv CR'$, but a different product is isolated. This can be the consequence of the work-up conditions (e.g., the saponification of the tetrahydropyran-2-ylmethyl group to $HOCH_2$). $1,8-(RC\equiv C)_2C_{10}H_6$ (from 1,8-diiodonaphthalene and 2 $RC\equiv CCu$) rearranges to benzo[k]fluoranthenes, and the relative yield of the two products is a function of the heating time (see Section 1.1.2.1.4.9). In some cases there is definite proof that $RC\equiv CR'$ is formed during the reaction as a real species.

If R'X contains a suitable structural unit, in most cases -CX=C(ZH)- (Z = NH, NR'', O, S, CO_2), ring closure reactions are observed (Reaction Types 8a to 8c); see Scheme I. Usually A means an anellated ring system and the product is a bi- or polycyclic compound.

I

For instance, the reaction of $RC\equiv CCu$ and 2-iodophenol (X=I, Z=O and A=CH=CH-CH=CH) gives a 2-substituted benzo[b]furan; see Scheme II, p. 74. The probable intermediate 2-hydroxytolane has never been isolated; see Section 1.1.2.1.5.1.

References on p. 75

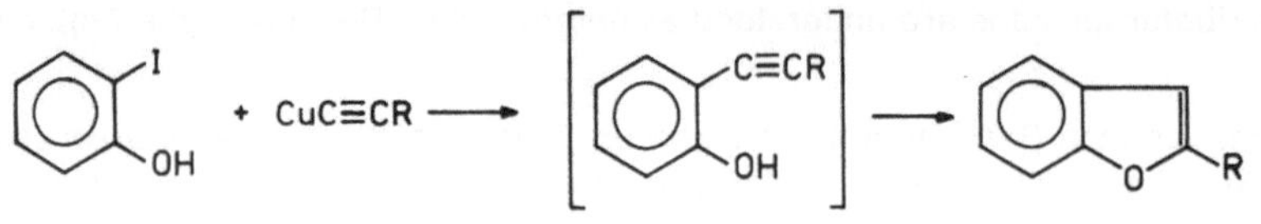

With Z=NH or NR'', in heterogeneous reaction systems the reaction runs analogously. From RC≡CCu and 2-iodoanilines indoles are formed. In contrast, with Z=NH or NR'' in homogeneous reaction systems, a mixture of RC≡CR' and the cyclization product (e.g., the indole) is formed; see Scheme III and Section 1.1.2.1.5.2.

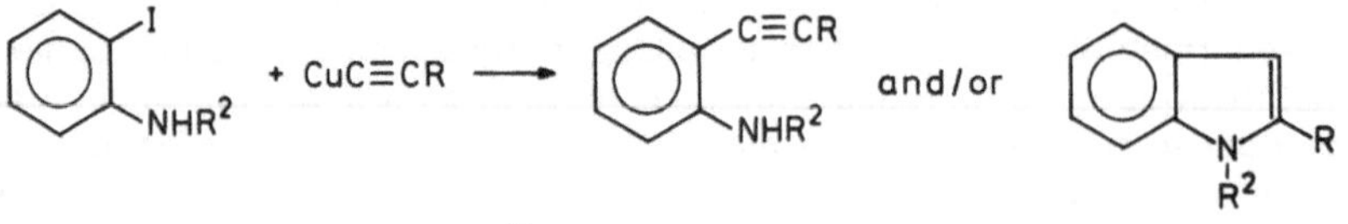

In Section 1.1.2.1.5.3 the formation of additional heterocyclic ring systems from RC≡CCu and R'X is listed as Reaction Type 8c. In almost all examples, A means an anellated benzene ring. The reaction is more important for Z=CO_2; phthalides and/or isocoumarines are thus produced; see Schemes IV to VI.

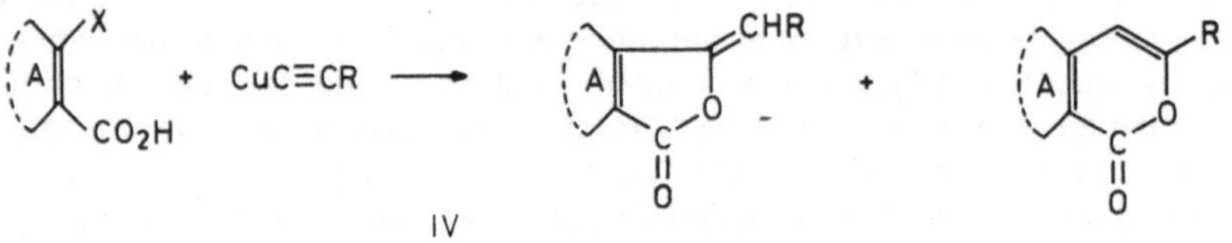

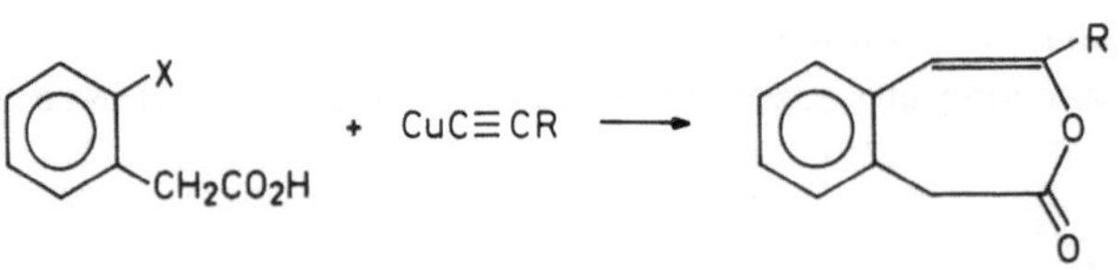

The synthesis of carbocyclic ring systems from RC≡CCu is listed under Reaction Type 8d in Section 1.1.2.1.5.4. Such rings can be formed intramolecularly, if R contains a leaving group like iodine. An intermolecular head-to-tail-substitution (2 to 4, or 6 molecules RC≡CCu per formed ring system) gives carbocycles. The tarry substances, that are normally the main products, should result from repeated intermolecular combination.

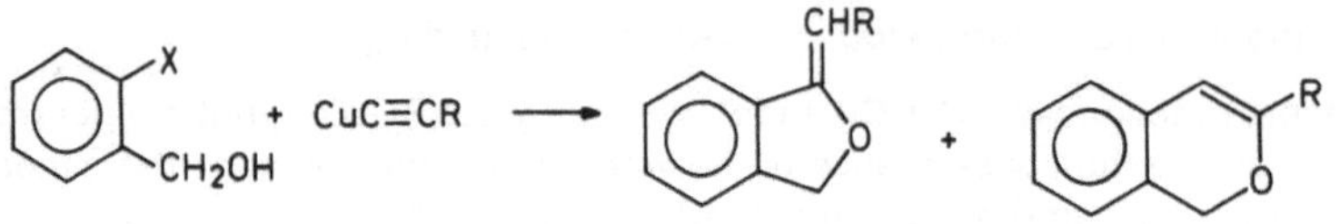

A further source of carbocycles from RC≡CCu are rearrangement reactions. Electrocyclic reactions starting from aromatic systems which are substituted by 2 C≡C groups are an important example. Reactions via arynes and dimerizations or trimerizations of unstable carbenes were also formulated (see Section 1.1.2.1.5.4).

References:

[1] W. J. Gensler, G. R. Thomas (J. Am. Chem. Soc. **73** [1951] 4601/4).
[2] W. J. Gensler, A. P. Mahadevan (J. Am. Chem. Soc. **77** [1955] 3076/9).
[3] G. Eglinton, W. McCrae (Advan. Org. Chem. **4** [1963] 225/328).
[4] M. S. Shvartsberg (Izv. Akad. Nauk SSSR Ser. Khim. **1970** 1144/9; Bull. Acad. Sci. USSR Div. Chem. Sci. **1970** 1079/82).
[5] M. S. Shvartsberg, A. A. Moroz, I. L. Kotlyarevskii (Izv. Akad. Nauk SSSR Ser. Khim. **1971** 1306/10; Bull. Acad. Sci. USSR Div. Chem. Sci. **1971** 1209/12).
[6] M. S. Shvartsberg, A. A. Moroz (Izv. Akad. Nauk SSSR Ser. Khim. **1971** 1582/5; Bull. Acad. Sci. USSR Div. Chem. Sci. **1971** 1488/90).
[7] A. Kasahara, T. Izumi, M. Maemura (Bull. Chem. Soc. Japan **50** [1977] 1021/2).
[8] J. S. Kiely, P. Boudjouk, L. L. Nelson (J. Org. Chem. **42** [1977] 2626/9).
[9] R. P. Duffley, R. Stevenson (J. Chem. Res. S **1978** 468).

1.1.2.1.4   Reactions with R′X to Form RC≡CR′

1.1.2.1.4.1   Reactions with R′X where X is Bonded to a sp³ Carbon

This reaction is referred to as "Reaction Type 7a" in Table 1 on pp. 17/41; X is usually halogen, sometimes benzothiazol-2-yloxy, CH_3CO_2, or $CH_3S(O)O$.

Simple primary, secondary, and tertiary alkyl halides do not react with RC≡CCu, but there exists one example of a catalyzed conversion. The reaction of $C_6H_5C≡CCu$ and CH_3I in tetrahydrofuran in the presence of $PdCl_2$ and $[N(C_4H_9)_4]I$ yields 46% $C_6H_5C≡CCH_3$ [18]. The same product has been obtained, when the complex "$C_6H_5C≡CCu·3P(C_4H_9-n)_3$" (see Section 1.1.2.4) and CH_3I were reacted [14].

In contrast to the simple alkyl compounds, halides of the allyl, benzyl, and phenacyl type react with RC≡CCu following the equation RC≡CCu + XR′ → RC≡CR′ + CuX; see Table 2, pp. 76/8. A catalyst is not necessary.

In many cases RC≡CCu is not added to the reaction mixture as a substance, but is formed in the batch from RC≡CH and Cu^I compounds. The precipitation of RC≡CCu occurs immediately. It is later consumed during the much slower substitution reaction. As discussed in Section 1.1.2.1.3, some mol% Cu^I, instead of the preformed RC≡CCu, is often sufficient to perform substitution reactions and sometimes even problems with side reactions can be cancelled. Nevertheless there are cases, where molar amounts of Cu^I are necessary [1]. Some reactions using in situ (visibly) formed RC≡CCu and even some "catalytic" reactions (with very low copper rates) are listed in Table 2. This has been exceptionally done, for they have in the case of Reaction Type 7a a high preparative value.

A detailed study and comparison using dry solid RC≡CCu, or stoichiometric amounts of Cu^I, RC≡CH, and R′X, or catalytic amounts of Cu^I, 1 mol RC≡CH, and 1 mol R′X is given in [11] for the compounds No. 8, 18, 69, and 90 (as in Table 1). Sometimes the yields depend more upon the nature of the auxiliary base than upon the method used. The following yields are reported, as shown for the reaction of $C_6H_5C≡CH$ (0.1 mol)/CuBr (1 g) with $CH_2=CHCH_2Br$: tertiary amines 0%, n-$C_4H_9NH_2$ 32%, K_2CO_3 68%. CuCl instead of CuBr is

References on p. 78

not recommended. Halogen exchange leads to the less reactive allyl chloride. Cupric acetate oxidizes $C_6H_5C\equiv CH$ to $C_6H_5(C\equiv C)_2C_6H_5$ [11].

The usual solvents are dimethylformamide, hexamethylphosphoric triamide, and tetrahydrofuran. N-methylpyrrolid-2-one, nitrobenzene, and mixtures of C_2H_5OH and H_2O were also used. Some reactions have been carried out by mixing the components without solvent.

The yields depend on the method of preparation of $RC\equiv CCu$. For instance, if $C_6H_5C\equiv CCu$ is prepared in hexamethylphosphoric triamide, the yield of the reaction with allyl bromide is 16%. When prepared in toluene, $C_6H_5C\equiv CCu$ does not react [11].

If the leaving group X is benzothiazol-2-yloxy, the enhanced reactivity of $C_6H_5C\equiv CCu$ and the high stereospecificity can be ascribed to the coordination of the organometallic species to the nitrogen atom in the heterocycle. A suitable solvent is tetrahydrofuran [16].

The reaction of $HOCH_2C\equiv CCu$ or $4\text{-}IC_6H_4C\equiv CCu$ and CH_2I_2 produces polymeric products, whose structures have not been reported [7].

Table 2
Reaction Type 7a (explanation see pp. 12/4):
$RC\equiv CCu + R'X \rightarrow RC\equiv CR' + CuX$ where X is Bonded to a sp^3 Carbon.

R in RC≡CCu	R′X	remarks (yield of RC≡CR′ in %) [a] and [b]: explanation see p. 78	Ref.
HOCH$_2$	CH$_2$=CHCH$_2$Br	(13)[a], (42)[b]	[5,7,11]
	CH$_2$=CHCH$_2$Cl	(12), (65)[b]	[4,6]
	(E)-CH$_3$CH=CHCH$_2$Cl	(65)[b] and rearrangement product	[6]
	CH$_2$=C(CH$_3$)CH$_2$Cl	(13), (13)[a], (65)[b]	[4,5,6]
	(E)-C$_6$H$_5$CH=CHCH$_2$Br	(8)[a]	[5]
CH$_3$CH(OH)	CH$_2$=CHCH$_2$Cl	(85)[b]	[6]
	CH$_2$=C(CH$_3$)CH$_2$Cl	(67)[b]	[6]
C$_2$H$_5$O	CH$_2$=CHCH$_2$Br	isolation as C$_2$H$_5$O$_2$C(CH$_2$)$_2$CH=CH$_2$ (39)[b]	[11]
(Z)-CH$_3$OCH=CH			
	CH$_2$=CHCH$_2$Br	no yield given	[15]
	C$_6$H$_5$CH$_2$Br	no yield given	[15]
(CH$_3$)$_2$C(OH)	CH≡CCH$_2$Cl	(32)[a]	[4,5]
	CH$_2$=CBrCH$_2$Br	only (CH$_3$)$_2$C(OH)C≡CCH$_2$CBr=CH$_2$ (50)[a]	[5]
	ClCH$_2$CH=CHCH$_2$Cl	(28)[a] monosubstituted and a few di-substituted product	[5]
	CH$_2$=CHCH$_2$Br	(53)[a]	[5]
	CH$_2$=CHCH$_2$Cl	(53)[a], (59)[b]	[4,6]
	CH$_2$=C(CH$_3$)CH$_2$Cl	(50)[a], (50)[b]	[4,5,6]
	(E)-C$_6$H$_5$CH=CHCH$_2$Br	(38), (45)[a]	[4,5]

References on p. 78

Table 2 [continued]

R in RC≡CCu	R′X	remarks (yield of RC≡CR′ in %) [a] and [b]: explanation see p. 78	Ref.
(CH$_3$)$_3$Si	(CH$_3$)$_2$C=C=CHBr	(33) with 4 mol % [Pd(P(C$_6$H$_5$)$_3$)$_4$] at 45°	[19]
	(CH$_3$)$_2$C=C=CHCO$_2$CH$_3$	(4) with 4 mol % [Pd(P(C$_6$H$_5$)$_3$)$_4$] at 45°	[19]
n-C$_4$H$_9$	n-C$_4$H$_9$C≡CCH$_2$Br	part of the n-C$_4$H$_9$C≡C moiety is introduced as n-C$_4$H$_9$C≡CMgBr	[2]
n-C$_5$H$_{11}$	CH≡CCH$_2$Br	(88) and n-C$_5$H$_{11}$C≡CCH=C=CH$_2$ (12)	[13]
	CH$_2$=CHCH$_2$Br	(13 to 96, strongly dependent on the auxiliary base and the solvent), (44[a]), (70)[b]	[11, 13]
	CH$_2$=CHCH$_2$Cl	(60)	[13]
	CH$_2$=CHCH$_2$I	(74)	[13]
	CH$_2$=CHCH(CH$_3$)OC$_7$H$_4$NS	rearrangement to (E)-n-C$_5$H$_{11}$C≡CCH$_2$CH=CHCH$_3$ (40)	[16]
	CH$_3$CH=CHCH$_2$Br	n-C$_5$H$_{11}$C≡CCH$_2$CH=CHCH$_3$ and n-C$_5$H$_{11}$C≡CCH(CH$_3$)CH=CH$_2$, total yield 76%	[13]
	CH$_2$=C(CH$_3$)CH$_2$Br	(80)	[13]
	CH$_2$=C(CH$_3$)CH$_2$Cl	(63)	[13]
	(CH$_3$)$_2$C=CHCH$_2$Br	(50)	[13]
	(CH$_3$)$_2$C=CHCH$_2$Cl	(60)	[13]
(C$_2$H$_5$)$_2$C(OH)	CH$_2$=CHCH$_2$Cl	(30)[a]	[5]
C$_6$H$_5$	CH$_2$=CHCH$_2$Br	(16) in hexamethylphosphoric triamide, (94) in the same solvent + 1 mol [N(C$_4$H$_9$)$_4$]I + 0.01 mol C$_6$H$_5$PdI[P(C$_6$H$_5$)$_3$]$_2$, (83) in nitrobenzene at 240°	[7, 9, 10, 17, 18]
	CH$_2$=CHCH$_2$Cl	(12, excess C$_6$H$_5$C≡CCu necessary), (30)[a]	[3, 6]
	CH$_2$=CHCH(CH$_3$)OC$_7$H$_4$NS	rearrangement to (E)-C$_6$H$_5$C≡CCH$_2$CH=CHCH$_3$ (70)	[16]
	CH$_2$=CHCH(C$_2$H$_5$)OC$_7$H$_4$NS	rearrangement to (E)-C$_6$H$_5$C≡CCH$_2$CH=CHC$_2$H$_5$ (68)	[16]
	CH$_3$CH=CHCH$_2$Cl, (E)/(Z)-mixture	(E)-C$_6$H$_5$C≡CCH$_2$CH=CHCH$_3$ (73) and C$_6$H$_5$C≡C(CH$_2$)$_2$CH=CH$_2$ (27), no (Z)-compound	[12]
	CH$_2$=C(CH$_3$)CH$_2$Cl	(20)[b]	[6]
	(CH$_3$)$_2$C=CHCH$_2$Br	(<5)	[11]

References on p. 78

Table 2 [continued]

R in RC≡CCu	R'X	remarks (yield of RC≡CR' in %) a) and b): explanation see below	Ref.
C_6H_5 [continued]	$C_6H_5CH_2Br$	(90) at 245° in N–methylpyrrolidine	[9]
	$C_6H_5COCH_2Br$	at 240° in nitrobenzene only 2,5–di-phenylfuran (54), at 140° in glycol only $C_6H_5COCH_3$ and tars	[8, 9]

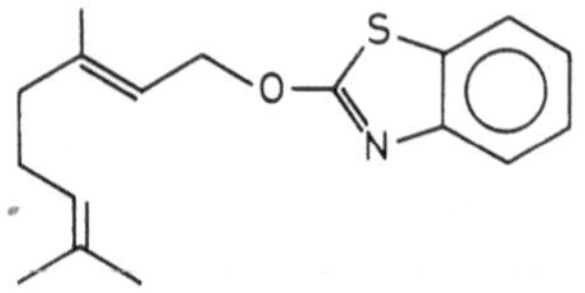

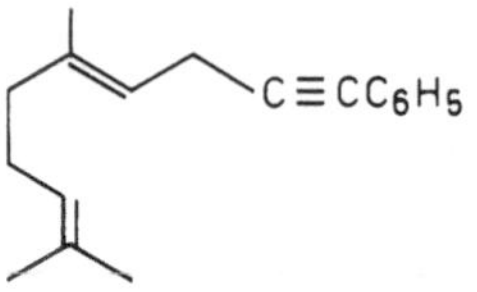

(60)   [16]

Explanations for Table 2:

a) RC≡CCu is formed in the reaction mixture without isolation from equimolar amounts RC≡CH and Cu^I.

b) RC≡CCu is formed in the reaction mixture without isolation from RC≡CH and less than equimolar amounts of Cu^I.

Yields not marked by a) or b) are obtained with pure RC≡CCu.

C_7H_4NS = benzothiazol-2-yl.

References:

[1] P. Kurtz, Bayer A.-G. (Ger. 1007767 [1954/57]; C.A. **1959** 14936).

[2] W. J. Gensler, A. P. Mahadevan (J. Am. Chem. Soc. **77** [1955] 3076/9).

[3] P. Kurtz, Bayer A.-G. (Brit. 775723 [1955/57]; C.A. **1957** 16511).

[4] J. Colonge, R. Falcotet (Compt. Rend. **242** [1956] 1484/6).

[5] J. Colonge, R. Falcotet (Bull. Soc. Chim. France **1957** 1166/9).

[6] P. Kurtz (Liebigs Ann. Chem. **658** [1962] 6/20).

[7] A. M. Sladkov, L. Y. Ukhin, V. V. Korshak (Izv. Akad. Nauk SSSR Ser. Khim. **1963** 2213/5; Bull. Acad. Sci. USSR Div. Chem. Sci. **1963** 2043/5).

[8] C. E. Castro, E. J. Gaughan, D. C. Owsley (J. Org. Chem. **31** [1966] 4071/8).

[9] K. Gump, S. W. Moje, C. E. Castro (J. Am. Chem. Soc. **89** [1967] 6770/1).

[10] J. Ipaktschi, H. A. Staab (Tetrahedron Letters **1967** 4403/8).

[11] M. Bourgain, J. F. Normant (Bull. Soc. Chim. France **1969** 2477/81).

[12] C. E. Castro, R. Havlin, U. K. Honvad, A. Malte, S. Moje (J. Am. Chem. Soc. **91** [1969] 6464/70).

[13] J. F. Normant, M. Bourgain, A.-M. Rone (Compt. Rend. C **270** [1970] 354/7).

[14] T. Tsuda, Y. Chujo, T. Saegusa (J. Chem. Soc. Chem. Commun. **1975** 963/4).

[15] G. A. Kraus, K. Frazier (Tetrahedron Letters **1978** 3195/8).

[16] V. Calò, L. Lopez, G. Marchese, G. Pesce (Tetrahedron Letters **1979** 3873/4).

[17] N. A. Bumagin, I. O. Kalinovskii, I. P. Beletskaya (Izv. Akad. Nauk SSSR Ser. Khim. **1981** 2836; Bull. Acad. Sci. USSR Div. Chem. Sci. **30** [1981] 2366).

[18] N. A. Bumagin, I. O. Kalinovskii, A. B. Ponomarev, I. P. Beletskaya (Dokl. Akad. Nauk SSSR **265** [1982] 1138/43; Dokl. Chem. Proc. Acad. Sci. USSR **262/267** [1982] 262/6).

[19] K. Ruitenberg, H. Kleijn, H. Westmijze, J. Meijer, P. Vermeer (Rec. Trav. Chim. **101** [1982] 405/9).

1.1.2.1.4.2 Reactions with Acyl Halides

This reaction, a valuable source in the preparation of ynones, is referred to as "Reaction Type 7 b" in Table 1 on pp. 17/41:

$$RC \equiv CCu + XCOR^1 \rightarrow RC \equiv CCOR^1 + CuX$$

X is in most cases Cl, sometimes Br. Aryloxymethyl substituted Cu acetylides tend to undergo rearrangement and substitution at oxygen. From $ArOCH_2C \equiv CCu$ and R^1COX the following products can be expected: $ArOCH=C=CHCOR^1$, ArO_2CR^1, and $(ArOCH_2C \equiv C)_2$ [15, 16]. $C_6H_5OC \equiv CCu$ and C_6H_5COCl give only $C_6H_5CO_2C_6H_5$ [15].

The reaction is usually carried out in solvent mixtures containing pyridine, dioxane, or $Cl(CH_2)_2Cl$ and is catalyzed by tertiary amines like $N(C_2H_5)_3$. It is often sufficient to run the reaction at room temperature. Dimethylformamide is not a suitable solvent except in mixtures. In pure pyridine (no addition of a further amine) the reaction of $C_6H_5C \equiv CCu$ and acid chlorides gives only a complex from which the initial products can be recovered. No reaction takes place in anisole [2].

The yields depend on the order of the charging of the reactants too. The addition of first 2 mol triethylamine and then of 2 mol pyridine to a suspension or solution of $C_6H_5C \equiv CCu$ and R^1COCl (R^1 = aryl) in benzene, dioxane, or $Cl(CH_2)_2Cl$ gives better yields (up to 55%) than a simultaneous addition of both amines. Using only one amine, $C_6H_5C \equiv CCu$ does not react at a molar ratio of acid chloride to amine of 1:2 or 1:4 [2]. Advantageously, the reaction is carried out with more than equimolar amounts $RC \equiv CCu$. In this way the R^1COX is quantitatively converted and the excess of insoluble $RC \equiv CCu$ can be easily filtered. Often the same reaction is carried out with and without addition of Li, Mg, or similar salts. The presence of LiX, MgX_2, etc., implies the formation of heterocuprates (see Section 1.1.2.3) and is normally faster than the reaction with pure $RC \equiv CCu$ [10]. Acetyl chloride and $C_6H_5C \equiv CCu$ do not react to form methyl phenylethynylketone [2, 3]. This compound is only produced after addition of LiI [5].

To avoid the "troublesome preparation" of $RC \equiv CCu$, a method has been proposed reacting $RC \equiv CH$, R^1COCl, CuI, $(C_2H_5)_2NH$, and $[P(C_6H_5)_3]_2PdCl_2$. $RC \equiv CCu$ is supposed to be the intermediate, but a catalytic reaction has not been excluded. Therefore, the examples are not entered into Table 3 [11]. This reaction has also been used to prepare macrocyclic ketones [13].

Whereas the reaction to ynones has a wide scope, an analogous reaction with sulfonyl halides to give $RC \equiv CSO_2R^1$ is not feasible [2, 3].

Table 3
Reaction Type 7 b (explanation see pp. 12/4):
$RC \equiv CCu + R'X \rightarrow RC \equiv CR' + CuX$ (R'X = acyl halide).

R in $RC \equiv CCu$	R'X	remarks (yield in %; refers to $RC \equiv CR'$, if not otherwise stated)	Ref.
(Z)–$CH_3OCH=CH$	C_6H_5COCl	no yield given	[12]
n–C_3H_7	CH_3COCl	(75), protect from light	[4, 7]
	$C_6H_5(CH_2)_2COCl$	(81)	[4]

Table 3 [continued]

R in RC≡CCu	R'X	remarks (yield in %; refers to RC≡CR', if not otherwise stated)	Ref.
(CH$_3$)$_3$Si	CH$_3$COCl	(30)	[9]
	(CH$_3$)$_2$CHCOCl	(48)	[9]
	n-C$_5$H$_{11}$COCl	(62)	[9]
	4-ClC$_6$H$_4$COCl	(61)	[9]
	C$_6$H$_5$COCl	(66)	[9]
	4-CH$_3$C$_6$H$_4$COCl	(48)	[9]
n-C$_4$H$_9$	CH$_3$COCl	(10)	[5]
	CH$_3$O$_2$C(CH$_2$)$_2$COCl	(56)	[10]
	CH$_3$O$_2$CCH(CH$_3$)CH$_2$COCl	(55)	[10]
	CH$_3$O$_2$CCH(C$_6$H$_5$)CH$_2$COCl	(44)	[10]
n-C$_5$H$_{11}$	CH$_3$COBr	(40)	[6]
	C$_6$H$_5$COCl	(58)	[6]
C$_6$H$_5$	CH$_3$O$_2$C(CH$_2$)$_2$COCl	(52)	[10]
	CH$_3$O$_2$C(CH$_2$)$_3$COCl	(56)	[10]
	CH$_3$O$_2$CCH(CH$_3$)CH$_2$COCl	(53)	[10]
	4-BrC$_6$H$_4$COCl	(58)	[2]
	3-O$_2$NC$_6$H$_4$COCl	(53)	[2]
	4-O$_2$NC$_6$H$_4$COCl	(52)	[2]
	C$_6$H$_5$COCl	(42)	[1, 2]
	CH$_3$O$_2$C(CH$_2$)$_4$COCl	(61)	[10]
	C$_6$H$_4$(COCl)$_2$-1,3	yields at a 2:1 ratio only 1,3-bis-(phenylethynyl)benzene (42)	[2]
	C$_6$H$_4$(COCl)$_2$-1,4	yields at a 2:1 ratio only 1,4-bis(phenylethynyl)benzene (45)	[2]
	C$_6$H$_5$CH=CHCOCl	(40)	[2]
	CH$_3$O$_2$CCH(C$_6$H$_5$)CH$_2$COCl	(40)	[10]
	COCl$_2$	yields in toluene/pyridine only C$_6$H$_5$C≡CCOCl	[2]
	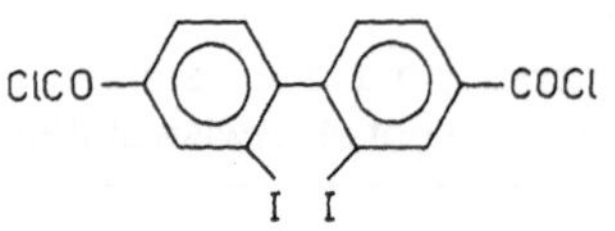 ClCO—⟨⟩—⟨⟩—COCl I I	a polymer containing free C$_6$H$_5$C≡CCu is reacted, no information about the incorporation of the supposedly formed ynone into the polymer is given; this reaction is claimed to be important to make polymers heat resistent	[14]

Table 3 [continued]

R in RC≡CCu	R'X	remarks (yield in % ; refers to RC≡CR', if not otherwise stated)	Ref.
C_6H_5O	C_6H_5COCl	yields only $C_6H_5CO_2C_6H_5$	[15]
$2,4,6-Br_3C_6H_2OCH_2$	C_6H_5COCl	yields only $2,4,6-Br_3C_6H_2O_2CC_6H_5$ (88)	[15]
$2,4,6-Cl_3C_6H_2OCH_2$	C_6H_5COCl	yields only $2,4,6-Cl_3C_6H_2O_2CC_6H_5$ (68)	[15]
$2-BrC_6H_4OCH_2$	C_6H_5COCl	yields only $(2-BrC_6H_4OCH_2C≡C)_2$ (40)	[15]
$4-BrC_6H_4OCH_2$	C_6H_5COCl	yields only $4-BrC_6H_4OCH=C=CHCOC_6H_5$ (39)	[15]
$4-O_2NC_6H_4OCH_2$	C_6H_5COCl	gives in benzene/$N(C_2H_5)_3$/pyridine at 60° $4-O_2NC_6H_4OCH=C=CHCOC_6H_5$ (52), $4-O_2NC_6H_4O_2CC_6H_5$ (20), and $(4-O_2NC_6H_4OCH_2C≡C)_2$ (28)	[16]
$C_6H_5OCH_2$	C_6H_5COCl	yields only $C_6H_5OCH=C=CHCOC_6H_5$ (30 [15], 81 [8])	[8, 15]
$4-CH_3C_6H_4OCH_2$	C_6H_5COCl	yields only $4-CH_3C_6H_4OCH=C=CHCOC_6H_5$ (37)	[15]
$4-CH_3OC_6H_4OCH_2$	C_6H_5COCl	reported to give $4-CH_3OC_6H_4OCH=C=CHCOC_6H_5$ (30) [8] or $(4-CH_3O-C_6H_4OCH_2C≡C)_2$ (72) [15] under similar conditions	[8, 15]
$2,4,6-(CH_3)_3C_6H_2$	C_6H_5COCl	(41)	[2]
$2,5-(CH_3)_2C_6H_3OCH_2$	C_6H_5COCl	yields only $2,5-(CH_3)_2C_6H_3OCH=C=CHCOC_6H_5$ (89)	[15]

References:

[1] A. M. Sladkov, I. R. Gol'ding (Zh. Org. Khim. 3 [1967] 1338; J. Org. Chem. [USSR] 3 [1967] 1298).

[2] A. M. Sladkov, I. R. Gol'ding (Khim. Atsetilena Tr. 3rd Vses. Konf., Dushanbe 1968 [1972], pp. 45/7; C.A. 79 [1973] No. 5431).

[3] M. Bourgain, J. F. Normant (Bull. Soc. Chim. France 1969 2477/81).

[4] C. E. Castro, R. Havlin, U. K. Honvad, A. Malte, S. Moje (J. Am. Chem. Soc. 91 [1969] 6464/70).

[5] J. F. Normant, M. Bourgain (Tetrahedron Letters 1970 2659/62).

[6] J. F. Normant, M. Bourgain, A.-M. Rone (Compt. Rend. C 270 [1970] 354/7).

[7] H. O. House, L. E. Huber, M. J. Umen (J. Am. Chem. Soc. 94 [1972] 8471/5).

[8] A. Kh. Filippova, G. S. Lyashenko, N. S. Vyazankin (Khim. Atsetilena Tr. 5th Vses. Konf., Tbilisi 1975, p. 104; C.A. 88 [1978] No. 190287).

[9] M. W. Logue, G. L. Moore (J. Org. Chem. 40 [1975] 131/2).

[10] R. Robin (Compt. Rend. C 282 [1976] 281/2).

[11] Y. Tohda, K. Sonogashira, N. Hagihara (Synthesis **1977** 777/8).

[12] G. A. Kraus, K. Frazier (Tetrahedron Letters **1978** 3195/8).

[13] Y. Onishi, M. Iyoda, M. Nakagawa (Tetrahedron Letters **22** [1981] 3641/4).

[14] S. Venkateso, U.S. Dept. of the Air Force (U.S. Appl. 219396 [1981]; C.A. **96** [1982] No. 7629).

[15] A. Kh. Filippova, G. S. Lyashenko, I. D. Kalikhman, V. Yu. Vitkovskii, G. G. Naumova, N. S. Vyazankin (Izv. Akad. Nauk SSSR Ser. Khim. **1983** 950/3; Bull. Acad. Sci. USSR Div. Chem. Sci. **1983** 861/3).

[16] L. V. Tumasheva, A. Kh. Filippova, G. S. Lyashenko, N. S. Vyazankin (Izv. Akad. Nauk SSSR Ser. Khim. **1984** 2797/8; Bull. Acad. Sci. USSR Div. Chem. Sci. **1984** 2563/4).

1.1.2.1.4.3 Reactions with Haloalkenes, Haloallenes, and Allenylesters

This reaction is referred to as Reaction Type 7c in Table 1 on pp. 17/41.

In contrast to halogen at sp^3-hybridized carbon, halogen in haloalkenes is readily substituted by an alkynyl group from $RC\equiv CCu$ forming the corresponding enynes and a cuprous halide. Depending on the reaction conditions, mono- and disubstitution is possible with 1,2-diiodoalkenes. Tetraiodoethene gives a tetrasubstituted product.

The reactions are normally carried out in pyridine (sometimes in dimethylformamide) at 40 to 130 °C, the reaction times are between a few minutes and 24 hours. They can be strongly catalyzed by $C_6H_5PdI[P(C_6H_5)_3]_2$ (see Section 1.1.2.1.4.5) [12]. Analogous reactions using $RC\equiv CH$, $R'X$, and catalytic amounts of both the Pd catalyst and CuI are also known [16]. The ease of replacement of the halogen follows the order $I > Br > Cl$. Fluorides do not react [1]. In the presence of $Pd[P(C_6H_5)_3]_4$ and $n\text{-}C_4H_9NH_2$, $RC\equiv CCu$ formed in situ from $RC\equiv CH$ and 5% CuI reacts with both (E)- and (Z)-ClCH=CHCl at room temperature to give $RC\equiv CCH=CHCl$ in more than 99% isomeric purity. The yields are almost quantitative $(R = CH_3SCH_2$, $n\text{-}C_5H_{11}$, pyran-2-yloxymethyl) except for $R = CH_3CO_2CH_2$ ((E)-form 65% and (Z)-form 72%) [14].

An analogous reaction is possible with the pseudo-halogen CN. The reaction of $C_6H_5C\equiv CCu$ and tetracyanoethylene gives after 1 h in tetrahydrofuran $(NC)_2C=C(CN)C\equiv CC_6H_5$ and resinous products [3].

The first reports about the straight forward reaction of $RC\equiv CCu$ and $(E)\text{-}R'C(I)=CHSO_2R''$ give no data about a possible change of the configuration at the double bond [4] or claim the retention of the (E)-structure without giving the molar ratio of the reactants [7]. Later the stereochemistry of the substitution reaction was studied in detail. The (E)-2-iodo-1-sulfonylalkenes on coupling with cuprous phenylacetylide gave the (E)-1-sulfonylbut-1-en-3-ynes in good yields, when a 1:1 ratio of reactants was used. However, when an excess of cuprous phenylacetylide was used, the reaction time doubled or quadrupled, and mixtures of the two 1-sulfonylbut-1-en-3-yne isomers were isolated with the thermodynamically more stable product, the (Z)-isomer, predominating. The (Z)-2-iodo-1-sulfonylalkenes were also coupled with cuprous phenylacetylide to give the (Z)-1-sulfonylbut-1-en-3-ynes in fair yields. The assignment of configuration to the 1-sulfonylbut-1-en-3-ynes was based upon NMR coupling constants and chemical shift data. In this work, however, no substituents are given [8]. The products from [4] should have the (E)-configuration for the molar ratio (as far as mentioned) is 1:1.

Table 4

Reaction Type 7c (explanation see pp. 12/4):

Reactions of RC≡CCu with Haloalkenes, Haloallenes, and Allenylesters.

If not otherwise stated, halogen has been substituted by RC≡C, and the reaction products are formed with a retention of configuration.

R in RC≡CCu	haloalkene, haloallene, or allenylester	remarks (yield in %)	Ref.
CH_3	(Z)-ICH=CHCO$_2$CH$_3$	isolated after hydrolysis to the carbonic acid (16)	[13]
C_2H_5	(E)-ICH=CHI	only (E)-C$_2$H$_5$C≡CCH=CHC≡CC$_2$H$_5$ (71)	[11]
$CH_3CH(OH)$	(Z)-ICH=CHCO$_2$CH$_3$	(57)	[13]
$CH_2=C(CH_3)$	(Z)-ICH=CHCO$_2$CH$_3$	(34)	[13]
(Z)-CH$_3$OCH=CH	(E)-ICH=CHC$_4$H$_9$-n	(1Z, 5E)-CH$_3$OCH=CHC≡CCH=CH-C$_4$H$_9$-n	[9]
n-C$_3$H$_7$	(E)-ICH=CHI	only (E)-n-C$_3$H$_7$C≡CCH=CHC≡CC$_3$H$_7$-n (75)	[11]
(CH$_3$)$_3$Si	(CH$_3$)$_2$C=C=CHBr	(CH$_3$)$_2$C=C=CHC≡CSi(CH$_3$)$_3$ (33) in presence of [Pd(P(C$_6$H$_5$)$_3$)$_4$]	[15]
	(CH$_3$)$_2$C=C=CHO$_2$CCH$_3$	(CH$_3$)$_2$C=C=CHC≡CSi(CH$_3$)$_3$ (4) in presence of [Pd(P(C$_6$H$_5$)$_3$)$_4$]	[15]
n-C$_4$H$_9$	(Z)-C$_2$H$_5$C(CH$_3$)=CHI	(70)	[6]
	(E)-n-C$_4$H$_9$C(C$_2$H$_5$)=CHI	(85)	[6]
	(E)-ICH=CHI	in boiling pyridine after 10 min 40% (E)-n-C$_4$H$_9$C≡CCH=CHI; after 2 h 60% (E)-n-C$_4$H$_9$C≡CCH=CHC≡CC$_4$H$_9$-n, no monosubstituted product	[2]
4-ClC$_6$H$_4$	(E)-ICH=CHI	mixture of the mono- and disubstituted products, no determination of the configuration	[2]
4-IC$_6$H$_4$	(E)-ICH=CHI	mixture of the mono- and disubstituted products, no determination of the configuration	[2]
C$_6$H$_5$	(E)-CH$_3$SO$_2$CH=C(I)CH$_3$	(53)	[7]
	(E)-CH$_3$SO$_2$CH=C(I)C$_6$H$_5$	(68)	[7]
	(E)-C$_2$H$_5$SO$_2$CH=C(I)CH$_3$	(50)	[7]
	(E)-C$_2$H$_5$SO$_2$CH=C(I)C$_6$H$_5$	(76)	[4, 7]
	(E)-C$_6$H$_5$CH=CHBr	(94) at 20°/0.4 h in hexamethylphosphoric triamide, catalyzed by C$_6$H$_5$PdI[P(C$_6$H$_5$)$_3$]$_2$ and [N(C$_4$H$_9$)$_4$]I	[12]
	C$_6$H$_5$CH=CHBr	(75)	[1]

Table 4 [continued]

R in RC≡CCu	haloalkene, haloallene, or allenylester	remarks (yield in %)	Ref.
C_6H_5 [continued]	(E)-$C_6H_5SO_2CH=CHI$ (63)		[7]
	(E)-$C_6H_5SO_2CH=C(I)CH_3$ (62)		[7]
	(E)-$C_6H_5SO_2CH=C(I)C_6H_5$ (68)		[7]
	(E)-4-$CH_3C_6H_4SO_2CH=CHI$ (76)		[7]
	(E)-4-$CH_3C_6H_4SO_2CH=C(I)CH_3$ (76)		[7]
	(E)-4-$CH_3C_6H_4SO_2CH=C(I)C_4H_9-n$ (72)		[7]
	(E)-4-$CH_3C_6H_4SO_2CH=C(I)C_6H_5$ (58)		[4, 7]
	(E)-4-$CH_3C_6H_4SO_2CH=C(I)C_6H_{11}$-cyclo (80), geometry not given		[4]
	(E)-4-$CH_3C_6H_4SO_2CH=C(I)C_6H_{13}-n$ (60)		[7]
	(E)-$ClCH=CHI$	60 to 90% (E)-$C_6H_5C≡CCH=CHCl$	[1, 5]
	(E)-$ICH=CHI$	in boiling pyridine after 10 min 30% (E)-$C_6H_5C≡CCH=CHI$; after 2 h in boiling pyridine or dimethylformamide 55 or 65% (E)-$C_6H_5C≡CCH=CHC≡CC_6H_5$, no monosubstituted product	[2, 5]
	$ICH=CHI$	only $C_6H_5C≡CCH=CHC≡CC_6H_5$ (90)	[1]
	$I_2C=CI_2$	40 to 90% $(C_6H_5C≡C)_2C=C(C≡CC_6H_5)_2$, best yields after 2 h at 80° in pyridine	[1, 5]
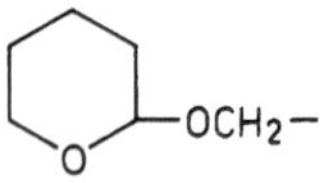	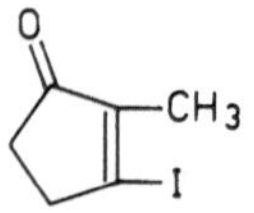	isolated as 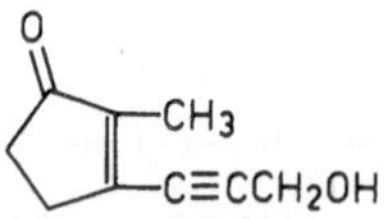 after treatment with toluenesulfonic acid (72)	[10]
$C_6H_5OCH_2$	(E)-$ClCH=CHI$	(E)-$C_6H_5OCH_2C≡CCH=CHCl$ (60 to 90)	[5]
	$ClCH=CHI$	$C_6H_5OCH_2C≡CCH=CHCl$ (70)	[1]
	$ICH=CHI$	only $C_6H_5OCH_2C≡CCH=CHC≡CCH_2OC_6H_5$ (40)	[1]

References:

[1] J. Burdon, P. L. Coe, C. R. Marsh, J. C. Tatlow (J. Chem. Soc. Chem. Commun. **1967** 1259/60).

[2] L. Yu. Ukhin, A. M. Sladkov, V. I. Gorshkov (Zh. Org. Khim. **4** [1968] 25/7; J. Org. Chem. [USSR] **4** [1968] 21/3).

[3] L. Yu. Ukhin, A. M. Sladkov, Zh. I. Orlova (Izv. Akad. Nauk SSSR Ser. Khim. **1969** 705/6; Bull. Acad. Sci. USSR Div. Chem. Sci. **1969** 637/8).

[4] W. E. Truce, G. C. Wolf (J. Org. Chem. **36** [1971] 1727/32).

[5] J. Burdon, P. L. Coe, C. R. Marsh, J. C. Tatlow (J. Chem. Soc. Perkin Trans. I **1972** 639/41).

[6] A. Commercon, J. F. Normant, J. Villiéras (J. Organometal. Chem. **93** [1975] 415/21).

[7] W. E. Truce, A. W. Borel, P. J. Marek (J. Org. Chem. **41** [1976] 401/2).

[8] A. W. Borel (Diss. Purdue Univ. 1976, pp. 1/366; Diss. Abstr. Intern. B **37** [1977] 5079; C.A. **87** [1977] No. 38724).

[9] G. A. Kraus, K. Frazier (Tetrahedron Letters **1978** 3195/8).

[10] R. H. Bradbury, T. L. Gilchrist, C. W. Rees (J. Chem. Soc. Perkin Trans I **1981** 3234/8).

[11] T. P. Lockhart, P. B. Comita, R. G. Bergman (J. Am. Chem. Soc. **103** [1981] 4082/90).

[12] N. A. Bumagin, I. O. Kalinovskii, A. B. Ponomarev, I. P. Beletskaya (Dokl. Akad. Nauk SSSR **265** [1982] 1138/43).

[13] G. Struve, S. Seltzer (J. Org. Chem. **47** [1982] 2109/13).

[14] V. Ratovelomanana, G. Linstrumelle (Tetrahedron Letters **22** [1981] 315/8).

[15] K. Ruitenberg, H. Kleijn, H. Westmijze, J. Meijer, P. Vermeer (Rec. Trav. Chim. **101** [1982] 405/9).

[16] N. A. Bumagin, A. B. Ponomarev, I. P. Beletskaya (Izv. Akad. Nauk SSSR Ser. Khim. **1984** 1561/6; Bull. Acad. Sci. USSR Div. Chem. Sci. **1984** 1433/8).

1.1.2.1.4.4 Reactions with Haloalkynes

This reaction is referred to as Reaction Type 7d in Table 1 on pp. 17/41. The reaction $RC{\equiv}CCu + XC{\equiv}CR^1 \rightarrow R(C{\equiv}C)_2R^1$ ($X = Br$, I) is very useful to prepare asymmetric diynes. Symmetric diynes $R(C{\equiv}C)_2R$ can be prepared more easily by oxidation of $RC{\equiv}CCu$ (see Reaction Type 2, Section 1.1.2.1.2).

Chloroalkynes do not react; in most cases, bromoalkynes are used. Iodoalkynes give principally the same results. The reaction conditions are very different. Dimethylformamide at 0 °C, H_2O/C_2H_5OH, and pyridine at room temperature as well as n-hexane, CH_2Cl_2, and boiling dimethylformamide are possible. The reaction has also been conducted without any solvent.

Formation of $R(C{\equiv}C)_2R$ as a side reaction is often observed. The also possible product $R^1(C{\equiv}C)_2R^1$ has rarely been reported. As outlined under "Preparation Method IV" in Section 1.1.2.1.2 (see p. 4), the Cu^+ liberated in the course of the reaction forms with $R^1C{\equiv}CBr$ the corresponding $R^1C{\equiv}CCu$ and Cu^{2+}. This for its part causes a side reaction to give $R(C{\equiv}C)_2R$ (see Reaction Type 2b, Section 1.1.2.1.2). To suppress this reaction, the amount of Cu^+ present must be reduced. This corresponds to the so-called "Chodkiewicz–Cadiot Coupling", which uses instead of $RC{\equiv}CCu$ a mixture of $RC{\equiv}CH$ and approximately 2 mol% Cu^+. With this method, the addition of a base to neutralize the protons formed during the reaction is necessary and the formation of Cu^+ from Cu^{2+} by reduction with $[NH_3OH]Cl$ to limit side reactions is preferable. The amine concentration should not be too high, for too much amine favors the formation of $R(C{\equiv}C)_2R$. During the Chodkiewicz–Cadiot coupling, $RC{\equiv}CCu$ is formed in a fast reaction between $RC{\equiv}CH$ and $Cu^+/$amine, whereas the substitution reaction with $R^1C\equiv CBr$ to give $R(C\equiv C)_2R^1$ proceeds more slowly [7, 13]. However, a clear proof of the reaction path has not yet been furnished. For this reason examples of this intensively studied catalytic reaction have not been entered in Table 5 [2, 3].

Table 5
Reaction Type 7d (explanation see pp. 12/4):
$RC\equiv CCu + XC\equiv CR^1 \rightarrow R(C\equiv C)_2R^1 + CuX$.
a) The acetylide is generated as a slurry in the reaction mixture and not isolated.
$C_5H_9O_2$ = tetrahydropyran-2-yloxy.

| R in RC≡CCu | $XC\equiv CR^1$ | | remarks | Ref. |
	R^1	X	(yield of $R(C\equiv C)_2R^1$ in %)	
CH_3	$HOCH_2$	Br	(60)	[5, 7]
$CH_3CH(OH)$	$HOCH_2$	Br	no yield given	[7]
$CH_3C\equiv C$	$HOCH_2$	Br	(66)	[5, 7]
$n-C_3H_7$	$(CH_3)_3Si$	Br	in pyridine at 25° (96) and $(CH_3)_3Si(C\equiv C)_2Si(CH_3)_3$ (1), no reaction in THF/hexane	[12]
$(CH_3)_2C(OH)$	$HOCH_2$	Br	(56)a)	[1]
	C_6H_5	Br	(73)a)	[1, 3]
	$(CH_3)_2C(OH)(C\equiv C)_n$	Br	n = 0 (95)a), n = 1 (66)a)	[1]
	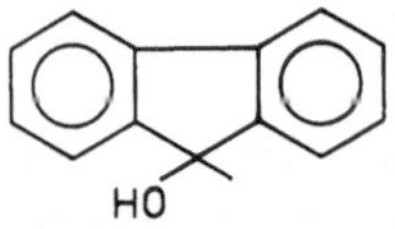 (9-hydroxyfluorene-9,9-diyl structure)	Br	(58)a)	[1]
$(CH_3)_3Si$	$n-C_3H_7$	Br	low yield, mixture with $(CH_3)_3Si(C\equiv C)_2$-$Si(CH_3)_3$ and $n-C_3H_7(C\equiv C)_2C_3H_7-n$	[12]
$(CH_3)_3C$	$(CH_3)_3Si$	Br	(93)a)	[12]
$(C_2H_5O)_2CH$	$HOCH_2$	Br	no yield given	[7]
$4-BrC_6H_4$	C_6H_5	Br	excess acetylide, no yield given	[4]
$4-ClC_6H_4$	C_6H_5	I	$R(C\equiv C)_2R^1:R^1(C\equiv C)_2R^1:R(C\equiv C)_2R =$ 4:1:2	[9]
$4-O_2NC_6H_4$	C_6H_5	Br	excess acetylide, no yield given	[4]
C_6H_5	$HOCH_2$	Br	(65)	[5, 7]
	$(CH_3)_2C(OH)$	Br	(87)a)	[3]
	C_6H_5	I	(96), $C_6H_5C\equiv CCl$ can be formed in situ from $C_6H_5C\equiv CCu$ and I_2	[5, 11]
	$4-CH\equiv CC_6H_4$	Br	large excess of acetylide, no yield given	[4]
	H	Br	isolated as the copper salt	[8, 10]

Table 5 [continued]

R in RC≡CCu	XC≡CR^1 R^1	X	remarks (yield of R(C≡C)$_2$R^1 in %)	Ref.
cyclohex-1-enyl	(CH$_3$)$_3$Si	Br	(85)[a]	[12]
cyclo-C$_6$H$_{11}$	(CH$_3$)$_3$Si	Br	(81)[a]	[12]
(Z)-n-C$_4$H$_9$OCH=CH	C$_6$H$_5$	Br	[a], no yield given	[6]
C$_5$H$_9$O$_2$CH$_2$	HOCH$_2$	Br	no yield given	[7]
n-C$_6$H$_{13}$	(CH$_3$)$_3$Si	Br	(83)[a]	[12]
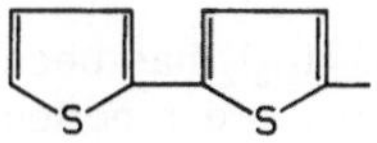	HOCH$_2$	Br	(71)	[5, 7]
naphth-1-yl	4-CH≡CC$_6$H$_4$	Br	large excess of acetylide, no yield given	[8, 10]
naphth-2-yl	4-CH≡CC$_6$H$_4$	Br	large excess of acetylide, no yield given	[8, 10]
1-C$_{10}$H$_7$C≡C	4-CH≡CC$_6$H$_4$	Br	large excess of acetylide, no yield given	[8, 10]
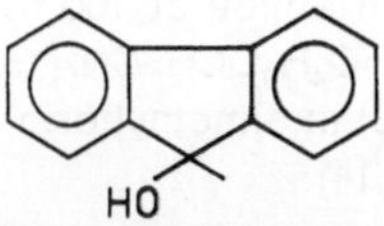	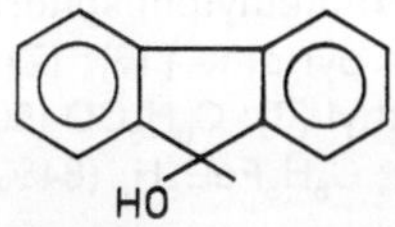	Br	(74)[a]	[1]
(C$_6$H$_5$)$_2$C(OH)	HOCH$_2$	Br	(50)[a]	[1]
	C$_6$H$_5$	Br	(60)[a]	[1]
	(C$_6$H$_5$)$_2$C(OH)(C≡C)$_n$	Br	n=0 (65)[a], n=1 (51)[a]	[1]
C$_5$H$_9$O$_2$(CH$_2$)$_9$	(CH$_3$)$_3$Si	Br	(70)[a]	[12]

References:

[1] W. Chodkiewicz, P. Cadiot (Compt. Rend. **241** [1955] 1055/7).
[2] W. Chodkiewicz, J. S. Alhuwalia, P. Cadiot, A. Willemart (Compt. Rend. **245** [1957] 322/4).
[3] W. Chodkiewicz (Ann. Chim. [Paris] [13] **2** [1957] 819/69).
[4] G. N. Gorshkova, M. A. Chubarova, A. M. Sladkov, L. Yu. Ukhin, V. I. Kasatochkin (Zh. Fiz. Khim. **38** [1964] 2516/20; Russ. J. Phys. Chem. **38** [1964] 1367/9).
[5] R. F. Curtis, J. A. Taylor (Tetrahedron Letters **1968** 2919/20).
[6] A. N. Volkov, Yu. M. Skvortsov, I. I. Danda, M. F. Shostakovskii (Zh. Org. Khim. **6** [1970] 897/902; J. Org. Chem. [USSR] **6** [1970] 903/7).
[7] R. F. Curtis, J. A. Taylor (J. Chem. Soc. C **1971** 186/8).
[8] Y. Okamoto, K. L. Chellappa, S. K. Kundu (J. Org. Chem. **37** [1972] 3185/7).
[9] A. M. Sladkov, I. R. Gol'ding (Dokl. 4th Vses. Konf. Khim. Atsetilena, Alma-Ata 1972, Vol. 1, pp. 59/64; C.A. **79** [1973] No. 78253).
[10] Y. Okamoto (AD-760772) [1973] 1/84; C.A. **80** [1974] No. 37537).

[11] J. S. Miller, S. T. Matsuo (J. Polym. Sci. Polym. Letters Ed. **18** [1980] 809/10).
[12] J. A. Miller, G. Zweifel (Synthesis **1983** 128/30).
[13] G. Eglinton, W. McCrae (Advan. Organic Chem. **4** [1963] 225/328).

1.1.2.1.4.5 Reactions with C_6H_5I

This reaction (called Reaction Type 7e in Table 1, pp. 17/41) is a very useful method to prepare $RC \equiv CC_6H_5$. Sensitive tolanes can best be prepared by this mild conversion:

$$RC \equiv CCu + IC_6H_5 \rightarrow RC \equiv CC_6H_5 + CuI$$

The reaction is in most cases accomplished by refluxing the components for 5 to 48 h in pyridine. These long reaction times demand a particularly careful exclusion of air to avoid the formation of $R(C \equiv C)_2R$ (see Section 1.1.2.1.2).

Recently a catalysis by 1 mol $[N(C_4H_9-n)_4]I$ and 0.01 mol $C_6H_5PdI[P(C_6H_5)_3]_2$ has been suggested. Under these conditions yields up to 95% at room temperature are reported [15, 16]. This reaction is also possible with $RC \equiv CH$, $R'X$, 1 mol% $C_6H_5PdI[P(C_6H_5)_3]_2$, 1 mol% CuI, and 2 equivalents $N(C_2H_5)_3$ [17].

The following $RC \equiv CCu$ were reacted with C_6H_5I to give $RC \equiv CC_6H_5$ (yields in parentheses):

R = (Z)-$C_2H_5OCH=CH$ (15 to 20% (Z)-product) [6]; $(CH_3)_3C$ (84%) [12]; (E)-$CH_3CH=CH$-$C \equiv C$ yields a (Z)/(E)-mixture at a 7:3 ratio [4]; tetrahydropyran-2-yloxy [7]; n-C_5H_{11} (36%) in hexamethylphosphoric triamide at 100 °C [11]; $(C_2H_5O)_2CH$ (71%) [7, 11]; C_6F_5 (74%) [9]; C_6H_5 (85 to 90%) in pyridine [1, 2], (65%) in hexamethylphosphoric triamide at 100 °C [11], (90%) for the catalyzed reaction (see above) in pyridine [16]; (Z)-n-$C_4H_9OCH=CH$ (15 to 20% (Z)-product) [6]; tetrahydropyran-2-yloxymethyl [3]; C_6H_5CO (90%) in dimethylformamide at 160 °C [5]; 2,5-d_2-$C_5H_5FeC_5H_2D_2$ (77%) [10]; $C_5H_5FeC_5H_4$ (84%) [14].

For R=C_6H_5, the rate constant $k_2 = 8 \times 10^{-6}$ $M^{-1} \cdot s^{-1}$ in pyridine at 100 °C, $\Delta G^{\neq}_{298} = 26$ kcal $\cdot$ mol^{-1}, $\Delta H^{\neq}_{298} = 7.2$ kcal $\cdot$ mol^{-1}, $\Delta S^{\neq} = -63$ cal $\cdot$ mol$^{-1} \cdot K^{-1}$, has been determined [8].

References:

[1]   C. E. Castro, R. D. Stevens (J. Org. Chem. **28** [1963] 2163).
[2]   R. D. Stevens, C. E. Castro (J. Org. Chem. **28** [1963] 3313/5).
[3]   R. E. Atkinson, R. F. Curtis, D. M. Jones, J. A. Taylor (J. Chem. Soc. Chem. Commun. **1967** 718/9).
[4]   E. R. H. Jones, S. Safe, V. Thaller (J. Chem. Soc. C **1967** 1038/44).
[5]   M. S. Shvartsberg, A. N. Kozhevnikova, I. L. Kotlyarevskii (Izv. Akad. Nauk SSSR Ser. Khim. **1967** 466/7; Bull. Acad. Sci. USSR Div. Chem. Sci. **1967** 460/1).
[6]   M. F. Shostakovskii, A. N. Volkov, A. N. Khudyakova, Yu. M. Skvortsov, I. I. Danda (Khim. Atsetilena Tr. 3rd Vses. Konf., Dushanbe 1968 [1972], pp. 94/7; C.A. **79** [1973] No. 18257).
[7]   R. E. Atkinson, R. F. Curtis, D. M. Jones, J. A. Taylor (J. Chem. Soc. C **1969** 2173/6).
[8]   C. E. Castro, R. Havlin, U. K. Honvad, A. Malte, S. Moje (J. Am. Chem. Soc. **91** [1969] 6464/70).
[9]   M. D. Rausch, A. Siegel, L. P. Klemann (J. Org. Chem. **34** [1969] 468/70).
[10] M. D. Rausch, A. Siegel (J. Organometal. Chem. **17** [1969] 117/25).

[11] J. F. Normant, M. Bourgain, A.-M. Rone (Compt. Rend. C **270** [1970] 354/7).
[12] J. J. Eisch, M. W. Foxton (J. Org. Chem. **36** [1971] 3520/6).
[13] J. S. Kiely, P. Boudjouk, L. L. Nelson (J. Org. Chem. **42** [1977] 2626/9).

[14] D. W. Slocum, M. D. Rausch, A. Siegel (Organometal. Polym. Symp. **1977/78** 39/51, 49).

[15] N. A. Bumagin, I. O. Kalinovskii, I. P. Beletskaya (Izv. Akad. Nauk SSSR Ser. Khim. **1981** 2836; Bull. Acad. Sci. USSR Div. Chem. Sci. **30** [1981] 2366).

[16] N. A. Bumagin, I. O. Kalinovskii, A. B. Ponomarev, I. P. Beletskaya (Dokl. Akad. Nauk SSSR **265** [1982] 1138/43; Dokl. Chem. Proc. Acad. Sci. USSR **262/267** [1982] 262/6).

[17] N. A. Bumagin, A. B. Ponomarev, I. P. Beletskaya (Izv. Akad. Nauk SSSR Ser. Khim. **1984** 1561/6; Bull. Acad. Sci. USSR Div. Chem. Sci. **1984** 1433/8).

1.1.2.1.4.6   Reactions with Monosubstituted Halobenzenes

This reaction is the most important method to synthesize compounds of the type $RC\equiv CC_6H_4R^1$, especially substituted tolanes. The reaction is in most cases accomplished by refluxing the reactants for 5 to 48 h in solvents like pyridine or dimethylformamide. Far milder conditions are sufficient, if the reaction is catalyzed by Pd compounds (see Section 1.1.2.1.4.5) [37]. For the analogous reaction using $RC\equiv CH$ and catalytic amounts of CuI, see [38].

Bromobenzene was not reacted with $RC\equiv CCu$; a few $R^1C_6H_4Br$ gave the corresponding tolanes. If the ortho-position in $R^1C_6H_4I$ bears a group containing active hydrogen (e.g., $R^1 = OH$, NH_2, NHR^2, CO_2H, etc.), a cyclization of the initially formed $RC\equiv CC_6H_4R^1$-2 is possible. The course of this reaction (formation of the alkyne or of the cyclization product, respecitvely, e.g., of a pyrrol) can depend on the solvent. Often mixtures of $RC\equiv CC_6H_4R^1$-2 and of a cyclic product are obtained. For these cyclization reactions, see Sections 1.1.2.1.3 (general remarks) and 1.1.2.1.5.1 to 1.1.2.1.5.4 (single reactions). For the reaction of $RC\equiv CCu$ with 2-substituted halobenzenes ("Reaction Type 7f") see Table 6, with 3-substituted halobenzenes ("Reaction Type 7g") see Table 7, and with 4-substituted halobenzenes ("Reaction Type 7h") see Table 8.

Table 6
Reaction Type 7f (explanation see pp. 12/4):
$RC\equiv CCu + 2\text{-}XC_6H_4R^1 \rightarrow 2\text{-}RC\equiv CC_6H_4R^1 + CuX$ (X = I or Br).
The yield of the cyclization product is only given if a mixture with $RC\equiv CC_6H_4R^1$ is obtained or if the reaction is kinetically characterized. For all other cyclizations, see Sections 1.1.2.1.5.1 to 1.1.2.1.5.4.
Rate constant k_2 in pyridine at 100 °C in $M^{-1} \cdot s^{-1}$. $C_5H_9O_2$ = tetrahydropyran-2-yloxy.

R in $RC\equiv CCu$	2-$R^1C_6H_4X$ R^1	X	remarks (yield of 2-$RC\equiv CC_6H_4R^1$ in %)	Ref.
C_2H_5	NH_2	I	in pyridine (28 + 12 cyclization product)	[5]
(Z)-$CH_3OCH=CH$	NO_2	Br	(87)	[32]
	NH_2	Br	tarry products only	[32]
n-C_3H_7	NH_2CO	I	(50)	[5]
	C_2H_5NH	I	in pyridine (5 + 50 cyclization product)	[5]
	NH_2	I	in pyridine (17 + 70 cyclization product)	[5]
$(CH_3)_2C(OH)$	CH_3O	I	(50), acetylide prepared in situ	[34]
	CH_3O_2C	I	(90), acetylide prepared in situ	[34]

Table 6 [continued]

R in RC≡CCu	2-R¹C₆H₄X R¹	X	remarks (yield of 2-RC≡CC₆H₄R¹ in %)	Ref.
n–C₄H₉	NH₂	I	in pyridine (20+35 cyclization product)	[5]
pyrid–2–yl	NH₂	I	in refluxing pyridine (50), at 125° in dimethylformamide no reaction	[5]
2–O₂NC₆H₄	NH₂CO	I	yields a mixture of	[10]

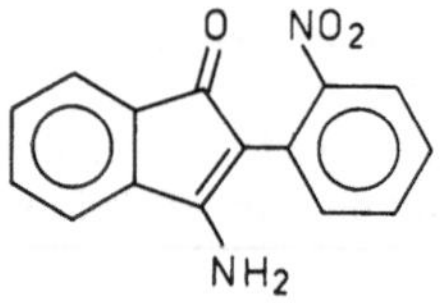

66% ... and

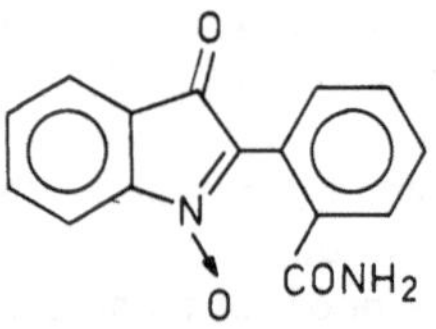

6% ... and

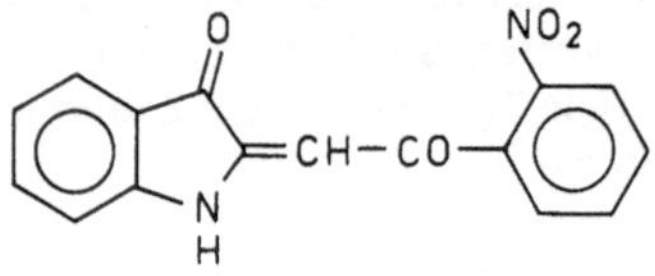

	CH₃	I	(67)	[10]
	CH₃O	I	(83)	[10]
	CH₃O₂C	I	(93)	[10]
	C₂H₅O₂C	I	(92)	[10]
	NO₂	I	(93)	[10]
C₆H₅	Br	I	(70 to 80)	[14]
	HO₂C	I	only cyclization, see Table 17, Section 1.1.2.1.5.3; $k_2 = 7 \times 10^{-2}$	[5, 11]
	NH₂CO	I	(47)	[5]
	CH₃O	I	(91), $k_2 = 1.7 \times 10^{-5}$	[3, 11]
	HOCH₂	I	(50)	[5]
	CH₃O₂C	I	(80), with isolation as 2–carboxytolane (95) reaction also possible with acetylide prepared in situ	[4, 19, 21, 34]
	CH₃CONH	I	no yield given	[26]
	C₂H₅	I	(70 to 80)	[14]
	(CH₃)₂CH	I	(70 to 80)	[14]

References on pp. 95/6

Table 6 [continued]

R in RC≡CCu	2-R^{1}C$_6$H$_4$X R^1	X	remarks (yield of 2-RC≡CC$_6$H$_4$R^1 in %)	Ref.
	(CH$_3$)$_3$C	I	(70 to 80)	[14]
	2-O$_2$NC$_6$H$_4$	I	2-nitro-2'-(phenylethynyl)biphenyl and 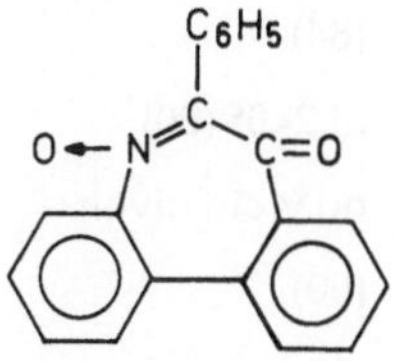	[18]

$$C_6H_5$$
$$O \leftarrow N \overset{\diagup}{=} C \overset{\diagdown}{} C = O$$

| | C$_6$H$_5$ | I | (56) | [6] |
| | 2-C$_6$H$_5$C≡CC$_6$H$_4$ | I | 2,2'-bis(phenylethynyl)biphenyl (65) and 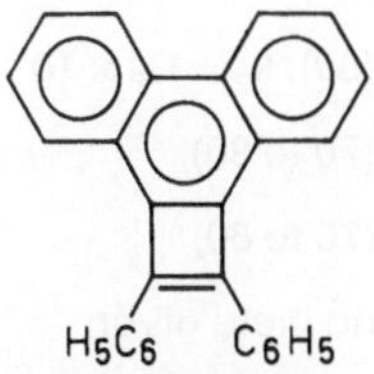 | [6] |

H$_5$C$_6$ C$_6$H$_5$

	Cl	I	in CH$_3$CO$_2$H (52)	[6]
	F	I	(70 to 80)	[14]
	I	I	1,2-bis(phenylethynyl)benzene (61)	[5]
	NH$_2$	I	(59) in pyridine (homogeneous solution) with k$_2$ = 3.4 × 10^{-5} in dimethylformamide (heterogeneous mixture) only cyclization	[5, 11]
	NO$_2$	I	(84), k$_2$ = 2.0 × 10^{-3}; ΔG$^{\ddagger}_{298}$ = 26 kcal·mol^{-1}, ΔH‡ = 16 kcal·mol^{-1}, ΔS$^{\ddagger}_{298}$ = −29 cal·mol^{-1}·K^{-1} in pyridine	[1, 3, 11, 12]
	HO	I	only cyclization, see Table 14, Section 1.1.2.1.5.1; k$_2$ = 8.6 × 10^{-3}	[11]
C$_5$H$_9$O$_2$CH$_2$	CF$_3$	I	no yield given	[29]
	CH$_3$O	I	no yield given	[29]
C$_6$H$_5$CO	CH$_3$	I	(82)	[8]
2-CH$_3$C$_6$H$_4$	CH$_3$	I	(90)	[7, 15]
4-CH$_3$OC$_6$H$_4$	NH$_2$CO	I	no yield given	[10]
naphth-1-yl	NH$_2$CO	I	no yield given	[10]

Table 7

Reaction Type 7 g (explanation see pp. 12/4):

$RC{\equiv}CCu + 3\text{-}XC_6H_4R^1 \rightarrow 3\text{-}RC{\equiv}CC_6H_4R^1 + CuX$ (X = I or Br).

Rate constant k_2 in pyridine at 100 °C in $M^{-1} \cdot s^{-1}$. $C_5H_9O_2$ = tetrahydropyran-2-yloxy.

R in $RC{\equiv}CCu$	3-$R^1C_6H_4X$ R^1	X	remarks (yield of 3-$RC{\equiv}CC_6H_4R^1$ in %)	Ref.
CH_3	C_6H_5CO	I	(84)	[31]
	C_6H_5CO	Br	at 240° (80)	
pyrid-3-yl	$OCHNHC(CH_3)_2$	I	no yield given	[25]
$(C_2H_5O)_2CH$	CH_3	I	(50)	[30]
C_6H_5	Br	I	(70 to 80)	[14]
	HO_2C	I	$k_2 = 2.9 \times 10^{-5}$	[11]
	CH_3	I	(37)	[27]
	CH_3O	I	(80), $k_2 = 1.4 \times 10^{-5}$	[1, 11]
	C_2H_5	I	(70 to 80)	[14]
	$(CH_3)_2CH$	I	(70 to 80)	[14]
	$OCHNHC(CH_3)_2$	I	no yield given	[25]
	$(CH_3)_3C$	I	(70 to 80)	[14]
	Cl	I	(53)	[27]
	F	I	(70 to 80)	[14]
	I	I	1,3-bis(phenylethynyl)benzene (42)	[5]
	NH_2	I	(70), $k_2 = 1.0 \times 10^{-5}$	[11, 24]
	NO_2	I	$k_2 = 9.1 \times 10^{-5}$; $\Delta G^{\ddagger}_{298} = 27$ kcal $\cdot$ mol^{-1}, $\Delta H^{\ddagger}_{298} =$ 15 kcal $\cdot$ mol^{-1}, $\Delta S^{\ddagger}_{298} = -36$ cal $\cdot$ mol$^{-1} \cdot K^{-1}$ in pyridine	[11, 13]
	HO	I	$k_2 = 1.3 \times 10^{-5}$	[11]
$C_5H_9O_2CH_2$	CF_3	I	(75 to 80)	[33]
	CH_3O	I	no yield given	[29]
C_6H_5CO	CH_3	I	(81)	[8]

References on pp. 95/6

Table 8

Reaction Type 7h (explanation see pp. 12/4):

$RC{\equiv}CCu + 4\text{-}XC_6H_4R^1 \rightarrow 4\text{-}RC{\equiv}CC_6H_4R^1 + CuX$ (X = I or Br).

[a] The reaction is carried out in hexamethylphosphoric triamide at room temperature in the presence of 1 mol $[N(C_4H_9\text{-}n)_4]I$ and 0.01 mol $C_6H_5PdI[P(C_6H_5)_3]_2$.

Rate constant k_2 in pyridine at 100 °C in $M^{-1} \cdot s^{-1}$. $C_5H_9O_2$ = tetrahydropyran-2-yloxy.

R in RC≡CCu	4-R¹C₆H₄X		remarks	Ref.
	R^1	X	(yield of 4-RC≡CC₆H₄R¹ in %)	
CH_3	NO_2	I	(<1)	[20]
cyclo-C_3H_5	NO_2	I	(33)	[20]
(Z)-$CH_3OCH{=}CH$	CH_3	I	(65)	[32]
	Cl	I	(79)	[32]
	NO_2	I	(87)	[32]
$(CH_3)_2CH$	NO_2	I	(1, 4)	[20]
n-C_4H_9	$OCHNHC(CH_3)_2$	I	no yield given	[25]
pyrid-3-yl	$OCHNHC(CH_3)_2$	I	(86)	[25]
$(C_2H_5O)_2CH$	CH_3	I	(<57)	[30]
	CH_3O	I	(<55)	[30]
	NO_2	I	no reaction after 48 h reflux in pyridine	[30]
4-BrC_6H_4	HO	I	no yield given	[9]
4-ClC_4H_4	HO	I	no yield given	[9]
4-FC_6H_4	$OCHNHC(CH_3)_2$	I	no yield given	[25]
4-$O_2NC_6H_4$	HO_2CCH_2	I	(38)	[16]
C_6H_5	Br	I	(70 to 80)	[14]
	NC-	I	(96)	[36, 37]
	HO_2C	I	(85), $k_2 = 4.6 \times 10^{-5}$	[3, 11, 14]
	HO_2C	Br	no reaction	[11]
	CH_3O	I	(94)[a], 6 h (83) or 10 h (99) in refluxing pyridine, $k_2 = 8.5 \times 10^{-6}$	[1, 3, 11, 36]
	CH_3S	I	(62)	[17]
	$CH_3S(O_2)$	I	(56)	[17]
	CH_3CO	I	(20 to 70), (94)[a]	[2, 36, 37]
	CH_3CO	Br	(89)[a]	[36, 37]
	CH_3O_2C	I	(94)[a]	[36, 37]

Table 8 [continued]

R in RC≡CCu	4–R¹C₆H₄X R¹	X	remarks (yield of 4–RC≡CC₆H₄R¹ in %)	Ref.
C₆H₅ [continued]	HO₂CCH₂	I	in dimethylformamide (67) or pyridine (60)	[16]
	CH₃CONH	I	no yield given	[26]
	C₂H₅	I	(70 to 80)	[14]
	(CH₃)₂N	I	(90)	[17]
	(CH₃)₂CH	I	(70 to 80)	[14]
	OCHNHC(CH₃)₂	I	(86)	[25]
	(CH₃)₃C	I	no yield given	[14]
	F	I	(70 to 80)	[14]
	I	I	16 h at 120° in pyridine (45) or at 125° in hexamethylphosphoric triamide (73), only 1,4-(C₆H₅C≡C)₂C₆H₄ is formed at 8 to 16 h refluxing in dimethylformamide also forms 4-C₆H₅C≡CC₆H₄I	[2, 5, 22]
	NH₂	I	(76), $k_2 = 1.4 \times 10^{-5}$	[2, 3, 11, 24]
	NO₂	I	6 h in boiling pyridine (74) or 20 to 60 min at 20° in presence of NaI and Pd compounds like C₆H₅PdI[P(C₆H₅)₃]₂, [(C₆H₅)₃As]₂PdCl₂, [(C₆H₅)₃Sb]₂PdCl₂, LiPdCl₃, (CH₃CN)₂PdCl₂, Fe(C₅H₄P(C₆H₅)₂)₂PdCl₂, PdCl₂ (98), or less effective [(C₆H₅)₃P]₂NiCl₂ $k_2 = 1.4 \times 10^4$; $\Delta G_{298}^{\ddagger} = 26$ kcal·mol⁻¹, $\Delta H_{298}^{\ddagger} = 13$ kcal·mol⁻¹, $\Delta S_{298}^{\ddagger} = -41$ cal·mol⁻¹·K⁻¹ in pyridine	[1 to 3, 11, 36, 37]
	OH	I	(82), $k_2 = 2.0 \times 10^{-5}$	[2, 3, 9, 11]
C₅H₉O₂CH₂	CH₃O₂C	I	no yield given	[28]
	(CH₃)₂NS(O₂)	I	no yield given	[29]
	4-CH₃OC₆H₄CH₂O₂C	I	(74)	[29]
	NO₂	I	no yield given	[28]
C₆H₅CO	CH₃	I	(77)	[8]
4-CH₃C₆H₄	OCHNHC(CH₃)₂	I	no yield given	[25]

Table 8 [continued]

R in RC≡CCu	4-R¹C₆H₄X		remarks	Ref.
	R¹	X	(yield of 4-RC≡CC₆H₄R¹ in %)	
4-CH₃C₆H₄	I	I	1,4-bis(4-tolylethynyl)benzene (35)	[23]
[continued]	HO	I	no yield given	[9]
4-CH₃OC₆H₄	OCHNHC(CH₃)₂	I	no yield given	[25]
	HO	I	no yield given	[9]
C₅H₅FeC₅H₄	CH₃O	I	(62)	[35]

References:

[1] C. E. Castro, R. D. Stevens (J. Org. Chem. **28** [1963] 2163).

[2] A. M. Sladkov, L. Y. Ukhin, V. V. Korshak (Izv. Akad. Nauk SSSR Ser. Khim. **1963** 2213/5; Bull. Acad. Sci. USSR Div. Chem. Sci. **1963** 2043/5).

[3] R. D. Stevens, C. E. Castro (J. Org. Chem. **28** [1963] 3313/5).

[4] C. E. Castro, R. D. Stevens (J. Am. Chem. Soc. **86** [1964] 4358/63).

[5] C. E. Castro, E. J. Gaughan, D. C. Owsley (J. Org. Chem. **31** [1966] 4071/8).

[6] S. A. Kandil, R. E. Dessy (J. Am. Chem. Soc. **88** [1966] 3027/34).

[7] H. Staab, F. Graf (Tetrahedron Letters **1966** 751/7).

[8] M. S. Shvartsberg, A. N. Kozhevnikova, I. L. Kotlyarevskii (Izv. Akad. Nauk SSSR Ser. Khim. **1967** 466/7; Bull. Acad. Sci. USSR Div. Chem. Sci. **1967** 460/1).

[9] H. Veschambre, G. Dauphin, A. Kergomard (Bull. Soc. Chim. France **1967** 2846/54).

[10] C. Bond, M. Hooper (J. Chem. Soc. C **1969** 2453/60).

[11] C. E. Castro, R. Havlin, U. K. Honvad, A. Malte, S. Moje (J. Am. Chem. Soc. **91** [1969] 6464/70).

[12] J. M. Appleton, B. D. Andrews, I. D. Rae, B. E. Reichert (Australian J. Chem. **23** [1970] 1667/77).

[13] C. W. Bird, A. F. Harmer (Org. Prep. Proced. **2** [1970] 79/81).

[14] G. Martelli, P. Spagnole, M. Tiecco (J. Chem. Soc. B **1970** 1413/8).

[15] H. A. Staab, F. Graf (Chem. Ber. **103** [1970] 1107/17).

[16] M. P. Briede, O. Neilands (U.S.S.R. 424856 [1970/74]; C.A. **81** [1974] No. 37397).

[17] J. J. Eisch, C. K. Hordis (J. Am. Chem. Soc. **93** [1971] 2974/80).

[18] C. C. Leznoff, R. J. Hayward (Can. J. Chem. **49** [1971] 3596/601).

[19] M. S. Shvartsberg, A. A. Moroz (Izv. Akad. Nauk SSSR Ser. Khim. **1971** 1582/5; Bull. Acad. Sci. USSR Div. Chem. Sci. **1971** 1488/90).

[20] C. E. Hudson, N. L. Bauld (J. Am. Chem. Soc. **94** [1972] 1158/63).

[21] M. S. Shvartsberg, A. A. Moroz, I. L. Kotlyarevskii (Izv. Akad. Nauk SSSR Ser. Khim. **1972** 981; Bull. Acad. Sci. USSR Div. Chem. Sci. **1972** 946).

[22] S. E. Wentworth (AD-757601 [1973] 1/8; C.A. **79** [1973] No. 18285).

[23] S. E. Wentworth, G. D. Mulligan (Polym. Prepr. Am. Chem. Soc. Div. Polym. Chem. **15** [1974] 697/700).

[24] G. G. I. Moore, J. K. Harrington, K. F. Swingle (J. Med. Chem. **18** [1975] 386/91).

[25] D. C. Remy, W. A. van Saun, E. L. Engelhardt, M. L. Torchiana, C. A. Stone (J. Med. Pharm. Chem. **18** [1975] 142/8).

[26] T. D. Roberts, L. Munchausen, H. Schechter (J. Am. Chem. Soc. **97** [1975] 3112/7).

[27] D. Seyferth, M. O. Nestle, A. T. Wehman (J. Am. Chem. Soc. **97** [1975] 7417/26).

[28] M. A. Harris, I. McMillan, J. H. C. Nayler, N. F. Osborne, M. J. Pearson, R. Southgate (J. Chem. Soc. Perkin Trans. I **1976** 1612/5).

[29] E. G. Brain, A. J. Eglington, B. G. James, J. H. C. Nayler, N. F. Osborne, M. J. Pearson, T. C. Smale, R. Southgate, P. Tolliday, M. J. Basker, L. W. Mizen, R. Sutherland (J. Med. Chem. **20** [1977] 1086/90).

[30] J. S. Kiely, P. Boudjouk, L. L. Nelson (J. Org. Chem. **42** [1977] 2626/9).

[31] B. Župančić, B. Jenko, LEK Tovarna Farmacevtskih in Kemičnih Izdelkov (Austrian 351511 [1979]; Brit. 1591426 [1978]; Yugosl. Appl. 76-2549 [1976]; Pol. 106809 [1976/78]; C.A. **91** [1979] No. 175017).

[32] G. A. Kraus, K. Frazier (Tetrahedron Letters **1978** 3195/8).

[33] M. M. Kwatra, D. Z. Simon, R. L. Salvador, P. D. Cooper (J. Med. Chem. **21** [1978] 253/7).

[34] M. S. Shvartsberg, A. A. Moroz, A. N. Kozhevnikova (Izv. Akad. Nauk SSSR Ser. Khim. **1978** 875/9; Bull. Acad. Sci. USSR Div. Chem. Sci. **1978** 756/60).

[35] D. Kaufmann, R. Kupper, T. Neal (J. Org. Chem. **44** [1979] 3076/7).

[36] N. A. Bumagin, I. O. Kalinovskii, I. P. Beletskaya (Izv. Akad. Nauk SSSR Ser. Khim. **1981** 2836; Bull. Acad. Sci. USSR Div. Chem. Sci. **1981** 2366).

[37] N. A. Bumagin, I. O. Kalinovskii, A. B. Ponomarev, I. P. Beletskaya (Dokl. Akad. Nauk SSSR **265** [1982] 1138/43; Dokl. Chem. Proc. Acad. Sci. USSR **262/267** [1982] 262/6).

[38] N. A. Bumagin, A. B. Ponomarev, I. P. Beletskaya (Izv. Akad. Nauk SSSR Ser. Khim. **1984** 1561/6; Bull. Acad. Sci. USSR Div. Chem. Sci. **1984** 1433/8).

1.1.2.1.4.7   Reactions with Di- and Polysubstituted Halobenzenes

This reaction is referred to as "Reaction Type 7i" in Table 1 on pp. 17/41 and reveals no general difference from the reactions of RC≡CCu and monosubstituted halobenzenes. If the di- or polysubstituted halobenzene contains acceptor groups, the reaction proceeds more easily; in contrast to monosubstituted benzenes, Br and Cl are also typical leaving groups. With $2,4,6-(O_2N)_3C_6H_3$ even H can be substituted by RC≡C; see Section 1.1.2.1.1, under No. 90. However, whereas $C_6H_5C≡CCu$ reacts with C_6Cl_5I, C_6F_5I, or C_6F_5Br to give $C_6H_5C≡CC_6X_5$ (X = Cl or F), it does not react with C_6Cl_6 [5].

Table 9
Reaction Type 7i (explanation see pp. 12/4):

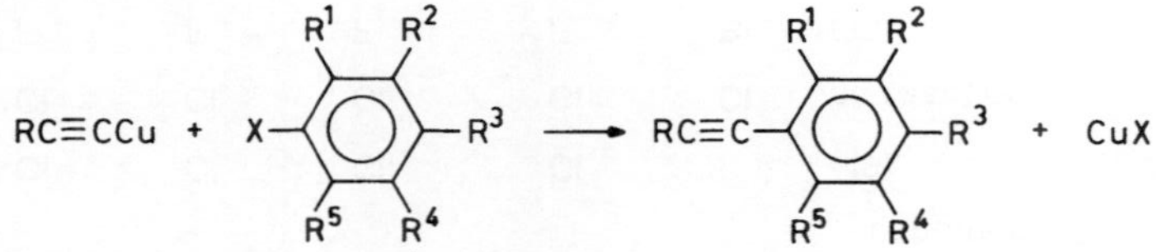

$$RC{\equiv}CCu + X{-}C_6H(R^1R^2R^3R^4R^5) \longrightarrow RC{\equiv}C{-}C_6H(R^1R^2R^3R^4R^5) + CuX$$

(X = I, Br or Cl)

a) The product is obtained as a mixture with the corresponding biphenyl. $C_5H_9O_2$ = tetrahydropyran-2-yloxy.

R in RC≡CCu	R^1	R^2	R^3	R^4	R^5	X	remarks (yield of RC≡CC$_6$R^{1}R^{2}R^{3}R^{4}R^5 in %)	Ref.
CH$_3$	CH$_3$	CH$_3$O$_2$C	H	H	H	I	(70 to 80)	[11]
C$_2$H$_5$	CH$_3$	CH$_3$O$_2$C	H	H	H	I	(70 to 80)	[11]
CH$_2$=C(CH$_3$)	CHO	HO	H	H	HO	I	frustulosine (antibiotic)	[17]
CH$_3$CO$_2$	CH$_3$CO$_2$	H	H	OCH$_2$O	H	I	mixture of RC≡CR' (12) and 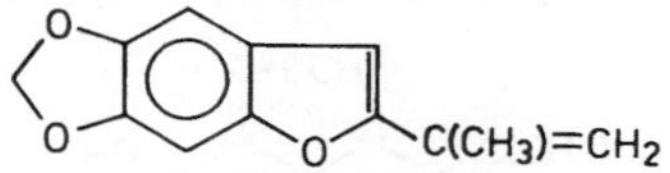 [benzofuran bearing C(CH$_3$)=CH$_2$, fused methylenedioxy] (48, formed after deacetylation)	[16]
CH$_3$CO$_2$	CH$_3$CO$_2$	H	CH$_3$O	CH$_3$O	H	I	1:1 mixture of RC≡CR' and of 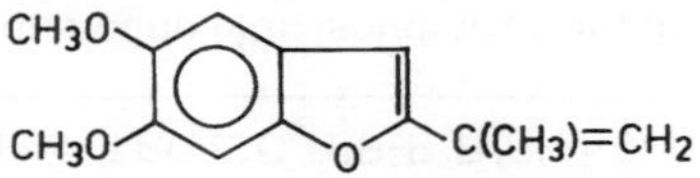 [benzofuran bearing C(CH$_3$)=CH$_2$, two CH$_3$O substituents] (formed after deacetylation)	[16]

References on p. 100

Table 9 [continued]

R in RC≡CCu	R¹	R²	R³	R⁴	R⁵	X	remarks (yield of RC≡CC₆R¹R²R³R⁴R⁵ in %)	Ref.
	CH₃CO₂	CH₃CO₂	CH₃CO₂	CH₃CO	H	Br	in quinoline (40), isolation after partial hydrolysis and ring closure as 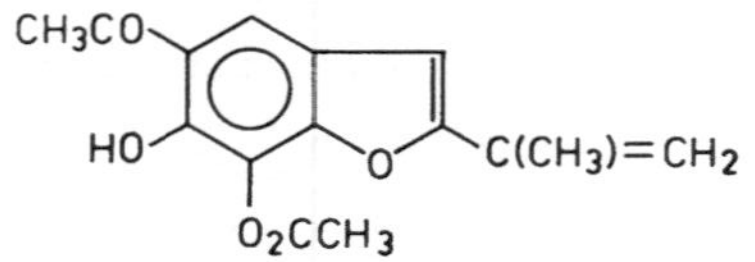 [benzofuran structure bearing CH₃CO–, HO–, –O₂CCH₃ and a 2-C(CH₃)=CH₂ group]	[9]
(Z)-CH₃OCH=CH	HO	HO	HO	CH₃CO	H	Br	only 2,3,4-trihydroxyacetophenone	[9]
	NO₂	H	CH₃O	H	H	I	(93)	[15]
	NH₂	H	NO₂	H	H	Br	only tarry products	[15]
C₂H₅O₂C	(CH₃)₃C	H	(CH₃)₃C	H	(CH₃)₃C	I	in N-methylpyrrolidone	[7]
n-C₃H₇	CH₃	CH₃O₂C	H	H	H	I	(70 to 80)	[11]
i-C₃H₇	CH₃	CH₃O₂C	H	H	H	I	(70 to 80)	[11]
n-C₄H₉	CH₃	CH₃O₂C	H	H	H	I	(70 to 80)	[11]
(E)-CH₃CH=CHC≡C	T or H	T or H	T or H	T or H	T or H	I	an (E)/(Z)-mixture (60)	[3]
C₆H₅	CH₃	H	H	CH₃	H	I	no yield given	[6]
	CH₃	H	CH₃	CH₃	CH₃	I	no yield given [6]; no reaction under similar conditions [5]	[5, 6]
	Cl	Cl	Cl	Cl	Cl	I	(49)	[5]
	Cl	Cl	Cl	Cl	Cl	Cl	no reaction	[5]
	F	F	F	F	F	Br	(33)	[5]
	F	F	F	F	F	I	in dimethylformamide (20) or in pyridine (55), yield decreases on longer heating	[3, 5, 10]

References on p. 100

				X	products (yields in %) / remarks	Ref.
F	F	F	F	I	$1,2\text{-}(C_6H_5C\equiv C)_2C_6F_4$ (25) + $C_6H_5C\equiv CC_6F_4H\text{-}2$ (40) + $(2\text{-}HC_6F_4)_2$ (2)	[10, 12]
NH_2	H	H	CH_3	–	(92)	[1]
NO_2	H	NO_2	H	Br	in hexamethylphosphoric triamide at 25° in presence of 1 mol $[N(C_4H_9\text{-}n)_4]I$ and 0.01 mol $C_6H_5PdI[P(C_6H_5)_3]_2$ (84)	[19, 20]
NO_2	H	NO_2	H	Cl	in dimethylformamide at 100° (34)	[1]
$C_{55}H_{37}O_9S_2$	H	H	$C_{25}H_{17}O_5S$, H (see p. 100, compound I)	–	no yield given	[8]
F	F	CF_3	F	–	a)	[13]
F	F	CH_3	F	–	a)	[13]
F	F	CH_3O	F	–	a)	[13]
F	F	$(CH_3)_2N$	F	–	a)	[13]
F	F	NH_2	F	–	no identifiable products	[13]
F	F	F	F	–	a), the optimization of the reaction is given	[13]
F	F	F	F	Br	no reaction	[13]
$C_5H_9O_2CH_2$	F	H	F	–	a), together with much $1,4\text{-}(C_5H_9O_2CH_2CH_2C\equiv C)_2C_6F_4$	[13]
4-$CH_3OC_6H_4$	NO_2	CH_3O	H	–		[4]
2,4-$(CH_3)_2C_6H_3$	CH_3	CH_3	CH_3	–	(60)	[14]
2,4-$(CH_3O)_2C_6H_3$	CH_3CO_2	OCH_2O	H	–	(30)	[18]

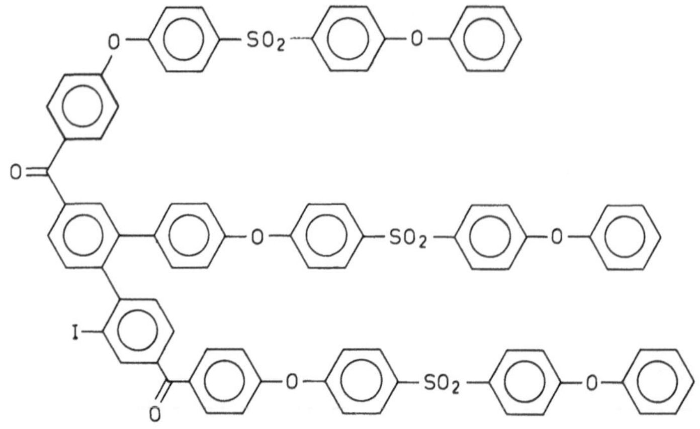

References:

[1] C. E. Castro, E. J. Gaughan, D. C. Owsley (J. Org. Chem. **31** [1966] 4071/8).

[2] R. Filler, E W. Heffern (J. Org. Chem. **32** [1967] 3249/51).

[3] E. R. H. Jones, S. Safe, V. Thaller (J. Chem. Soc. C **1967** 1038/44).

[4] C. Bond, M. Hooper (J. Chem. Soc. C **1969** 2453/60).

[5] M. D. Rausch, A. Siegel, L. P. Klemann (J. Org. Chem. **34** [1969] 468/70).

[6] G. Martelli, P. Spagnole, M. Tiecco (J. Chem. Soc. B **1970** 1413/8).

[7] H. E. Zimmerman, J. R. Dodd (J. Am. Chem. Soc. **92** [1970] 6507/14).

[8] A. Banihashemi, C. S. Marvel (J. Polym. Sci. Polym. Chem. Ed. **15** [1977] 2653/65).

[9] F. Bohlmann, U. Bühmann (Chem. Ber. **105** [1972] 863/73).

[10] J. Burdon, P. L. Coe, C. R. Marsh, J. C. Tatlow (J. Chem. Soc. Perkin Trans. I **1972** 763/9).

[11] J. I. DeGraw, V. H. Brown, W. T. Colwell, N. E. Morrison (J. Med. Chem. **17** [1974] 762/4).

[12] R. S. Dickson, L. J. Michel (Australian J. Chem. **28** [1975] 1943/55).

[13] I. Barrow, A. E. Pedler (Tetrahedron **32** [1976] 1829/34).

[14] S. Staicu, I. G. Dinulescu, F. Chiraleu, M. Avram (J. Organometal. Chem. **117** [1976] 385/94).

[15] G. A. Kraus, K. Frazier (Tetrahedron Letters **1978** 3195/8).

[16] F. G. Schreiber, R. Stevenson (Org. Prep. Proced. Intern. **10** [1978] 137/42).

[17] R. C. Ronald, J. M. Lansinger (J. Chem. Soc. Chem. Commun. **1979** 124/5).

[18] R. T. Scannel, R. Stevenson (J. Heterocycl. Chem. **17** [1980] 1727/8).

[19] N. A. Bumagin, I. O. Kalinovskii, I. P. Beletskaya (Izv. Akad. Nauk SSSR Ser. Khim. **1981** 2836; Bull. Acad. Sci. USSR Div. Chem. Sci. **1981** 2366).

[20] N. A. Bumagin, I. O. Kalinovskii, A. B. Ponomarev, I. P. Beletskaya (Dokl. Akad. Nauk SSSR **265** [1982] 1138/43; Dokl. Chem. Proc. Acad. Sci. USSR **262/267** [1982] 262/6).

1.1.2.1.4.8 Reactions with Compounds Containing More than One Iodobenzene Moiety

This reaction is referred to as "Reaction Type 7j" in Table 1 on pp. 17/41. The group RC≡C can be easily introduced not only into more simple benzene derivatives, but also into complex carbocyclic systems. The products are valuable intermediates in organic chemistry. This method has been applied especially to 2,2′- and 4,4′-diiodo biphenyls, diphenylmethanes, and stilbenes. In most cases all iodine atoms are substituted by RC≡C. Only the reaction of $C_6H_5C≡CCu$ and 2,2′-diiodobiphenyl in boiling pyridine yields 2-phenylethynyl-2'-iodobiphenyl [1]. The only 3,3'-substituted compound reacted is 3,3'-diiodobenzophenone, which gives 86% 3,3'-bis(phenylethynyl)benzophenone with excess $C_6H_5C{\equiv}CCu$ [8]. The compounds I and III (n not given) and $C_6H_5C{\equiv}CCu$ react in boiling pyridine (reaction time up to 72 h) to form the corresponding $C_6H_5C{\equiv}C$ substituted products II or IV, respectively, in yields between 95 and 100% [10].

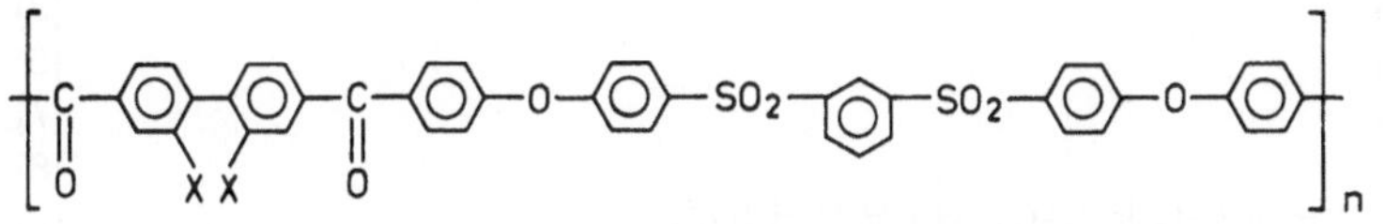

I: X = I

II: X = C≡CC$_6$H$_5$

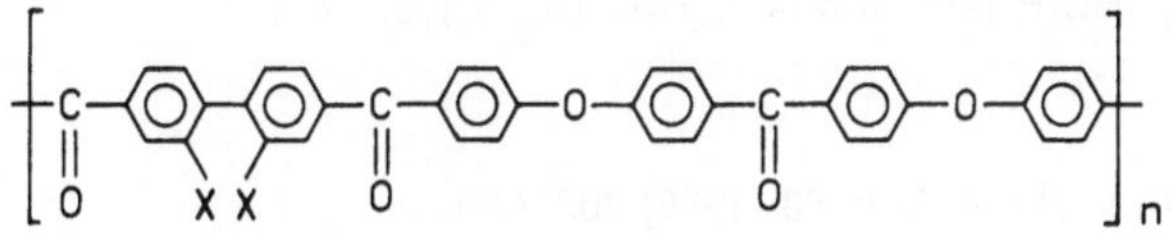

III: X = I

IV: X = C≡CC$_6$H$_5$

Reactions of the Type

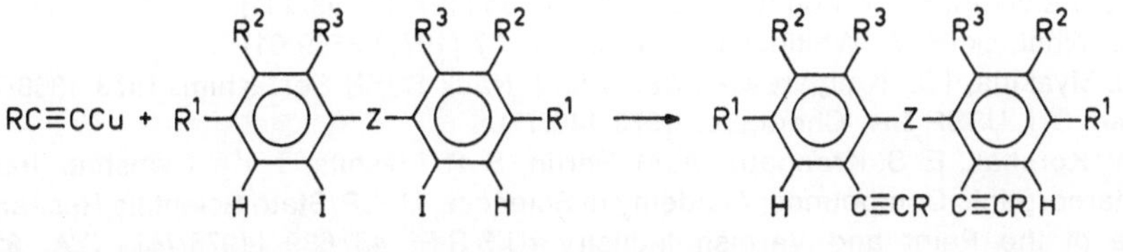

(belongs to Reaction Type 7j, see explanation on pp. 12/4; $C_5H_9O_2$ = tetrahydropyran-2-yloxy):

R	Z	R¹	R²	R³	remarks (yield in %)	Ref.
$C_2H_5O_2C$	–	H	H	CH_3O_2C	(5) and 16% V	[4]
C_6H_5	–	CH_3O_2C	H	H	(58)	[10]
	–	NH_2	NO_2	H	(57)	[9]
	–	COCl	H	H	see Section 1.1.2.1.5	[12]
	C≡C	H	H	H	(61)	[5]
$C_5H_9O_2CH_2$	C(CH₃)=C (E)-compound	H	H	H	isolated as VI (57)	[11]
$2\text{-}CH_3C_6H_4$	CH=CH (E)-compound	H	H	H	no yield given	[3]

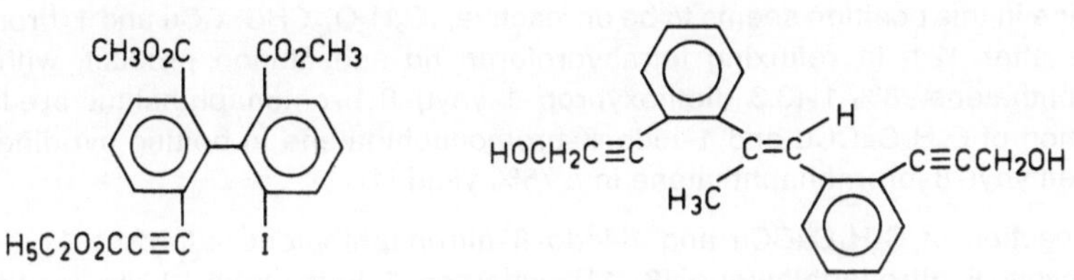

V

VI

References on p. 102

Reactions of the type $2C_6H_5C{\equiv}CCu + 4\text{-}IC_6H_4ZC_6H_4I\text{-}4' \rightarrow 4\text{-}C_6H_5C{\equiv}CC_6H_4ZC_6H_4C{\equiv}CC_6H_5\text{-}4'$ (belongs to Reaction Type 7j, see explanation on pp. 12/4):

Z	remarks (yield in %)	Ref.
–	(96)	[8]
CO	(81)	[8]
CH_2	(71)	[6]
$C(CF_3)_2$	(54)	[7]
$1,4\text{-}C_6H_4$	(72)	[8]
$CH(C_6H_4I\text{-}4)$	mixture of $4\text{-}C_6H_5C{\equiv}CC_6H_4CH(C_6H_4I\text{-}4')_2$,	[2]
	$(4\text{-}C_6H_5C{\equiv}CC_6H_4)_2CHC_6H_4I\text{-}4'$, and $(4\text{-}C_6H_5C{\equiv}CC_6H_4)_3CH$ (41)	

References:

[1] S. A. Kandil, R. E. Dessy (J. Am. Chem. Soc. **88** [1966] 3027/34).
[2] N. I. Popova, G. I. Skubnevskaya, Yu. N. Molin, I. L. Kotlyarevskii (Izv. Akad. Nauk SSSR Ser. Khim. **1969** 2424/30; Bull. Acad. Sci. USSR Div. Chem. Sci. **1969** 2271/5).
[3] H. A. Staab, R. Bader (Chem. Ber. **103** [1970] 1157/67).
[4] M. S. Newman, M. W. Logue (J. Org. Chem. **36** [1971] 1398/401).
[5] B. J. Whitlock, H. W. Whitlock (J. Org. Chem. **37** [1972] 3559/61).
[6] N. I. Myakina, I. L. Kotlyarevskii (Izv. Akad. Nauk SSSR Ser. Khim. **1973** 1368/70; Bull. Acad. Sci. USSR Div. Chem. Sci. **1973** 1405/7).
[7] V. V. Korshak, E. S. Krongauz, A. M. Berlin, B. R. Livshits, T. Kh. Dymshits, Institute of Heteroorganic Compounds, Academy of Sciences, USSR/State Scientific Research Institute of the Paint and Varnish Industry (U.S.S.R. 457689 [1973/75]; C.A. **82** [1975] No. 139702).
[8] A. M. Berlin, N. M. Kofman, E. S. Krongauz, A. N. Novikov, I. R. Gol'ding (Izv. Akad. Nauk SSSR Ser. Khim. **1975** 441/2; Bull. Acad. Sci. USSR Div. Chem. Sci. **1975** 369/70).
[9] F. L. Hedberg, F. E. Arnold (J. Polym. Sci. Polym. Chem. Ed. **14** [1976] 2607/19).
[10] A. Banihashemi, C. S. Marvel (J. Polym. Sci. Polym. Chem. Ed. **15** [1977] 2653/65).
[11] H. A. Staab, P. Günthert (Chem. Ber. **110** [1977] 619/30).
[12] S. Venkateso, U.S. Dept. of the Air Force (U.S. Appl. 219396 [1981]; C.A. **96** [1982] No. 7629).

1.1.2.1.4.9 Reactions with Iodonaphthalenes

This reaction is referred to as "Reaction Type 7k" in Table 1 on pp. 17/41.

Monoiodonaphthalenes. $RC{\equiv}CCu$ and 1-iodonaphthalene in refluxing pyridine give $1\text{-}RC{\equiv}CC_{10}H_7$ in good yields (yields in parentheses): R = tetrahydropyran-2-yloxy [6], $(C_2H_5O)_2CH$ [6], $2\text{-}O_2NC_6H_4$ (88%) [7], C_6H_5 (65 to 75%) [1, 12], tetrahydropyran-2-yloxymethyl [3], $C_5H_5FeC_5H_4$ (83%) [9, 14].

Bromine in this position seems to be unreactive. $(C_2H_5O)_2CHC{\equiv}CCu$ and 1-bromonaphthalene give after 12 h in refluxing tetrahydrofuran no substitution product, with 1-iodo-8-bromonaphthalene 98% 1-(3,3-diethoxyprop-1-ynyl)-8-bromonaphthalene are formed [13]. The reaction of $C_6H_5C{\equiv}CCu$ and 1-iodo-8-bromonaphthalene in boiling pyridine gives only 1-phenylethynyl-8-bromonaphthalene in a 75% yield [1].

The reaction of $C_6H_5C{\equiv}CCu$ and 1-iodo-8-nitronaphthalene affords a 17% yield of 1-phenylethynyl-8-nitronaphthalene [8, 11], whereas 1-iodonaphth-2-ole in dimethylformamide at 152° C is dehalogenated to naphth-2-ole [2].

References on pp. 104/5

Diiodonaphthalenes. RC≡CCu and 1,5–diiodonaphthalenes react in boiling pyridine to form 1,5–$(RC≡C)_2C_{10}H_6$, where R=CH_3 (yield 36%) [8] or R=C_6H_5 (yield 10%) [4, 8].

In contrast to this simple course, the reaction of RC≡CCu and 1,8–diiodonaphthalenes usually gives mixtures of mono- and disubstitution products I and II. If R=aryl, 7–arylben-zo[k]fluoranthenes III can also be formed, probably by thermal rearrangement of the disubsti-tution product IV. This reaction can be understood as an intramolecular [2+2+2]-cycloaddi-tion. The yield of III increases with increasing reaction time and is sensitive to electronic and steric effects. With R^2=CH_3O as a donor the yield is considerably enhanced. It is reduced, however, if R^1, R^2, and/or R^3 indicate CH_3 instead of H. Surprisingly, III also forms from 2,4,6–$(CH_3)_3C_6H_2C≡CCu$. No information is given about the fate of the lost CH_3 group [5, 12].

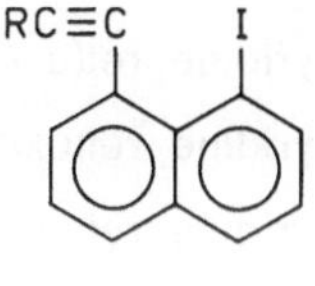

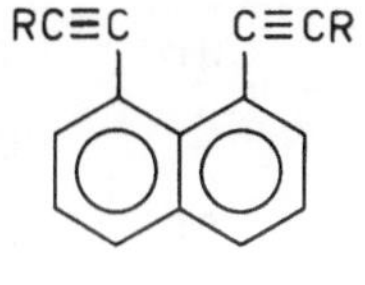

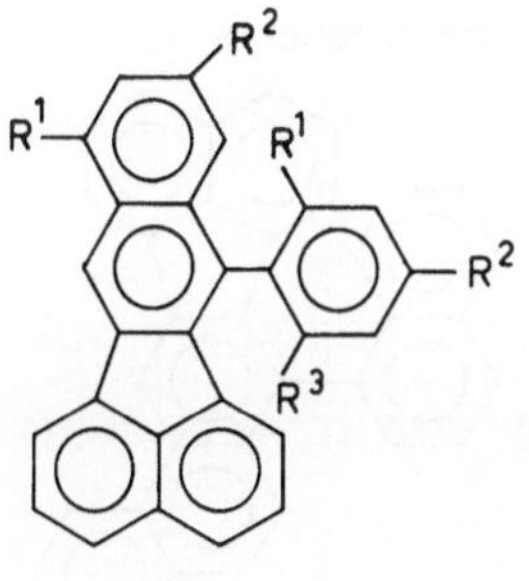

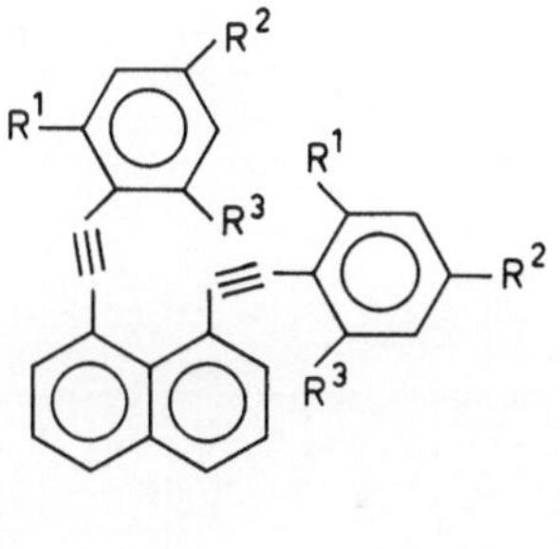

Table 10
Reactions of RC≡CCu and 1,8–Diiodonaphthalene (belongs to Reaction Type 7k, see explana-tion on pp. 12/4).
For abbreviations and dimensions, see p. X.

R	products (see above, yields in %)			molar ratio acetylide: $C_{10}H_6I_2$, conditions, remarks	Ref.
	I	II	III		
CH_3	—	62	—	—	[8]
C_2H_5	24	13	—	2:1, pyridine, 80°/12 h	[8]
	12	63	—	6:1, pyridine, 70 to 80°/72 h	[8]
n–C_4H_9	formed	—	—	2:1, pyridine, 25°	[8]
	formed	39	—	5:2, pyridine, 80°/24 h	[8]
$(C_2H_5O)_2CH$	91	—	—	—	[13]

Table 10 [continued]

R	products (see p. 103, yield in %)			molar ratio acetylide: $C_{10}H_6I_2$, conditions, remarks	Ref.
	I	II	III		
C_6H_5	—	formed	—	—	[4]
	—	72	—	pyridine, 20°/72 h	[8]
	—	57	18	pyridine, reflux/5 h	[5, 12]
	—	40	35	pyridine, reflux/10 h	[5, 12]
	—	16	59	pyridine, reflux/25 h	[5, 12]
$2\text{-}CH_3C_6H_4$	—	48	23	pyridine, reflux/10 h	[5, 12]
$4\text{-}CH_3C_6H_4$	—	41	17	pyridine, reflux/10 h	[5, 12]
$4\text{-}CH_3OC_6H_4$	—	—	65	pyridine, reflux/10 h	[5, 12]
$2,4\text{-}(CH_3)_2C_6H_3$	—	51	21	pyridine, reflux/10 h	[5, 12]
$2,4,6\text{-}(CH_3)_3C_6H_2$	—	80	5	loses one CH_3 group upon rearrangement	[5, 12]
naphth-1-yl	—	—	formed	the product is 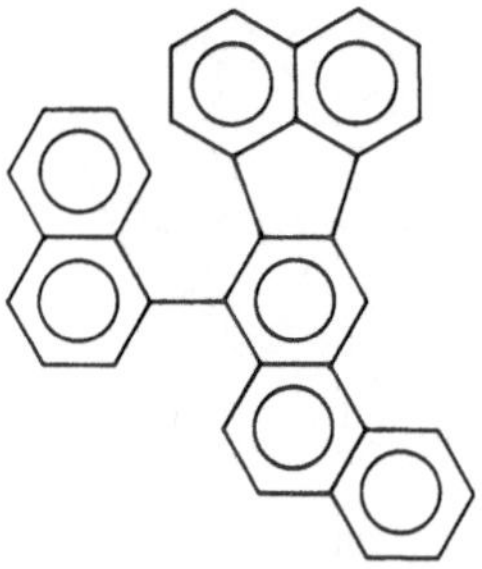	[5]

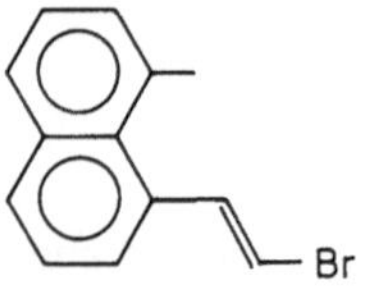	14	—	—	pyridine, 90°/2 h	[10]
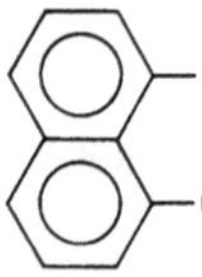	26	—	—	pyridine, 90°/2 h	[10]

References:

[1] R. E. Dessy, S. A. Kandil (J. Org. Chem. **30** [1965] 3857/60).
[2] C. E. Castro, E. J. Gaughan, D. C. Owsley (J. Org. Chem. **31** [1966] 4071/8).
[3] R. E. Atkinson, R. F. Curtis, D. M. Jones, J. A. Taylor (J. Chem. Soc. Chem. Commun. **1967** 718/9).
[4] B. Bossenbroek, H. Shechter (J. Am. Chem. Soc. **89** [1967] 7111/2).
[5] J. Ipaktschi, H. A. Staab (Tetrahedron Letters **1967** 4403/8).

[6] R. E. Atkinson, R. F. Curtis, D. M. Jones, J. A. Taylor (J. Chem. Soc. **1969** 2173/6).

[7] C. Bond, M. Hooper (J. Chem. Soc. C **1969** 2453/60).

[8] B. Bossenbroek, D. C. Sanders, H. M. Curry, H. Shechter (J. Am. Chem. Soc. **91** [1969] 371/9).

[9] M. D. Rausch, A. Siegel (J. Org. Chem. **34** [1969] 1974/6).

[10] R. H. Mitchell, F. Sondheimer (Tetrahedron **26** [1970] 2141/50).

[11] C. C. Leznoff, R. J. Hayward (Can. J. Chem. **49** [1971] 3596/601).

[12] H. A. Staab, J. Ipaktschi (Chem. Ber. **104** [1971] 1170/81).

[13] J. S. Kiely, P. Boudjouk, L. L. Nelson (J. Org. Chem. **42** [1977] 2626/9).

[14] D. W. Slocum, M. D. Rausch, A. Siegel (Organometal. Polym. Symp., New Orleans 1977 [1978], pp. 39/51, 49).

1.1.2.1.4.10   Reactions with Other Carbocycles R'X

This reaction is referred to as "Reaction Type 7l" in Table 1 on pp. 17/41.

Azulenes. The reaction of 1,3-diiodoazulene (Ia) with a tenfold excess of $C_6H_5C≡CCu$ in boiling dimethylformamide produces 15% Ib and 81% Ic [7]. The reaction of the dirhodano-azulene Id and $C_6H_5C≡CCu$ affords upon cleavage of the S-C bond 2.5% Ie (3 min in boiling dimethylacetamide). Under the same conditions the monorhodanoazulene If with $C_6H_5C≡CCu$ gives Ig [8].

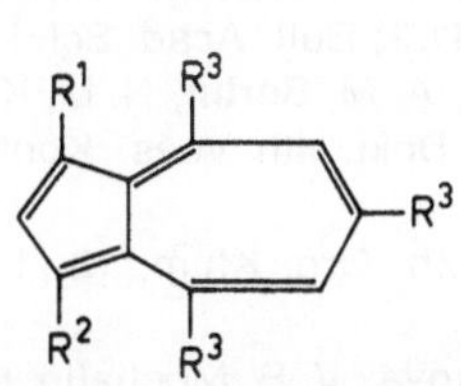

I

	R^1	R^2	R^3
Ia	I	I	H
Ib	I	$C_6H_5C≡C$	H
Ic	$C_6H_5C≡C$	$C_6H_5C≡C$	H
Id	NCS-	NCS-	CH_3
Ie	$C_6H_5C≡CS$	$C_6H_5C≡CS$	CH_3
If	NCS-	H	CH_3
Ig	$C_6H_5C≡CS$	H	CH_3

Phenanthrenes. The reaction of $C_6H_5C≡CCu$ and II (R=I) (30 h in boiling pyridine or 14 h in boiling dimethylformamide) gives 65% [5] II (R=$C_6H_5C≡C$) [5, 6].

II   III

Fluorenes. The reactions of $C_6H_5C≡CCu$ with III (X=H_2 or O, R=I) give 39% III (X=H_2, R = $C_6H_5C≡C$) [5] or 93% III (X=O, R = $C_6H_5C≡C$) [5, 6].

References on p. 106

Ferrocenes and Ruthenocenes. $C_6H_5C{\equiv}CCu$ and bromoferrocene react in dimethylform-amide to form $C_5H_5FeC_5H_4C{\equiv}CC_6H_5$ in a 48% yield [1]. The analogous reaction with iodoferro-cene in pyridine yields 84% of the same product [2, 3]. $C_5H_5FeC_5H_4C{\equiv}CCu$ and iodoferrocene react to form $C_5H_5FeC_5H_4C{\equiv}CC_5H_4FeC_5H_5$ (yield 84%) [2, 10]. The reaction of $C_5H_5RuC_5H_4C{\equiv}$ CCu and iodoferrocene gives 60% $C_5H_5RuC_5H_4C{\equiv}CC_5H_4FeC_5H_5$ [4].

$C_6H_5C{\equiv}CCu$ and $Fe(C_5H_4I)_2$ react in boiling pyridine to form $Fe(C_5H_4C{\equiv}CC_6H_5)_2$; the yield is 40% after 30 min [3] and 57% after 8 h [2]. The formation of 46% $Fe(C_5H_4C{\equiv}CC_6H_5)_2$ from $Fe(C_5H_4I)_2$, $C_6H_5C{\equiv}CH$, CuI, and $Pd(P(C_6H_5)_3)_2Cl_2$ in $HN(C_2H_5)_2$ at 80° C/6 h can be ex-plained as a reaction of $C_6H_5C{\equiv}CCu$ formed in situ [9]. $C_5H_5FeC_5H_4C{\equiv}CCu$ and $Fe(C_5H_4I)_2$ react to form 55% $Fe(C_5H_4C{\equiv}CC_5H_4FeC_5H_5)_2$ [3].

References:

[1] A. N. Nesmeyanov, V. A. Sazonova, V. N. Drozd (Dokl. Akad. Nauk SSSR **154** [1964] 158/9; Dokl. Chem. Proc. Acad.Sci. USSR **154/159** [1964] 30/2).

[2] M. D. Rausch, A. Siegel, L. P. Klemann (J. Org. Chem. **31** [1966] 2703/4).

[3] M. Rosenblum, N. Brawn, J. Papenmeier, M. Applebaum (J. Organometal. Chem. **6** [1966] 173/80).

[4] M. D. Rausch, A. Siegel (J. Org. Chem. **34** [1969] 1974/6).

[5] A. M. Berlin, N. M. Kofman, E. S. Krongauz, A. N. Novikov, I. R. Gol'ding (Izv. Akad. Nauk SSSR Ser. Khim. **1975** 441/2; Bull. Acad. Sci. USSR Div. Chem. Sci. **1975** 369/70).

[6] V. V. Korshak, E. S. Krongauz, A. M. Berlin, N. M. Kofman, A. P. Travnikova (Khim. Di-karbonil'nykh Soedin. Tezisy Dokl. 4th Vses. Konf., Riga 1975 [1976], pp. 76/7; C.A. **87** [1977] No. 39874).

[7] V. A. Nefedov, I. K. Tarygina (Zh. Org. Khim. **12** [1976] 1763/9; J. Org. Chem. [USSR] **12** [1976] 1730/5).

[8] Yu. N. Porshnev, T. N. Dvornikova, V. B. Mochalin (Zh. Org. Khim. **12** [1976] 2019/21; J. Org. Chem. [USSR] **12** [1976] 1966/7).

[9] A. Kasahara, T. Izumi, M. Maemura (Bull. Chem. Soc. Japan **50** [1977] 1021/2).

[10] D. W. Slocum, M. D. Rausch, A. Siegel (Organometal. Polym. Symp., New Orleans 1977 [1978], pp. 39/51, 49).

1.1.2.1.4.11 Reactions with 5–Membered Heterocycles R′X

This reaction is referred to as "Reaction Type 7 m" in Table 1 on pp. 17/41.

Ribofuranosides. The reaction of $C_6H_5C{\equiv}CCu$ and I (5d in benzene at room temperature) gives 8% of the ribofuranoside II and 32% of the ribofuranose III [26, 29, 32]. Unexpectedly, I attacks $HOCH_2C{\equiv}CCu$ (or $CuOCH_2C{\equiv}CH$, bonding unclear) at the oxygen. After 2d in ben-zene at room temperature 21% of the ribofuranoside IV and 5% of the ribose V are formed [29, 32].

Furans and Thiophenes. The introduction of the group $RC{\equiv}C$ into a thiophene ring has been widely used because of the importance of alkynylthiophenes as natural products and as precursors in synthesis. Certain (polyalkynyl)thiophenes are also used to prepare bi- or polythiophenes, which are plant metabolites, too. The synthesis using $RC{\equiv}CCu$ is general-ly applicable to 2–iodofurans and 2–iodothiophenes (Tables 11 and 12).

This reaction is sometimes catalyzed by Cu powder (mechanism unknown). In many cases the iodine in 2–position is partly substituted by H and a mixture of $RC{\equiv}CR'$ and R′H is formed. The product pattern of the reaction of $RC{\equiv}CCu$ with diiodothiophenes depends on the molar ratio of the reactants [3, 9, 10, 13, 16, 20, 21, 27]. In the synthesis of monosubsti-tuted species sometimes an excess of the diiodothiophene is used [3, 9]. In the presence

References on pp. 113/4

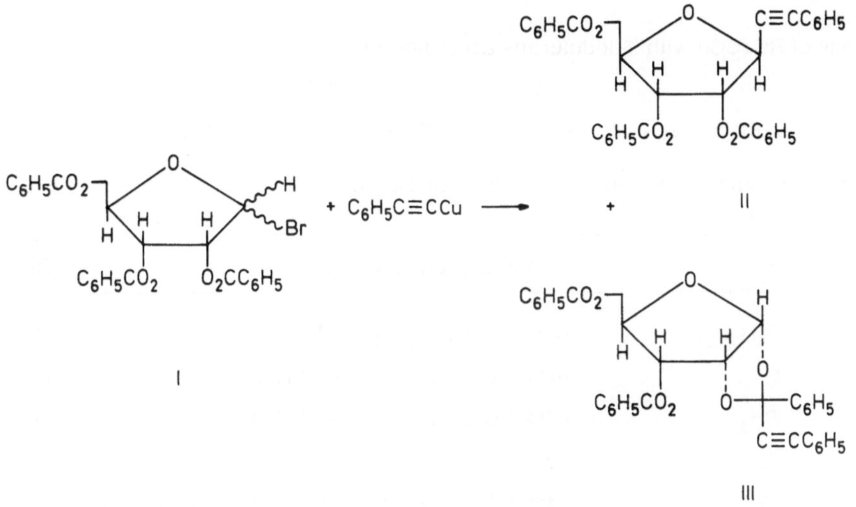

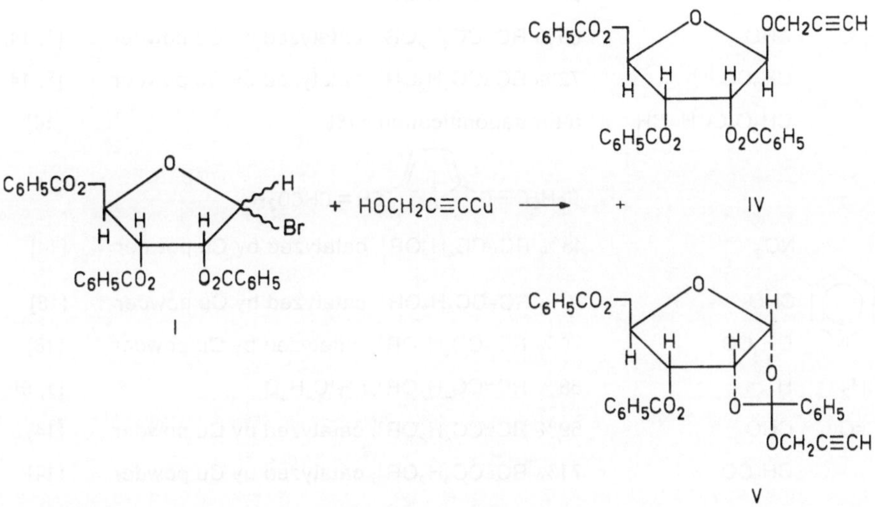

$C_6H_5CO_2$ ... O ... $C\equiv CC_6H_5$
$C_6H_5CO_2$ ... $O_2CC_6H_5$

II

$C_6H_5CO_2$ ... O ... H
+ $C_6H_5C\equiv CCu$ ⟶   +
$C_6H_5CO_2$ ... Br

I

$C_6H_5CO_2$ ... O ... H
$C_6H_5CO_2$ ... O—C_6H_5
$C\equiv CC_6H_5$

III

$C_6H_5CO_2$ ... O ... $OCH_2C\equiv CH$
$C_6H_5CO_2$ ... $O_2CC_6H_5$

IV

$C_6H_5CO_2$ ... O ... H
+ $HOCH_2C\equiv CCu$ ⟶   +
$C_6H_5CO_2$ ... Br

I

$C_6H_5CO_2$ ... O ... H
$C_6H_5CO_2$ ... O—C_6H_5
$OCH_2C\equiv CH$

V

of Pd compounds in dimethylformamide the reaction seems to proceed almost quantitatively even under very mild conditions, but until now only one example of this reaction has been published [35]. Usually the reaction is run in boiling pyridine (4 to 8 h); slowly heating to the reflux temperature enhances the yield [13, 16]. This reaction can be strongly catalyzed by Pd compounds [35] and an analogous reaction using $RC\equiv CH$, R'X, and catalytic amounts of both the Pd catalyst and CuI is also known [36].

3-Iodothiophene and $C_6H_5C\equiv CCu$ react to form the expected 3-phenylethynylthiophene [22]. The reaction of $RC\equiv CCu$ and 3-iodo-4-carboxythiophenes yields only thieno[3,4-c]-pyran-4-ones after ring closure; see Section 1.1.2.1.5.3 [19]. The attempts to react $RC\equiv CCu$ and 2-bromothiophenes were not successful (e.g., $CH_3C\equiv CCu$ with 5-bromothiophene-2-carboxaldehyde) or gave low yields (e.g., $HOCH_2C\equiv CCu$ with 2-bromothiophene, 14% yield in boiling pyridine) [9].

Table 11
Reactions of RC≡CCu with 2-Iodofurans according to

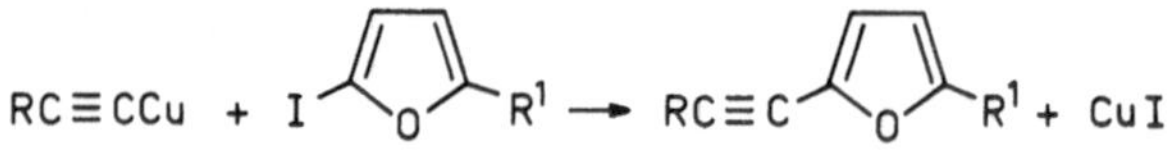

$$RC\equiv CCu \ + \ I\underset{O}{\overset{}{\bigcirc}}R^1 \longrightarrow RC\equiv C\underset{O}{\overset{}{\bigcirc}}R^1 + CuI$$

(belongs to Reaction Type 7m, explanation see pp. 12/4).
C_4H_3O = fur-2-yl.

R	R^1	remarks, products, yields	Ref.
CH$_3$	C$_2$H$_5$O$_2$C	60% RC≡CC$_4$H$_2$OR^1+R^{1}C$_4$H$_3$O	[1, 9]
HOCH$_2$	H	67% RC≡CC$_4$H$_2$OR^1+R^{1}C$_4$H$_3$O	[1, 9]
	CH$_3$	65% RC≡CC$_4$H$_2$OR^1+R^{1}C$_4$H$_3$O	[1, 9]
	(E)-CH$_3$O$_2$CCH=CH	60% RC≡CC$_4$H$_2$OR^1 in presence of [Cu(NH$_3$)$_4$]Cl	[17]
n-C$_4$H$_9$	CHO	74% RC≡CC$_4$H$_2$OR^1, catalyzed by Cu powder	[14]
C$_6$H$_5$	H	82% RC≡CC$_4$H$_2$OR^1	[23]
	CHO	65% RC≡CC$_4$H$_2$OR^1, catalyzed by Cu powder	[7, 14, 18]
	CH$_3$CO	72% RC≡CC$_4$H$_2$OR^1, catalyzed by Cu powder	[7, 14, 18]
	CH$_3$O$_2$CCH=CH	after saponification 64% 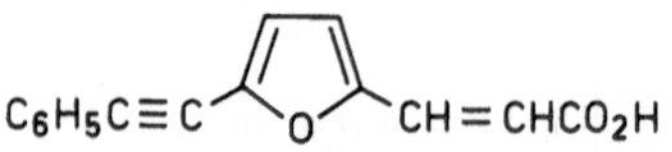	[14]

$$C_6H_5C\equiv C\underset{O}{\overset{}{\bigcirc}}CH=CHCO_2H$$

R	R^1	remarks, products, yields	Ref.
	NO$_2$	48% RC≡CC$_4$H$_2$OR^1, catalyzed by Cu powder	[14]
CH$_3$–⟨N⟩	CHO	70% RC≡CC$_4$H$_2$OR^1, catalyzed by Cu powder	[18]
	CH$_3$CO	70% RC≡CC$_4$H$_2$OR^1, catalyzed by Cu powder	[18]
C$_6$H$_5$CH$_2$	H	58% RC≡CC$_4$H$_2$OR^1+R^{1}C$_4$H$_3$O	[1, 9]
C$_6$H$_5$C≡C	CHO	69% RC≡CC$_4$H$_2$OR^1, catalyzed by Cu powder	[14]
	CH$_3$CO	71% RC≡CC$_4$H$_2$OR^1, catalyzed by Cu powder	[14]

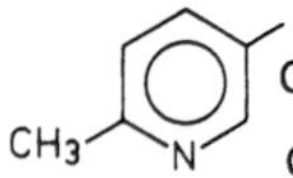

References on pp. 113/4

Table 12
Reactions of RC≡CCu with 2-Iodothiophenes according to

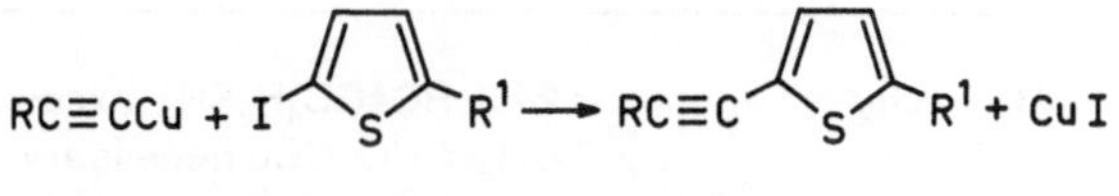

RC≡CCu + I⟨S⟩R¹ ⟶ RC≡C⟨S⟩R¹ + CuI

(belongs to Reaction Type 7m, explanation see pp. 12/4).
C_4H_3S = thien-2-yl, $C_5H_9O_2$ = tetrahydrodropyran-2-yloxy.

R	R¹	remarks, products, yields	Ref.
CH₃	CHO	61% RC≡CC₄H₂SR¹+R¹C₄H₃S	[1, 9]
	(E)-CH₃CH=CH	RC≡CC₄H₂SR¹	[21]
	C₂H₅CO	RC≡CC₄H₂SR¹	[21]
	C₄H₃S	RC≡CC₄H₂SR¹+R¹C₄H₃S	[11]
	(C₂H₅O)₂CH	RC≡CC₄H₂SR¹	[21]
	(E)-CH₂=CHCH=CHC≡C	RC≡CC₄H₂SR¹	[3]
	C₆H₅	51% RC≡CC₄H₂SR¹+R¹C₄H₃S	[1]
	5-(CH₃O)₂CHC≡CC₄H₂S	R¹C₄H₃S the only product	[28]
	I	18% RC≡CC₄H₂SR¹ after 6 h in pyridine at 120° [13] 20 to 60% RC≡CC₄H₂SR¹ after 4 h reflux in pyridine [16] "heated" in pyridine for 7.5 h gives 13% R¹C₄H₃S, 5,5′-diiodo-2,2′-bithienyl and RC≡CC₄H₂SR¹ [21] (in all cases with a molar ratio RC≡CCu:thiophene = 1:1)	[13, 16, 21]
HOCH₂	H	67 to 79% RC≡CC₄H₂SR¹+ R¹C₄H₃S	[1, 9]
	C₄H₃S	RC≡CC₄H₂SR¹+R¹C₄H₃S	[11]
CH₂=CH	H	48% RC≡CC₄H₂SR¹+R¹C₄H₃S 32% RC≡CC₄H₂SR¹ with acetylide prepared in situ from CH₂=CHC≡CMgBr and CuCl	[1, 6]
	CH₃	53% RC≡CC₄H₂SR¹+R¹C₄H₃S	[1]
	C₂H₅O₂C	43% RC≡CC₄H₂SR¹+R¹C₄H₃S	[1]
	C₄H₃S	32% RC≡CC₄H₂SR¹ with acetylide prepared in situ from CH₂=CHC≡CMgBr and CuCl	[2, 6, 11]

Table 12 [continued]

R	R^1	remarks, products, yields	Ref.
$CH_2=CH$ [continued]	$CH_3(C\equiv C)_2$	<24% $RC\equiv CC_4H_2SR^1$, excess $CH_2=CHC\equiv CCu$ necessary	[16]
	(E)-$CH_3CH=CHC\equiv C$	$RC\equiv CC_4H_2SR^1$	[3]
	$5\text{-}CH_3C_4H_2S$	$RC\equiv CC_4H_2SR^1 + R^1C_4H_3S$	[11]
	$(CH_3O)_2CHC\equiv C$	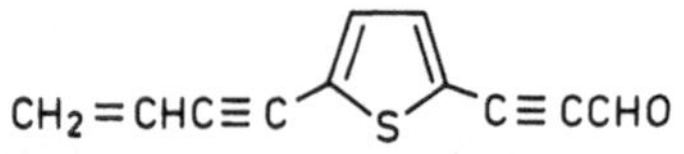 $CH_2=CHC\equiv C$ —[thiophene]— $C\equiv CCHO$ after saponification, excess $CH_2=CHC\equiv CCu$ necessary	[16]
	$5\text{-}CH_3O_2CC_4H_2S$	$RC\equiv CC_4H_2SR^1 + R^1C_4H_3S$	[11]
	I	29% $RC\equiv CC_4H_2SR^1 + R^1C_4H_3S$, 120°/4 h in pyridine, molar ratio 1:1, the acetylide is not totally converted	[13]
CH_3CO	C_4H_3S	$RC\equiv CC_4H_2SR^1 + R^1C_4H_3S$	[11]
CH_3O_2C	$CH_3C\equiv C$	35% $RC\equiv CC_4H_2SR^1$	[13]
$ClCH_2CH(OH)$	C_4H_3S	the active Cl causes side reactions, no product identified	[4]
$HO(CH_2)_2$	H	51% $RC\equiv CC_4H_2SR^1 + R^1C_4H_3S$	[1]
	I	<15% $RC\equiv CC_4H_2SR^1$, 4 h reflux in pyridine, molar ratio = 1:1	[16]
$CH_3C\equiv C$	$5\text{-}HO(CH_2)_2C\equiv CC_4H_2S$	<15% $RC\equiv CC_4H_2SR^1$	[16]
	$C_5H_9O_2CH(CH_2O_2CCH_3)C\equiv C$	<33% $RC\equiv CC_4H_2SR^1$	[16]
	I	<24% $RC\equiv CC_4H_2SR^1$, 4 h reflux in pyridine, molar ratio = 1:1	[16]
$CH_3CH=CH$	$C_5H_9O_2CH_2C\equiv C$	(E)/(Z)-mixture of 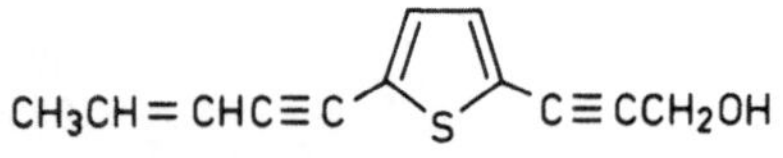 $CH_3CH=CHC\equiv C$ —[thiophene]— $C\equiv CCH_2OH$ after saponification	[16]
$C_2H_5O_2C$	C_4H_3S	$RC\equiv CC_4H_2SR^1$	[20]
$(CH_3O)_2CH$	$CH_3C\equiv C$	68% $RC\equiv CC_4H_2SR^1$	[25]
	I	72% $RC\equiv CC_4H_2SR^1$, 7 h reflux in pyridine, molar ratio $RC\equiv CCu$:thiophene = 2:1	[27]

References on pp. 113/4

Table 12 [continued]

R	R[1]	remarks, products, yields	Ref.
(CH$_3$O)$_2$CH [continued]	5-IC$_4$H$_2$S	yields 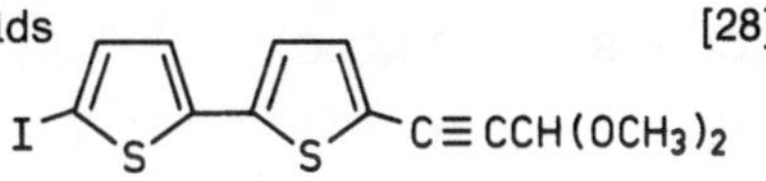 at a molar ratio RC≡CCu:thiophene = 1:1, 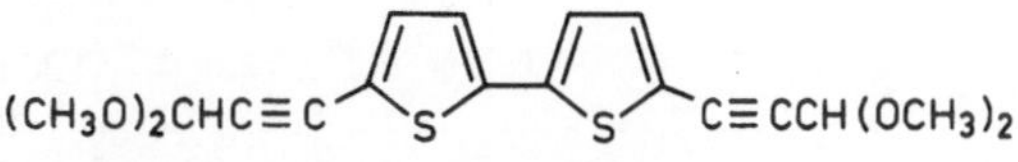 at a molar ratio of 2:1 (no yields given)	[28]
CH$_2$=CHC≡C	CH$_3$C≡C	gives RC≡CC$_4$H$_2$SR[1] with excess CH$_2$=CHC≡CCu	[16]
	C$_5$H$_9$O$_2$CH$_2$C≡C	<11% RC≡CC$_4$H$_2$SR[1], if slowly reacted under ice-cooling, excess CH$_2$=CHC≡CCu necessary	[16]
C$_5$H$_9$O$_2$	I	RC≡CC$_4$H$_2$SR[1] at a 1:1 molar ratio, 5 h reflux in pyridine	[20]
(C$_2$H$_5$O)$_2$CH	H	RC≡CC$_4$H$_2$SR[1]	[9]
	CH$_3$	85% RC≡CC$_4$H$_2$SR[1] + R[1]C$_4$H$_3$S	[1, 9]
	C$_4$H$_3$S	after saponification, R[1]C$_4$H$_3$S is also formed	[11]
	I	RC≡CC$_4$H$_2$SR[1] with a molar ratio RC≡CCu:thiophene = 8:13, 8 h reflux in pyridine	[3, 9]
C$_6$H$_5$	H	RC≡CC$_4$H$_2$SR[1]; 75% in boiling pyridine after 10 h, 44% with C$_6$H$_5$C≡CCu prepared in situ from C$_6$H$_5$C≡CMgBr and CuCl, 94% after 1 h at 20° in dimethylformamide in presence of C$_6$H$_5$PdI[P(C$_6$H$_5$)$_3$]$_2$/[N(C$_4$H$_9$-n)$_4$]I	[6, 8, 35]
	C$_4$H$_3$S	51% RC≡CC$_4$H$_2$SR[1] with C$_6$H$_5$C≡CCu prepared in situ from C$_6$H$_5$C≡CMgBr and CuCl	[6]
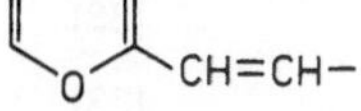	H	60% RC≡CC$_4$H$_2$SR[1] + R[1]C$_4$H$_3$S	[1]
	C$_2$H$_5$O$_2$C	55% RC≡CC$_4$H$_2$SR[1] + R[1]C$_4$H$_3$S	[1]

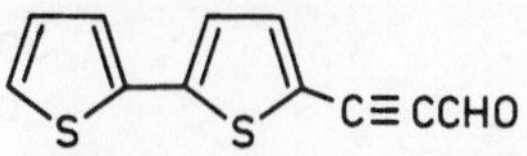

References on pp. 113/4

Table 12 [continued]

R	R[1]	remarks, products, yields	Ref.
4-CH$_3$O$_2$CC$_4$H$_2$S	CH$_3$C≡C	46% RC≡CC$_4$H$_2$SR[1]	[25]
C$_5$H$_9$O$_2$CH$_2$	H	RC≡CC$_4$H$_2$SR[1]	[9, 10]
	C$_4$H$_3$S	RC≡CC$_4$H$_2$SR[1]	[10]
	C$_6$H$_5$	60% RC≡CC$_4$H$_2$SR[1]	[15]
	I	20 to 60% RC≡CC$_4$H$_2$SR[1], 4 h reflux in pyridine, molar ratio = 1:1	[10, 16]
(E)-C$_5$H$_9$O$_2$CH$_2$CH=CH	H	17% 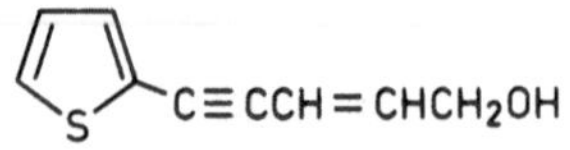 after saponification; 58% of the corresponding aldehyde after saponification and MnO$_2$ oxidation	[5, 30]
	CH$_3$CO$_2$CH$_2$	54% RC≡CC$_4$H$_2$SR[1]	[24]
	CH$_2$=CHC≡C	57% RC≡CC$_4$H$_2$SR[1]	[13]
C$_5$H$_9$O$_2$CH(CH$_2$O$_2$CCH$_3$)	C$_4$H$_3$S	54% RC≡CC$_4$H$_2$SR[1]	[12]
	I	<33% RC≡CC$_4$H$_2$SR[1], 4 h reflux in pyridine	[16]
C$_5$H$_5$FeC$_5$H$_4$	H	80% RC≡CC$_4$H$_2$SR[1]	[8, 31]

Pyrazoles. The halogen of 4-iodopyrazoles R'I can be substituted by RC≡C, using RC≡CCu as a reagent; see Table 13. For reactions involving a ring closure, see Reaction Type 8, Section 1.1.2.1.5.3.

Table 13
Reactions of the Type RC≡CCu +

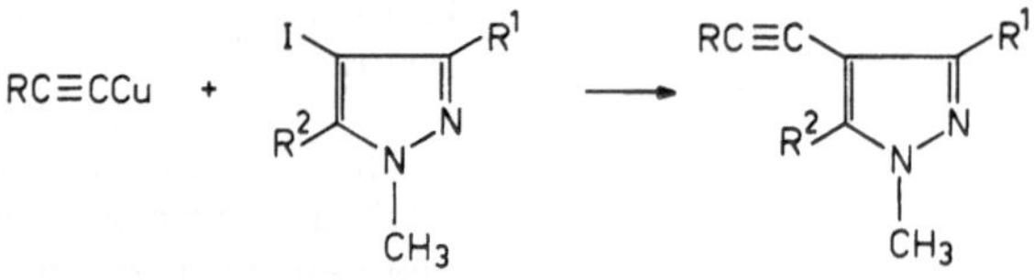

(belongs to Reaction Type 7 m, explanation see pp. 12/4).

R	R[1]	R[2]	yield	Ref.
CH$_3$OCH$_2$	CH$_3$	HO$_2$C	59%	[33]
	HO$_2$C	H	60%	[33]
n-C$_3$H$_7$	CH$_3$	HO$_2$C	53%	[33]
	HO$_2$C	H	62%	[33]

Table 13 [continued]

R	R^1	R^2	yield	Ref.
C$_6$H$_5$	CH$_3$	HO$_2$C	62%	[33]
	HO$_2$C	H	58%	[33]
	H	CH$_3$	80% 84% with C$_6$H$_5$C≡CCu in situ prepared from C$_6$H$_5$C≡CH, CuI, (C$_2$H$_5$)$_2$NH and catalyzed by Pd(P(C$_6$H$_5$)$_3$)$_2$Cl$_2$	[34]

Thiazoles. The reaction of 2-iodothiazole and C$_6$H$_5$≡CCu affords a 58% yield of 2-phenyl-ethynylthiazole [23].

References:

[1] R. Atkinson, R. F. Curtis, G. T. Phillips (Chem. Ind. [London] **1964** 2101/2).
[2] R. E. Atkinson, R. F. Curtis, G. T. Phillips (Tetrahedron Letters **1964** 3159/62).
[3] R. E. Atkinson, R. F. Curtis (Tetrahedron Letters **1965** 297/300).
[4] R. E. Atkinson, R. F. Curtis, G. T. Phillips (J. Chem. Soc. C **1966** 1101/3).
[5] F. Bohlmann, K.-M. Kleine, C. Arndt (Chem. Ber. **99** [1966] 1642/7).
[6] D. Brown, J. Cymerman-Craig, N. H. Dyson, J. W. Westley (J. Chem. Soc. C **1966** 89/91).
[7] R. I. Katkevich, S. P. Korshunov, L. I. Vereshchagin (Zh. Vses. Khim. Obshchestva **11** [1966] 705/6; C.A. **66** [1967] No. 55300).
[8] M. D. Rausch, A. Siegel, L. P. Klemann (J. Org. Chem. **31** [1966] 2703/4).
[9] R. E. Atkinson, R. F. Curtis, J. A. Taylor (J. Chem. Soc. C **1967** 578/82).
[10] R. E. Atkinson, R. F. Curtis, D. M. Jones, J. A. Taylor (J. Chem. Soc. Chem. Commun. **1967** 718/9).

[11] R. E. Atkinson, R. F. Curtis, G. T. Phillips (J. Chem. Soc. C **1967** 2011/5).
[12] F. Bohlmann, C. Zdero, W. Gordon (Chem. Ber. **100** [1967] 1193/9).
[13] F. Bohlmann, P.-H. Bonnet, H. Hofmeister (Chem. Ber. **100** [1967] 1200/5).
[14] S. P. Korshunov, R. I. Katkevich, L. I. Vereshchagin (Zh. Org. Khim. **3** [1967] 1327/31; J. Org. Chem. [USSR] **3** [1967] 1288/91).
[15] F. Bohlmann, C. Zdero (Chem. Ber. **101** [1968] 3243/54).
[16] F. Bohlmann, P. Blaszkiewicz, E. Bresinsky (Chem. Ber. **101** [1968] 4163/9).
[17] C. H. Fawcett, D. M. Spencer, R. L. Wain, A. G. Fallis, E. R. H. Jones, M. Le Quan, C. B. Page, V. Thaller, D. C. Shubrook, P. M. Whitham (J. Chem. Soc. C **1968** 2455/62).
[18] S. P. Korshunov, R. I. Katkevich, O. T. Shashlova, L. I. Vereshchagin (Zh. Org. Khim. **4** [1968] 676/80; J. Org. Chem. [USSR] **4** [1968] 659/62).
[19] S. A. Mladenović, C. E. Castro (J. Heterocycl. Chem. **5** [1968] 227/30).
[20] R. E. Atkinson, R. F. Curtis, D. M. Jones, J. A. Taylor (J. Chem. Soc. C **1969** 2173/6).

[21] R. F. Curtis, J. A. Taylor (J. Chem. Soc. C **1969** 1813/8).
[22] G. Martelli, P. Spagnole, M. Tiecco (J. Chem. Soc. B **1970** 1413/8).
[23] T. Teitei, P. J. Collin, W. H. F. Sasse (Australian J. Chem. **25** [1972] 171/82).
[24] F. Bohlmann, P.-D. Hopf (Chem. Ber. **106** [1973] 3621/5).
[25] F. Bohlmann, W. Skuballa (Chem. Ber. **106** [1973] 497/504).
[26] K. Arakawa, T. Miyasaka, N. Hamamichi (Chem. Letters **1974** 1305/8; C.A. **82** [1975] No. 73382).
[27] F. Bohlmann, J. Kocur (Chem. Ber. **107** [1974] 2115/9).

[28] F. Bohlmann, J. Kocur (Chem. Ber. **108** [1975] 2149/52).

[29] K. Arakawa, T. Miyasaka, N. Hamamichi (Nucl. Acids Spec. Publ. No. 2 [1976] 1/4; C.A. **86** [1977] No. 140379).

[30] F. Bohlmann, C. Hühn (Chem. Ber. **110** [1977] 1183/5).

[31] D. W. Slocum, M. D. Rausch, A. Siegel (Organometal. Polym. Symp., New Orleans 1977 [1978], pp. 39/51, 49).

[32] N. Hamamichi, T. Miyasaka, K. Arakawa (Chem. Pharm. Bull. [Tokyo] **26** [1978] 898/907).

[33] S. F. Vasilevskii, E. M. Rubinshtein, M. S. Shvartsberg (Izv. Akad. Nauk SSSR Ser. Khim. **1978** 1175/7; Bull. Acad. Sci. USSR Div. Chem. Sci. **1978** 1021/3).

[34] A. N. Sinyakov, M. S. Shvartsberg (Izv. Akad. Nauk SSSR Ser. Khim. **1979** 1126/8; Bull. Acad. Sci. USSR Div. Chem. Sci. **1979** 1053/4).

[35] N. A. Bumagin, I. O. Kalinovskii, A. B. Ponomarov, I. P. Beletskaya (Dokl. Akad. Nauk SSSR **265** [1982] 1138/43; Dokl. Chem. Proc. Acad. Sci. USSR **262/267** [1982] 262/6).

[36] N. A. Bumagin, A. B. Ponomarev, I. B. Beletskaya (Izv. Akad. Nauk SSSR Ser. Khim. **1984** 1561/6; Bull. Acad. Sci. USSR Div. Chem. Sci. **1984** 1433/8).

1.1.2.1.4.12 Reactions with 6-Membered Heterocycles R′X

This reaction is referred to as "Reaction Type 7n" in Table 1 on pp. 17/41.

Pyridines. The reaction of $RC{\equiv}CCu$ and 2-iodopyridine in boiling pyridine gives 2-alkinyl-pyridines. This reaction is much faster than the reaction of $RC{\equiv}CCu$ with iodobenzenes. For instance, $C_6H_5C{\equiv}CCu$ and 2-iodopyridine give 86% substitution product after 1 h [3], whereas under the same conditions with 2-iodobenzene a yield of 87% is obtained only after 10 h [1]. The reaction of $C_6H_5C{\equiv}CCu$ and 2-iodopyridine can be catalyzed by $Fe(C_5H_4P(C_6H_5)_2)_2 \cdot PdCl_2$ and $[N(C_4H_9\text{-}n)_4]I$ in hexamethylphosphoric triamide. After 2 h at 20° C, 67% $C_5H_4NC{\equiv}CC_6H_5$ $(C_5H_4N=\text{pyrid-2-yl})$ is obtained. The catalysis by $C_6H_5PdI[P(C_6H_5)_3]_2$ needs 5 h at 50° C to give a yield of 81% [8].

In boiling pyridine the following $RC{\equiv}CCu$ were reacted with 2-iodopyridine to give 2 $RC{\equiv}CC_5H_4N$ (yields in parentheses): $R=n\text{-}C_4H_9$ (95%) [3]; $4\text{-}O_2NC_6H_4$ (52%) [3]; C_6H_5 (25 to 86%) [2, 3], with $C_6H_5C{\equiv}CCu$ formed in situ yields up to 92% are obtained [5]; tetrahydro-pyran-2-yloxymethyl [6]; $4\text{-}CH_3OC_6H_4$ (68%) [3]; $4\text{-}C_6H_5C_6H_4$ (54%) [3]; $4\text{-}C_6H_5C_6H_4C{\equiv}C$ (7%) at 120° C, unstable product [3].

The reaction of 2-iodo-3-hydroxypyridines and $RC{\equiv}CCu$, in most cases, gives under ring closure furo[3,2-b]pyridines and is described in Section 1.1.2.1.5.1. Only $2\text{-}O_2NC_6H_4C{\equiv}CCu$ and 2-iodo-3-hydroxypyridine are said to give 6.5% 2-(2-nitrophenylethynyl)-3-hydroxypyridine [4]. This reaction product shows no $C{\equiv}C$ band in the IR spectrum and the correct structure is possibly I.

I   II   III

3-Iodopyridine and $C_6H_5C{\equiv}CCu$ react in boiling pyridine to give 47% 3-phenylethynylpyridine [2]. The reaction of 3-iodo-4-methylpyridine and II gives III [6].

4-Iodopyridine is reacted with in situ formed $HOCH_2C{\equiv}CCu$ using a mixture of equimolar amounts of $HOCH_2C{\equiv}CH$ and 4-iodopyridine, 1 mol% $C_6H_5PdI[P(C_6H_5)_3]_2$, 1 mol% CuI, and

2 equivalents $N(C_2H_5)_3$ in pyridine. The yield of 4-(3-hydroxyprop-1-ynyl)pyridine is 83% [9].

Pyrimidines. The reaction of II and IVa yields Va. An attempted analogous reaction of II and IVb failed, supposedly because of the steric hindrance by the $(CH_3)_3Si$ group [7].

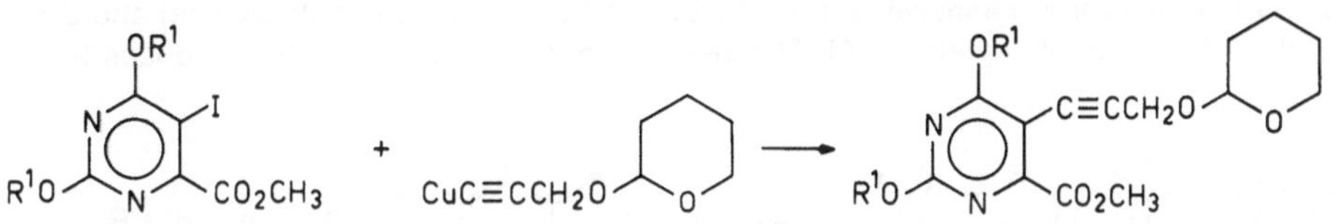

$$a : R^1 = CH_3$$
$$b : R^1 = (CH_3)_3 Si$$

References:

[1] R. D. Stevens, C. E. Castro (J. Org. Chem. **28** [1963] 3313/5).
[2] C. E. Castro, E. J. Gaughan, D. C. Owsley (J. Org. Chem. **31** [1966] 4071/8).
[3] I. L. Kotlyarevskii, V. N. Andrievskii, M. S. Shvartsberg (Khim. Geterotsikl. Soedin. **1967** 308/9; Chem. Heterocycl. Compounds [USSR] **1967** 236/7).
[4] C. Bond, M. Hooper (J. Chem. Soc. C **1969** 2453/60).
[5] M. S. Shvartsberg (Izv. Akad. Nauk SSSR Ser. Khim. **1970** 1144/9; Bull. Acad. Sci. USSR Div. Chem. Sci. **1970** 1079/82).
[6] E. G. Brain, A. J. Eglington, B. G. James, J. H. C. Nayler, N. F. Osborne, M. J. Pearson, T. C. Smale, R. Southgate, P. Tolliday, M. J. Basker, L. W. Mizen, R. Sutherland (J. Med. Chem. **20** [1977] 1086/90).
[7] R. S. Bhatt, N. G. Kundu, T. L. Chwang, C. Heidelberger (J. Heterocycl. Chem. **18** [1981] 771/4).
[8] N. A. Bumagin, I. O. Kalinovskii, A. B. Ponomarev, I. P. Beletskaya (Dokl. Akad. Nauk SSSR **265** [1982] 1138/43; Dokl. Chem. Proc. Acad. Sci. USSR **262/267** [1982] 262/6).
[9] N. A. Bumagin, A. B. Ponomarev, I. P. Beletskaya (Izv. Akad. Nauk SSSR Ser. Khim. **1984** 1561/6; Bull. Acad. Sci. USSR Div. Chem. Sci. **1984** 1433/8).

1.1.2.1.4.13 Reactions with Condensed Heterocycles R′X

This reaction is referred to as "Reaction Type 7o" in Table 1 on pp. 17/41.

Benzimidazolium Salts. These compounds represent a Mannich cation type structure and their reactions with $C_6H_5C\equiv CCu$ are therefore treated in Section 1.1.2.1.7.

Phthalides. The reaction of $C_6H_5C\equiv CCu$ and 6-iodophthalide (Ia) gives Ib in a 61% yield [1, 2].

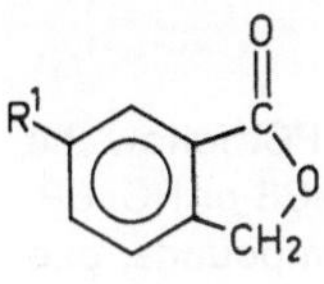

$$a : R^1 = I$$
$$b : R^1 = C_6H_5C\equiv C$$

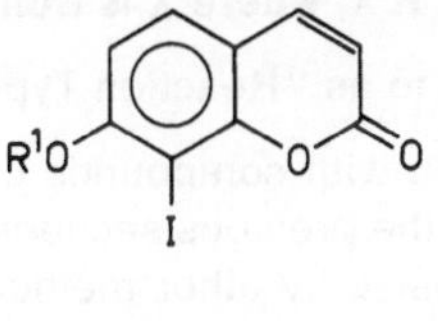

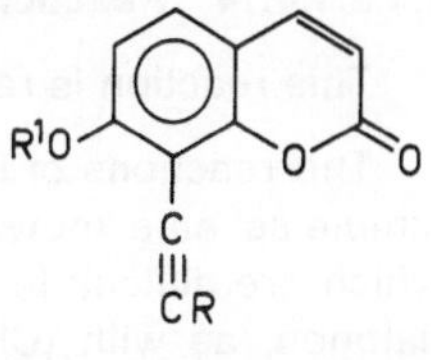

References on p. 116

Coumarines II react with RC≡CCu to give III (figures see p. 115): R = 1-methyl-1-(tetrahy-dropyran-2-yloxy)ethyl, $R^1 = CH_3$ [7]; R = CH_2=C(CH_3), $R^1 = CH_3OCH_2$, yield 65% [3].

The reaction of 7-hydroxy-8-iodocoumarines and RC≡CCu yields 2H-furo[2,3-h]-1-ben-zopyran-2-ones under cyclization and is described in Section 1.1.2.1.5.1.

Dibenzofuranes and -phenazines. The reaction of 3,6-diiododibenzofuran (IVa) and $C_6H_5C≡$ CCu gives IVb in a 94% yield [5, 6]. The reaction of $C_6H_5C≡CCu$ and Va produces 91% Vb [5].

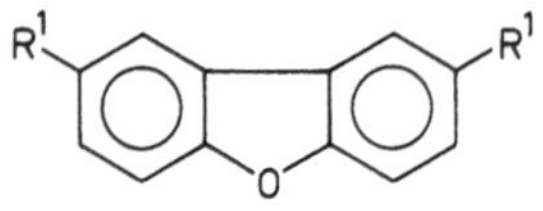

IV

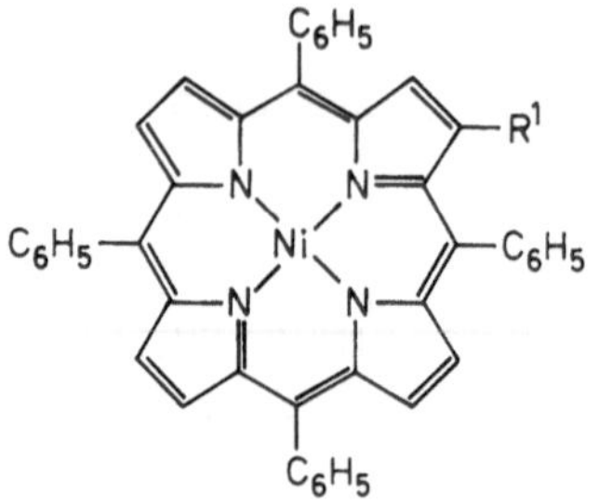

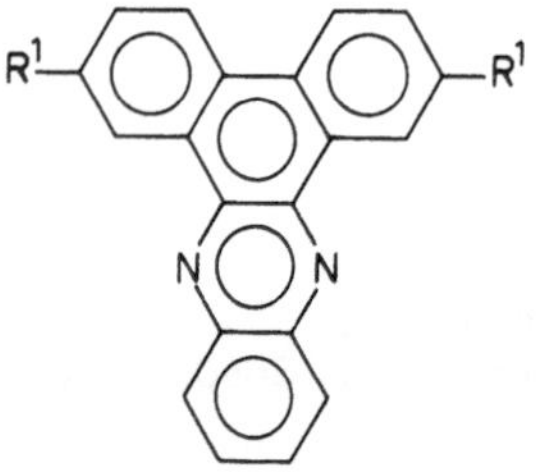

a : R^1 = I

b : R^1 = $C_6H_5C≡C$

V

a : R^1 = I

b : R^1 = $C_6H_5C≡C$

VI

a : R^1 = Br

b : R^1 = n–$C_4H_9C≡C$

Porphines. 8-Bromo-5,10,15,20-tetraphenylporphine nickel (VIa) and n-C_4H_9C≡CCu in refluxing pyridine yield VIb [4].

References:

[1] O. Ya. Neiland, M. P. Briede (Khim. Atsetilena Tr. 3rd Vses. Konf., Dushanbe 1968 [1972], pp. 160/3; C.A. **79** [1973] No. 18464).
[2] M. P. Briede, Ya. O. Neiland (Zh. Org. Khim. **6** [1970] 1701/6; J. Org. Chem. [USSR] **6** [1970] 1706/10).
[3] H. Franke (Diss. T.U. Berlin 1971).
[4] H. J. Callot (Tetrahedron Letters **1973** 4987/90).
[5] A. M. Berlin, N. M. Kofman, E. S. Krongauz, A. N. Novikov, I. R. Gol'ding (Izv. Akad. Nauk SSSR Ser. Khim. **1975** 441/2; Bull. Acad. Sci. USSR Div. Chem. Sci. **1975** 369/70).
[6] V. V. Korshak, E. S. Krongauz, A. M. Berlin, N. M. Kofman, A. P. Travnikova (Khim. Dikar-bonil'nykh Soedin. Tezisy Dokl. 4th Vses. Konf., Riga 1975 [1976], pp. 76/7; C.A. **87** [1977] No. 39874).
[7] R. D. H. Murray, I. T. Forbes (Tetrahedron Letters **1977** 3077/8).

1.1.2.1.4.14   Reactions with R'X, where X is Bonded to S, P, Si, or Sn

This reaction is referred to as "Reaction Type 7p" in Table 1 on pp. 17/41.

The reactions of RC≡CCu with compounds containing groups like SCl or PCl follow the simple scheme known from the previous sections and yield structures like RC≡CS or RC≡CP, which are difficult to synthesize by other methods. From polyhalogenated compounds, one halogen, as with $(CH_3)_2SnBr_2$, or all halogens, as with PCl_3, can be substituted in this way.

If X is bonded to transition metals, the simple substitution of X by RC≡C has only been occasionally observed. However, in most cases the reaction is much more complicated. Usually mixtures containing one or more transition metal complexes are formed. The reaction product pattern depends on the reaction conditions and on the ratio of the reactants, and redox reactions occur. In most cases there is no information about the reaction path. Therefore, all reactions of compounds containing groups M–X (M = transition metal) are treated separately in Section 1.1.2.1.10.

Sulfur Compounds. RC≡CCu and 2,4-$(O_2N)_2C_6H_3SCl$ react in CH_3CN to form 2,4-$(O_2N)_2C_6H_3SC≡CR$. The yields are 70% with R = n-C_3H_7 and 80% with R = C_6H_5 [2].

Phosphorus Compounds. The reaction of $C_6H_5C≡CCu$ and an excess of PCl_3 yields the white $P(C≡CC_6H_5)_3$ [1]. The reaction of RC≡CCu and compound I in tetrahydrofuran/ether/dimethylsulfide gives 60% II, isolated as a CuI complex. With R = CH_3OCH_2 the asymmetric induction is 31%, with R = C_6H_5 30% [5].

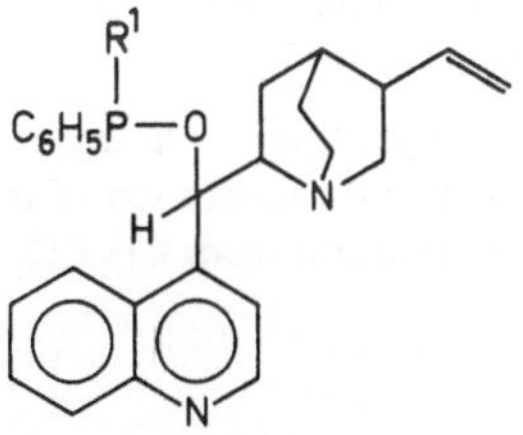

I: R^1 = Cl

II: R^1 = RC≡C

Silicon Compounds. The reactions of RC≡CCu (R = n-C_4H_9 or C_6H_5) with chlorosilanes $(CH_3)_nSiCl_{4-n}$ (n = 1, 2, or 3) afford a 60 to 80% yield of RC≡CSi$(CH_3)_nCl_{3-n}$. The acetylides RC≡CCu are best prepared in situ. The components are heated 120 h at 150° C [3].

Tin Compounds. 4-$CH_3C_6H_4C≡CCu$ and $(CH_3)_2SnBr_2$ give after 17 d in ether 4-$CH_3C_6H_4C≡CSn(CH_3)_2Br$ in a 22% yield [4].

References:

[1] A. M. Sladkov, L. Y. Ukhin, V. V. Korshak (Izv. Akad. Nauk SSSR Ser. Khim. **1963** 2213/5; Bull. Acad.Sci. USSR Div. Chem. Sci. **1963** 2043/5).
[2] C. E. Castro, R. Havlin, U. K. Honvad, A. Malte, S. Moje (J. Am. Chem. Soc. **91** [1969] 6464/70).
[3] G. Derleis, J. Dunogues, R. Calas, P. Lapouyade (J. Organometal. Chem. **80** [1974] C45/C46).
[4] G. Van Koten, C. A. Schaap, J. G. Noltes (J. Organometal. Chem. **99** [1975] 157/70).
[5] W. Chodkiewicz (J. Organometal. Chem. **194** [1980] C25/C28).

1.1.2.1.5   Reactions with R′X to Form Ring Closure Products of RC≡CR′

1.1.2.1.5.1   Ring Closure Giving Furans

This reaction of RC≡CCu with R′X is referred to as "Reaction Type 8a" in Table 1 on pp. 17/41. The formation of phthalides, which have a benzo[c]furan skeleton, is treated in Section 1.1.2.1.5.3 (Reaction Type 8c).

If R′X contains a structural element –CX=C(OH)– with X = I or Br, its reaction with RC≡CCu gives a furan instead of an enynol, which might be expected according to the reaction path outlined in Section 1.1.2.1.4.3. However, the addition of the OH group of the structural element $-(RC{\equiv}C)C{=}C(OH)-$ to the $C{\equiv}C$ bond to yield a furan ring is favored. The only exception is said to be the formation of 2-(2-nitrophenylethynyl)-3-hydroxypyridine [6]. This claim is doubtful. The compound shows no $C{\equiv}C$ band in the IR spectrum, thus the furan structure is more probable here too.

Simple furans are formed from phenacylbromides and $RC{\equiv}CCu$ according to Scheme I:

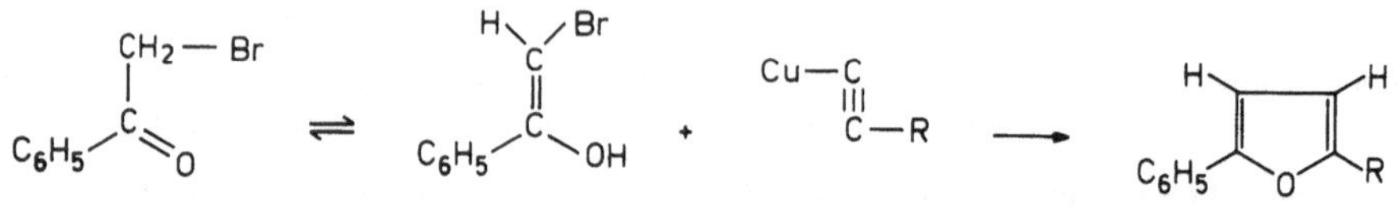

I

The reaction conditions are important. By heating the components 5 min at 140 °C with $R=n-C_3H_7$ the yield is 29%; with $R=C_6H_5$ 54% yield of the furan is obtained after 5 min in $C_6H_5NO_2$ at 240 °C [4]. Upon heating $C_6H_5C{\equiv}CCu$ and $C_6H_5COCH_2Br$ 16 h in glycol at 140 °C only $C_6H_5COCH_3$ (47%) and tars were obtained [3].

More important is the synthesis of benzo[b]furans (in many cases plant products), see Table 14. The nature of the solvent is not of great importance. The OH group is sometimes formed in the reaction mixture, e.g., by deacetylation from CH_3CO_2.

Table 14
Formation of Benzo[b]furans from Halophenols according to

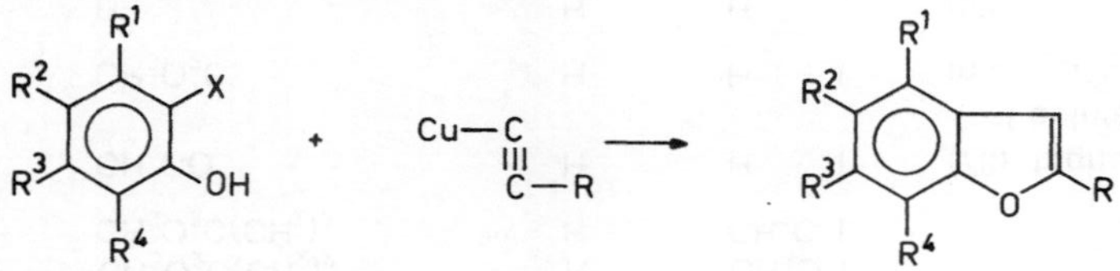

Reaction scheme:

halophenol (ring bearing R^1, R^2, R^3, R^4, X, OH) $+$ $Cu-C{\equiv}C-R$ $\longrightarrow$ benzo[b]furan (ring bearing R^1, R^2, R^3, R^4, R, O)

(belongs to Reaction Type 8a, explanation see pp. 12/4).

R	R^1	R^2	R^3	R^4	X	remarks (yield in %)	Ref.
$CH_2=C(CH_3)$	H	H	CH_3O	H	Br	(26)	[31]
	H	H	$C_6H_5CO_2$	H	—	(18)	[31]
	H	CH_3O	CH_3O	H	—	OH group by deacetylation	[24]
	H	CH_3CO	H	H	—	OH group by deacetylation	[24]
	H	CH_3CO	OH	H	—	dehydrotremetone	[19]
	H	CH_3CO	OH	H	Br	euparin	[19]
	H	CH_3CO	OH	H	—	euparin (32)	[31]
	H	CH_3CO	OH	CH_3O	Br	methoxyeuparin	[19]
	OH	CH_3O	H	H	—	—	[26]
	OH	CH_3O_2C	H	H	—	(42)	[26]
	CH_3CO_2	H	$n\text{-}C_5H_{11}$	H	—	OH group by deacetylation	[27]
$(Z)\text{-}CH_3OCH=CH$	H	H	H	H	—	gives an (E)/(Z)-mixture (35:65)	[23]
$n\text{-}C_3H_7$	H	H	H	H	—	(60)	[3]
	H	Br	H	H	Br	(40)	[3]
pyrid-2-yl	H	H	H	H	Br	(50)	[3]
	H	Br	H	H	Br	(38)	[3]
C_6H_5	H	H	H	H	Br	(53)	[3]
	H	H	H	H	—	in pyridine (85); in dimethylformamide (88); in dimethylformamide/N-methylpiperidine (41); with $C_6H_5C{\equiv}CCu$ prepared in situ in dimethylformamide (16)	[1 to 3, 33]

Annotation (bridging R^2/R^3): OCH_2O

References on p. 125

Table 14 [continued]

R	R¹	R²	R³	R⁴	X	remarks (yield in %)	Ref.
C₆H₅ [continued]	H	Br	H	H	Br	(53)	[3]
	H	OHC	H	–	–	(26)	[14]
	H	CH₃O₂CCH₂	H	–	–	(44)	[14]
	H	CH₃O₂CCH(NHCOCH₃)CH₂	H	–	–	(32)	[14]
2,4,6-Cl₃C₆H₂OCH₂	H	H	H	H	–	(86)	[9, 11]
2,4-Cl₂C₆H₃OCH₂	H	H	H	H	–	(84)	[9, 11]
[benzodioxol-5-yl structure, H₂C⟨O—O⟩]	H	CH₃O₂C(CH₂)₂	H	H	–	(90)	[15, 16]
	H	CH₃O₂C(CH₂)₂	H	CH₃O	Br	methyl egonoate (23)	[16]
2-BrC₆H₄OCH₂	H	H	H	H	–	(77)	[9]
4-BrC₆H₄OCH₂	H	H	H	H	–	(67)	[9]
2-ClC₆H₄OCH₂	H	H	H	H	–	(61)	[9, 11]
4-ClC₆H₄OCH₂	H	H	H	H	–	(76)	[9, 11]
4-CH₃OC₆H₄	H	CH₃O₂CCH₂	H	–	–	in dimethylformamide (16)	[14]
	H	CH₃O₂CCH(NHCOCH₃)CH₂	H	–	–	in dimethylformamide (18)	[14]
3,4-(CH₃O)₂C₆H₃	H	CH₃O₂C(CH₂)₂	H	H	–	(100)	[16]
	H	CH₃O₂C(CH₂)₂	H	CH₃O	Br	–	[16]
	H	CH₃O₂C(CH₂)₂	H	CH₃O	I	–	[16]
[tetrahydropyranyl structure], —O—C(CH₃)₂—	H	CH₃CO	H	H	–	(78), higher yields in presence of activated Cu	[29]
	H	CH₃O₂C	H	H	–	in CH₃CO₂H/H₂O	[22]
naphth-1-yl	H	H	H	H	–	(52)	[17]

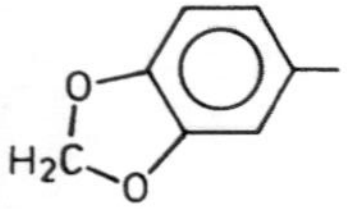

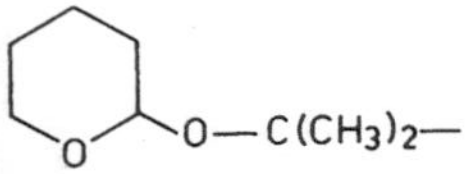

References on p. 125

R^1	R^2	R^3	R^4	R^5	product (yield %)	Ref.
$C_5H_5FeC_5H_4$	H	H	H	H	(85)	[7]
$2,4\text{-}(CH_3O)_2(3\text{-}CH_3O_2C)C_6H_2$	H	$C_6H_5CO_2$	H	—	isolation with $R^3 = OH$ as pterofuran (78)	[18]
$3\text{-}i\text{-}C_3H_7(4\text{-}CH_3O)C_6H_3$	$CH_3O_2CCH_2$	H	—	—	in dimethylformamide (63)	[14]
$3\text{-}i\text{-}C_3H_7(4\text{-}CH_3O)C_6H_3$	$CH_3O_2CCH(NHCOCH_3)CH_2$	H	—	—	in dimethylformamide (40)	[14]
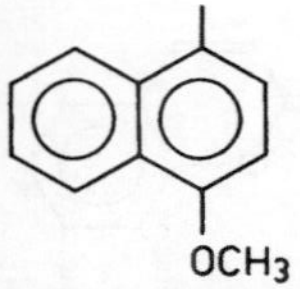 (structure with OCH_3 and CH_3)	H	H	H	—	(52)	[17]
$2\text{-}CH_3O(4\text{-}C_6H_5CH_2O)C_6H_3$	H	CH_3O	H	Br	vignafuran benzyl ether (34)	[18]

References on p. 125

For the synthesis of furo[3,2–b]pyridines from 2–iodo–3–hydroxypyridines, see Table 15. A typical run is carried out by heating the components 9 to 14 h in pyridine or dimethylformamide at 120 °C.

Table 15
Formation of Furo[3,2–b]pyridines from 2–Iodo–3–hydroxypyridines according to

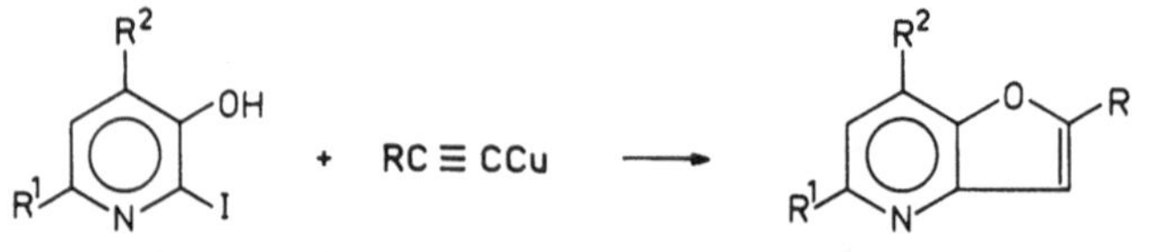

(belongs to Reaction Type 8a, explanation see pp. 12/4).

R	R^1	R^2	yield in %	Ref.
HO(CH$_2$)$_2$	H	H	61	[5]
n–C$_3$H$_7$	H	H	82	[5]
	H	CH$_3$	–	[10]
	CH$_3$	H	–	[10]
n–C$_4$H$_9$	H	H	88	[5]
	H	CH$_3$	–	[10]
	CH$_3$	H	–	[10]
C$_6$H$_5$	H	H	84	[5, 12]
n–C$_6$H$_{13}$	H	H	92	[5]

Analogously the reaction of 3,5–diiodo–4–hydroxypyridines II and RC≡CCu yields 7–iodo-furo[3,2–c]pyridines III. The reaction was carried out with the following R (yields in parentheses): R=HO(CH$_2$)$_2$ (49%), n–C$_3$H$_7$ (24%), pyrid–2–yl (37%) [5], C$_6$H$_5$ (86%) [3, 5], n–C$_6$H$_{13}$ (37%) [5].

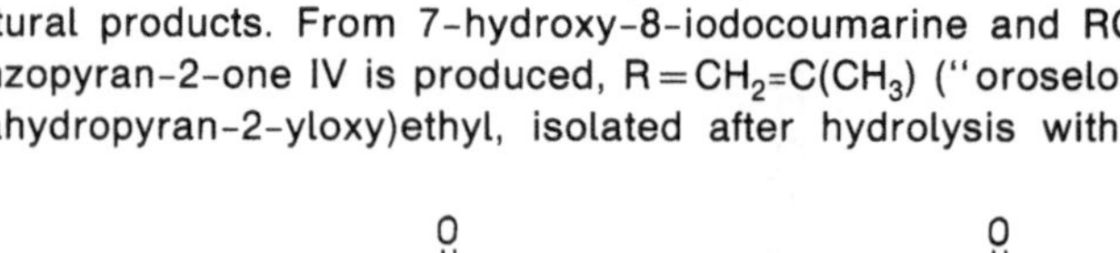

II   III

The reactions of RC≡CCu and halogenated hydroxycoumarines have also been used to synthesize natural products. From 7–hydroxy–8–iodocoumarine and RC≡CCu the 2H–furo[2,3–h]–1–benzopyran–2–one IV is produced, R=CH$_2$=C(CH$_3$) ("oroselone") [20, 25] or 1–methyl–1–(tetrahydropyran–2–yloxy)ethyl, isolated after hydrolysis with R=HOC(CH$_3$)$_2$ ("oroselol") [21].

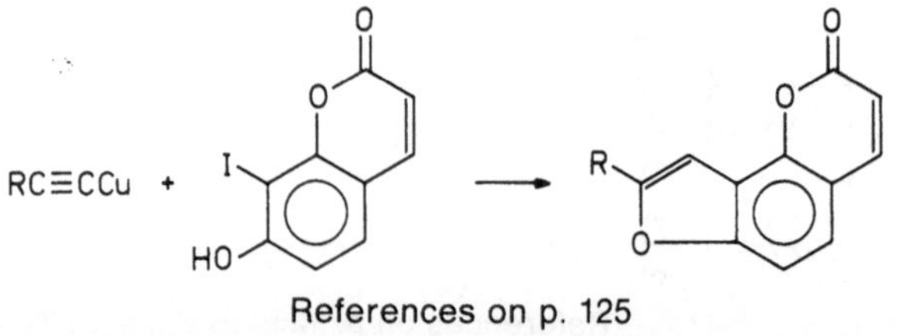

IV

References on p. 125

The reaction of 6-bromo-7-hydroxycoumarine and RC≡CCu in hot pyridine gives psoralenes V, which also are natural products. The yield is 65% with R=CH$_2$=C(CH$_3$) [25] and not given with R=(CH$_2$=CHCH$_2$)$_2$NC(CH$_3$)$_2$ [30].

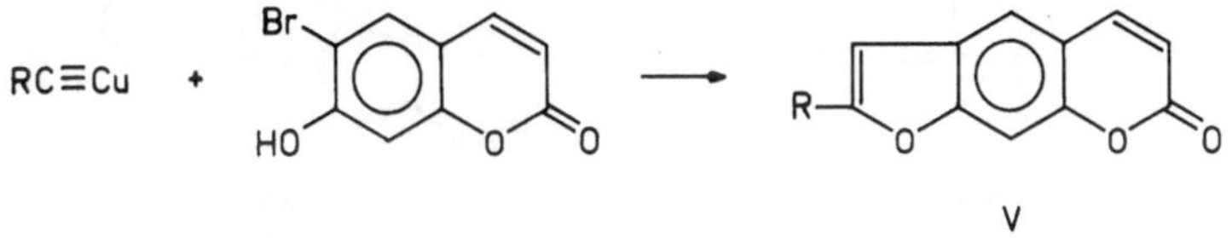

V

The reaction of 2-iodoestrone VI and RC≡CCu did not yield the expected ethynylestrones VII, but after cyclization 5'-substituted 17-keto-estra-1(10),4-dieno[3,2-b]furanes VIII were obtained [8, 13]. The following VIII were prepared (yields in parentheses): R=HO(CH$_2$)$_2$ (60%), n-C$_3$H$_7$ (69%), n-C$_4$H$_9$ (66%), C$_6$H$_5$ (71%), n-C$_6$H$_{13}$ (63%) [13]. The O-acylated 2-iodoestrone X reacts with RC≡CCu more slowly to form the same heterocycle, whereas the O-methylated estrone IX is inert towards RC≡CCu [13].

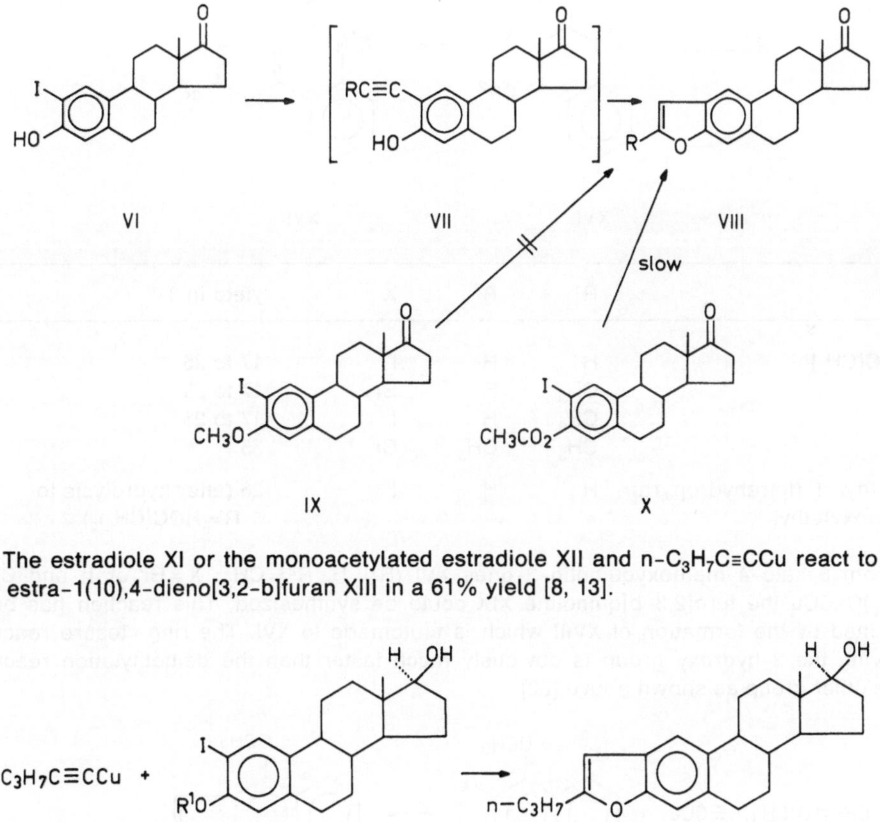

VI   VII   VIII

slow

IX   X

The estradiole XI or the monoacetylated estradiole XII and n-C$_3$H$_7$C≡CCu react to form the estra-1(10),4-dieno[3,2-b]furan XIII in a 61% yield [8, 13].

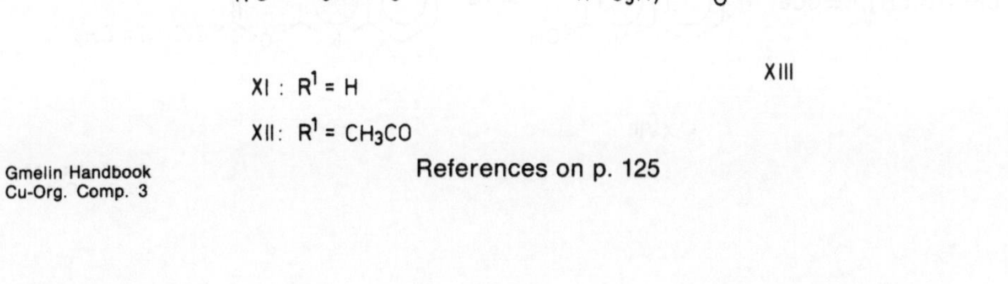

n−C$_3$H$_7$C≡CCu +

XIII

XI : R^1 = H

XII: R^1 = CH$_3$CO

The acetoxyiodoazulene XIV and $C_6H_5C\equiv CCu$ give XV in a 69% yield [28].

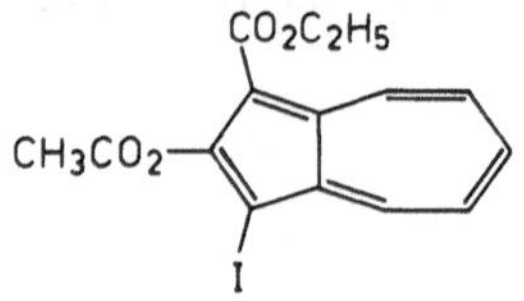

XIV 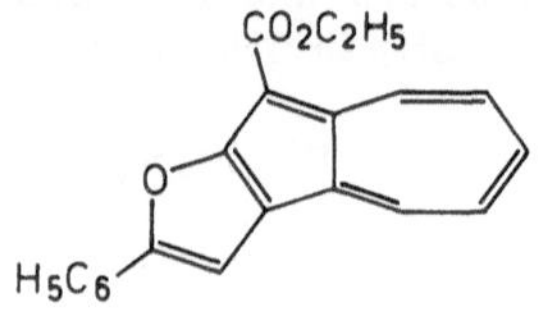   XV

Benzo[c]furans are byproducts of the synthesis of coumarines and similar heterocycles from $RC\equiv CCu$ and aromatic 2-halobenzoic acids or 2-halobenzyl alcohols. These reactions are described in Section 1.1.2.1.5.3 (Reaction Type 8c), see pp. 128/9.

The reaction of 3-halo-4-hydroxyquinolin-2-ones XVI ($R^2=H$) and $RC\equiv CCu$ in boiling pyridine gives 2-substituted 4-oxofuro[3,2-c]quinolines XVII. Surprisingly, the methoxy compound XVI ($R^1=R^2=CH_3$) also undergoes ring closure. Obviously, the demethylation by the pyridine solvent is fast enough [32]:

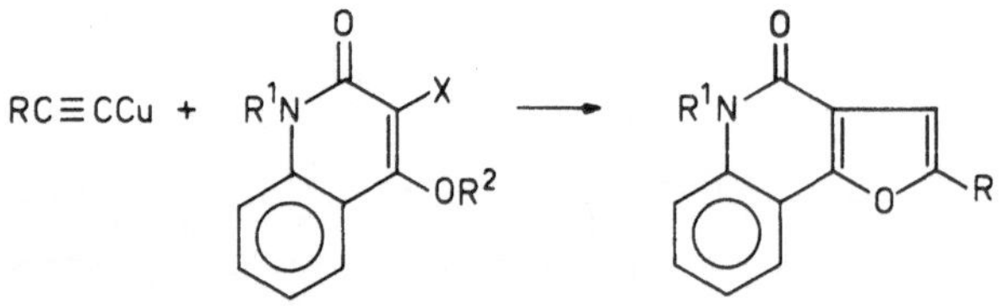

XVI   XVII

R	R^1	R^2	X	yield in %
$CH_2=C(CH_3)$	H	H	I	17 to 25
	CH_3	H	Br	17 to 25
	CH_3	H	I	17 to 25
	CH_3	CH_3	Br	35
1-methyl-1-(tetrahydropyran-2-yloxy)ethyl	H	H	I	26 (after hydrolysis to $R=HOC(CH_3)_2$)

From 3-halo-4-methoxyquinolin-2-ones XVI ($R^1=H$, $R^2=CH_3$, $X=Br$ or I) and $CH_2=C(CH_3)C\equiv CCu$ the furo[2,3-b]quinoline XIX could be synthesized. This reaction has been explained by the formation of XVIII which is tautomeric to XVI. The ring closure reaction involving the 2-hydroxy group is obviously much faster than the demethylation reaction of the ether group as shown above [32].

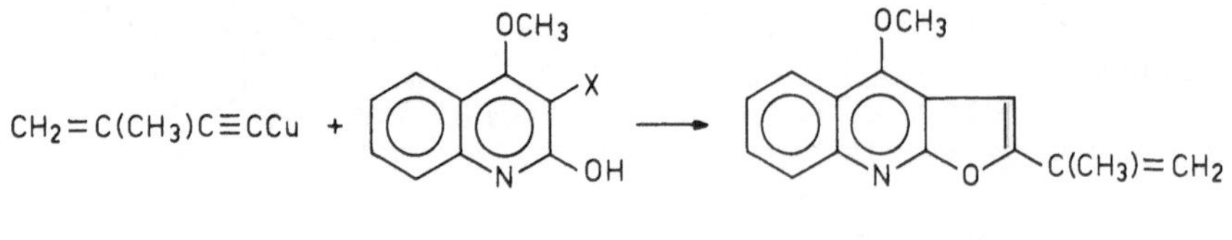

XVIII   XIX

References:

[1] C. E. Castro, R. D. Stevens (J. Org. Chem. **28** [1963] 2163).
[2] R. D. Stevens, C. E. Castro (J. Org. Chem. **28** [1963] 3313/5).
[3] C. E. Castro, E. J. Gaughan, D. C. Owsley (J. Org. Chem. **31** [1966] 4071/8).
[4] K. Gump, S. W. Moje, C. E. Castro (J. Am. Chem. Soc. **89** [1967] 6770/1).
[5] S. A. Mladenović, C. E. Castro (J. Heterocycl. Chem. **5** [1968] 227/30).
[6] C. Bond, M. Hooper (J. Chem. Soc. C **1969** 2453/60).
[7] M. D. Rausch, A. Siegel (J. Org. Chem. **34** [1969] 1974/6).
[8] M. Stefanović, L. Krstić, S. Mladenović (Tetrahedron Letters **1971** 3311/2).
[9] A. G. Makhsumov, O. V. Afanas'eva, Sh. U. Abdullaev, U. A. Abidov (U.S.S.R. 389090 [1971/73]; C.A. **79** [1973] No. 136978).
[10] P. I. Abramenko, V. G. Zhiryakov (Zh. Vses. Khim. Obshchestva **17** [1972] 695/6).

[11] O. V. Afanas'eva, A. G. Makhsumov, Sh. U. Abdullaev (Dokl. 4th Vses. Konf. Khim. Atsetilena, Alma-Ata 1972, Vol. 1, pp. 244/7; C.A. **79** [1973] No. 31756).
[12] D. C. Owsley, C. E. Castro (Org. Syn. **52** [1972] 128/31).
[13] M. Stefanović, L. Krstić, S. Mladenović (Glasnik Hem. Društva [Beograd] **38** [1973] 463/8; C.A. **82** [1975] No. 43645).
[14] M. T. Cox, J. J. Holohan (Tetrahedron **31** [1975] 633/5).
[15] F. G. Schreiber, R. Stevenson (Chem. Letters **1975** 1257/8).
[16] F. G. Schreiber, R. Stevenson (J. Chem. Soc. Perkin Trans. I **1976** 1514/8).
[17] M. Brenner, C. Brush (Tetrahedron Letters **1977** 419/22).
[18] R. P. Duffley, R. Stevenson (J. Chem. Soc. Perkin Trans. I **1977** 802/4).
[19] F. G. Schreiber, R. Stevenson (J. Chem. Soc. Perkin Trans. I **1977** 90/2).
[20] F. G. Schreiber (Diss. Brandeis Univ. from [21]).

[21] R. P. Duffley, R. Stevenson (Syn. Commun. **8** [1978] 175/80).
[22] R. P. Duffley, R. Stevenson (J. Chem. Res. S **1978** 468).
[23] G. A. Kraus, K. Frazier (Tetrahedron Letters **1978** 3195/8).
[24] F. G. Schreiber, R. Stevenson (Org. Prep. Proced. Intern. **10** [1978] 137/42).
[25] F. G. Schreiber, R. Stevenson (J. Chem. Res. S **1978** 92).
[26] G. Batu, R. Stevenson (J. Org. Chem. **44** [1979] 3948/9).
[27] R. T. Scannel, R. Stevenson (J. Chem. Res. S **1983** 36).
[28] T. Morita, T. Nakadate, K. Takase (Heterocycles **15** [1981] 835/8).
[29] G. E. Schneiders, R. Stevenson (Syn. Commun. **10** [1980] 699/705).
[30] G. E. Schneiders, R. Stevenson (J. Chem. Res. S **1982** 182).

[31] R. T. Scannel, R. Stevenson (J. Chem. Soc. Perkin Trans. I **1983** 2927/31).
[32] J. L. Gaston, R. J. Greer, M. F. Grundon (J. Chem. Res. S **1985** 135).
[33] C. E. Castro, R. Havlin, U. K. Honvad, A. Malte, S. Moje (J. Am. Chem. Soc. **91** [1969] 6464/70).

1.1.2.1.5.2   Ring Closure Giving Pyrroles or Thiophenes

This reaction of RC≡CCu with R'X is referred to as "Reaction Type 8b" in Table 1 on pp. 17/41.

If R'X contains a structural element $-C(I)=C(NHR^1)-$ or $-CBr=C(SH)-$, respectively, the expected RC≡CR' with the structural elements $-(RC≡C)C=C(NHR^1)-$ or $-(RC≡C)C=C(SH)-$ are not generally obtained. The proton at N or S can add to the C≡C bond and a pyrrol or a thiophene ring is formed. R'X is in most cases a 2-iodoaniline or 2-bromothiophenol and the formed heterocycle is then an indole or a benzo[b]thiophene, respectively.

Formation of Indoles. In homogeneous reaction systems, usually with pyridine as a solvent, the reaction of RC≡CCu with halo-aminophenols R'X (X = halogen) yields RC≡CR' or a mixture of RC≡CR' and the cyclization product [3], see Sections 1.1.2.1.3 and 1.1.2.1.4.6. In heterogeneous reaction systems, however, only indoles can be isolated, see Table 16. Best yields are obtained by allowing RC≡CCu and R'X to stand in a suitable solvent such as dimethylformamide, especially in the presence of certain copper salts. Obviously, a Cu^I polymer is essential for catalysis. Monomeric or low molecular weight aggregates of Cu^I compounds, which are obtained in warm pyridine or upon prolonged warming in dimethylformamide, are not capable of rapid catalysis. For further explanation of this fact and for a suggested mechanism, see [1, 5].

Table 16
Formation of Indoles from Iodoanilines according to

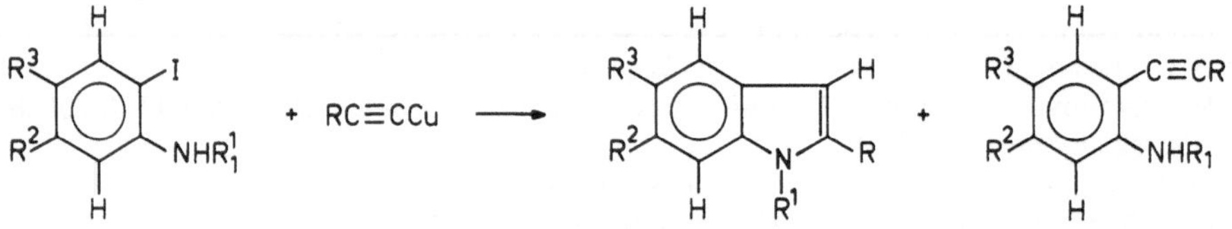

(Reaction Type 8b, explanation see pp. 12/4). For abbreviations and dimensions see p. X.

R	R1	R2	R3	yield of I in %	yield of II in %	remarks	Ref.
C_2H_5	H	H	H	12	28	pyridine, 120°/8 h	[3]
n-C_3H_7	H	H	H	89	–	pyridine, 120°/8 h	[3]
	H	H	H	70	17	pyridine, 120°/8 h	[3]
	C_2H_5	H	H	50	5	pyridine	[3]
n-C_4H_9	H	H	H	35	20	pyridine, 120°/8 h	[3]
$CH_2=CHC(CH_3)_2$	H	H	H	36	–	dimethylformamide, 140°/12 h	[6]
	H	Br	H	100	–	dimethylformamide, 130°/10 h, Br partially exchanged against I (Br:I = 5:4)	[7]
	H	$(CH_3)_2C=CHCH_2$	H	94	–	dimethylformamide, 125°/5 h	[7]
	H	I	H	42	–	dimethylformamide, 125°/9 h, gives a mixture of I and 37%	[7]

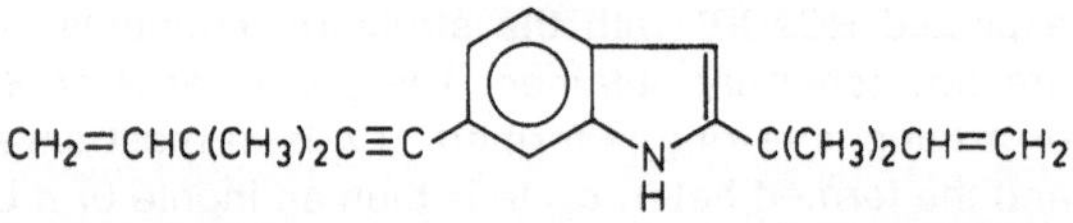

References on p. 128

Table 16 [continued]

R	R¹	R²	R³	yield of I in %	yield of II in %	remarks	Ref.
C_6H_5	H	H	H	>80	—	dimethylformamide/ 175°, pyridine/ 6 h reflux, dimethylsulfoxide or CH_3CO_2H	[1 to 3]
	C_2H_5	H	H	50	—	dimethylformamide, 120°/22 h	[3]
	H	H	CH_3	90	—	dimethylformamide, 120°/22 h	[3]
	H	H	OH	57	—	dimethylformamide, 120°/28 h	[3]

The acetylamino-iodo-azulene III and $C_6H_5C\equiv CCu$ give the condensed pyrrole IV in a 54% yield in boiling pyridine (which probably first causes the deacetylation) [8].

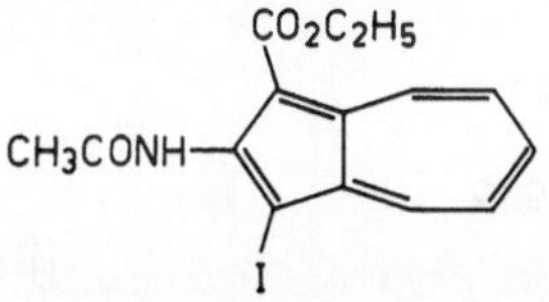

III 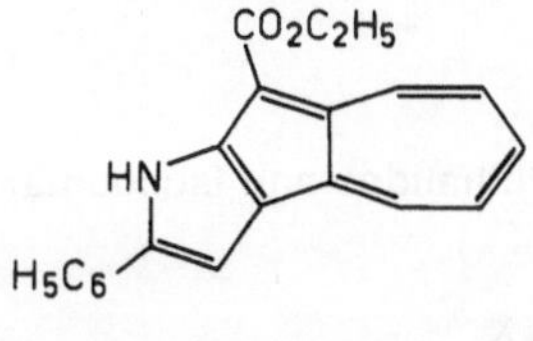   IV

Formation of Benzo[b]thiophenes. Cyclization V is carried out by adding a very dilute pyridine solution of the thiophenol into a suspension or solution of the acetylide in pyridine. With concentrated solutions, however, Cyclization VI takes place. The following benzo[b]thiophenes were prepared in this way (yields in parentheses): R=n-C_3H_7 (80%), n-C_4H_9 (80%), and C_6H_5 (90%). Some acetylides are less stable under the reaction conditions and therefore are generated in situ from $RC\equiv CH$, CuI, and N-ethylpiperidine. The following benzo[b]thiophenes were prepared in this way: R=$HOCH_2$ (10%) and $C_2H_5O_2C$ (35%) [4].

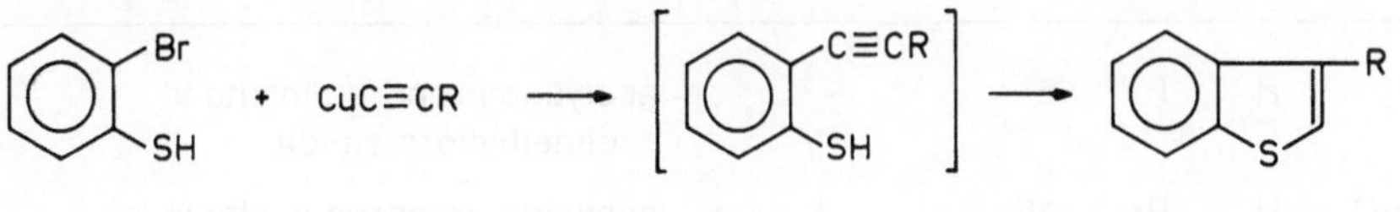

V

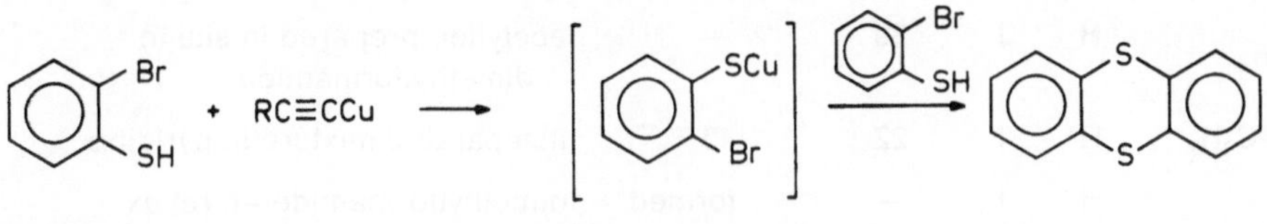

VI

References on p. 128

References:

[1] C. E. Castro, R. D. Stevens (J. Org. Chem. **28** [1963] 2163).
[2] R. D. Stevens, C. E. Castro (J. Org. Chem. **28** [1963] 3313/5).
[3] C. E. Castro, E. J. Gaughan, D. C. Owsley (J. Org. Chem. **31** [1966] 4071/8).
[4] A. M. Malte, C. E. Castro (J. Am. Chem. Soc. **89** [1967] 6770).
[5] C. E. Castro, R. Havlin, U. K. Honvad, A. Malte, S. Moje (J. Am. Chem. Soc. **91** [1969] 6464/70).
[6] H. Plieninger, H. Sirowej (Chem. Ber. **104** [1971] 2027/9).
[7] S. Inoue, N. Takamatsu, Y. Kishi (Yakugaku Zasshi **97** [1977] 564/8; C.A. **87** [1977] No. 184744).
[8] T. Morita, T. Nakadate, K. Takase (Heterocycles **15** [1981] 835/8).

1.1.2.1.5.3 Ring Closure Giving Other Heterocycles

This reaction is referred to as "Reaction Type 8c" in Table 1 on pp. 17/41.

Phthalides and Isocoumarines. The reaction of $RC\equiv CCu$ and $2\text{-}XC_6H_4CO_2H$ ($X=Cl$, Br, or I) gives phthalides I or isocoumarines II, or mixtures thereof; in no case was $2\text{-}RC\equiv CC_6H_4CO_2H$ observed. For examples, see Table 17. The components are heated in a suitable solvent, such as pyridine or dimethylformamide. No work was done about the dependence of the I:II ratio on the substitution type and the reaction conditions.

Table 17
Formation of Phthalides and Isocoumarines according to

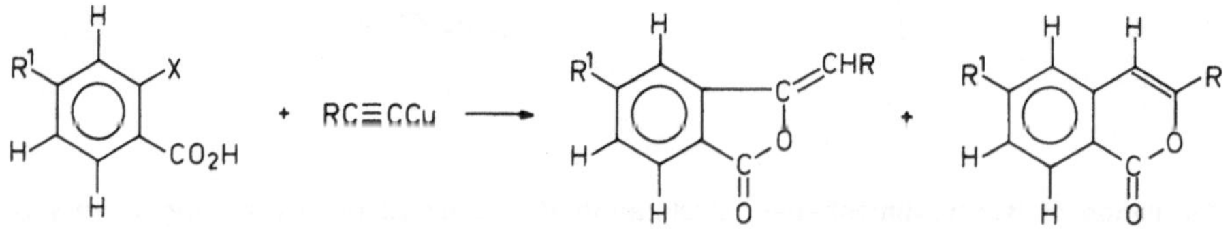

(belongs to "Reaction Type 8c", explanation see p. 12/4).

R	R^1	X	yield of I in %	yield of II in %	remarks	Ref.
$HOCH_2$	H	I	6	—	acetylide prepared in situ in dimethylformamide	[3]
$C_2H_5O_2C$	H	Br	15	—	acetylide prepared in situ in dimethylformamide	[3]
	H	I	39	—	acetylide prepared in situ in dimethylformamide	[3]
$n\text{-}C_3H_7$	H	I	22	40	inseparable mixture in pyridine	[3, 8]
	H	I	—	formed	dimethylformamide/4d reflux	[8]
$n\text{-}C_4H_9$	H	I	—	formed	dimethylformamide/4d reflux	[8]

References on pp. 130/1

Table 17 [continued]

R	R¹	X	yield of I in %	yield of II in %	remarks	Ref.
C_6H_5	H	Br	53	–	acetylide prepared in situ in dimethylformamide	[3]
	H	Cl	65	–	pyridine	[3]
	H	Cl	39	–	dimethylformamide	[3]
	H	I	14	–	pyridine in previous papers [1, 2] wrong structure and different yield (94%) given	[3]
	H	I	90	–	acetylide prepared in situ in dimethylformamide	[3]
	Cl	Cl	69	–	pyridine	[3]

1H-2-Benzopyrans and 1H,3H-Benzo[c]furans. The type of the reaction product from $RC\equiv CCu$ and $2\text{-}IC_6H_4CH_2OH$ depends on the group R.

In boiling pyridine with $R=n\text{-}C_3H_7$ 50% IV is formed, whereas with $R=C_6H_5$ 80% V can be isolated. Possibly $2\text{-}RC\equiv CC_6H_4CH_2OH$ (III) is the intermediate [5].

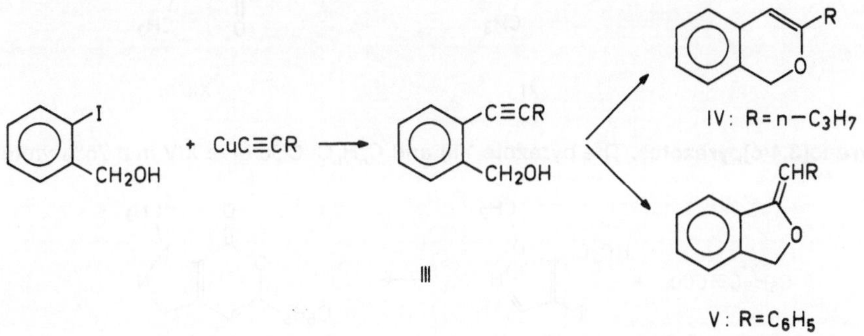

III

IV: $R=n\text{-}C_3H_7$

V: $R=C_6H_5$

Benzoxepines. The reaction of $n\text{-}C_3H_7C\equiv CCu$ and $2\text{-}IC_6H_4CH_2CO_2H$ affords a 60% yield of VI [5].

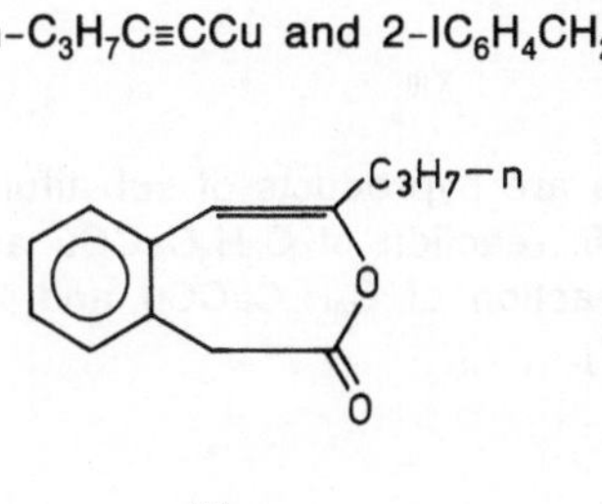

VI

Thieno[3,4-c]pyrans. The reaction of $RC\equiv CCu$ and VII in dimethylformamide at 125 °C yields 4H-thieno[3,4-c]pyran-4-ones VIII, $R=n\text{-}C_3H_7$ (85%), $n\text{-}C_4H_9$ (78%), C_6H_5 (57%), $n\text{-}C_6H_{13}$ (30%) [4].

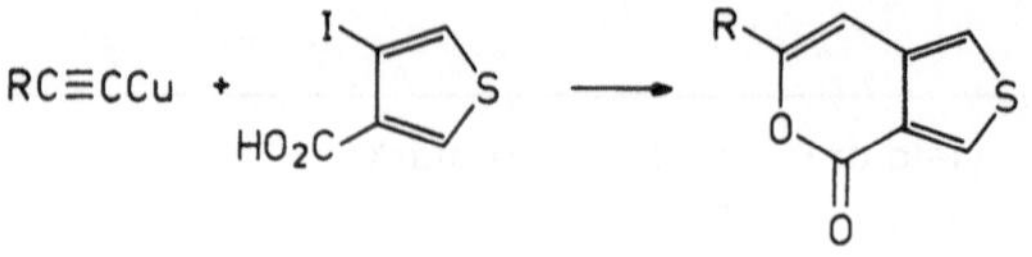

VII   VIII

Pyrano[4,3-c]pyrazoles. The reaction of RC≡CCu and 5-iodo-1,3-dimethylpyrazole-4-carboxylic acid IX in boiling pyridine yields X. The yields are 66% with $R=n-C_3H_7$ and 79% with $R=C_6H_5$ [9].

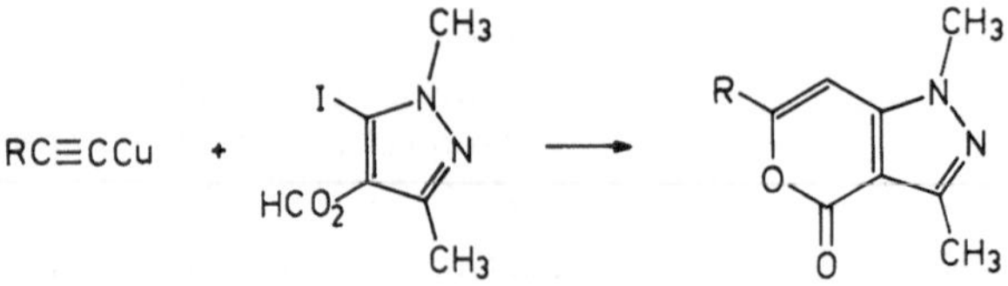

IX   X

The reaction of RC≡CCu and 3-iodo-1,5-dimethylpyrazole-4-carboxylic acid XI gives XII, $R=n-C_5H_{11}$ (67%) and C_6H_5 (67%) [9].

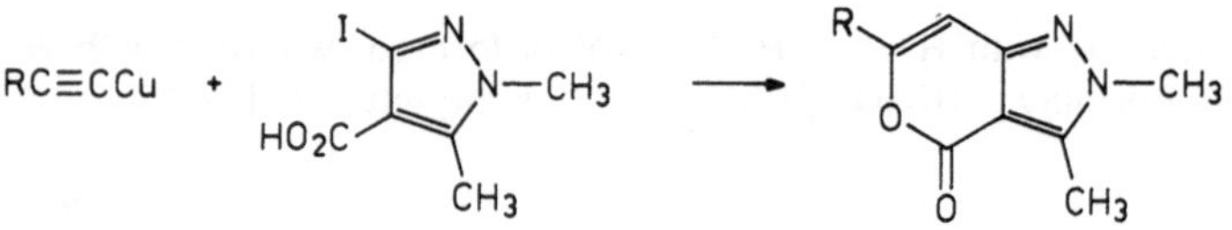

XI   XII

Pyrano[3,4-c]pyrazoles. The pyrazole XIII and $C_6H_5C≡CCu$ give XIV in a 75% yield [10].

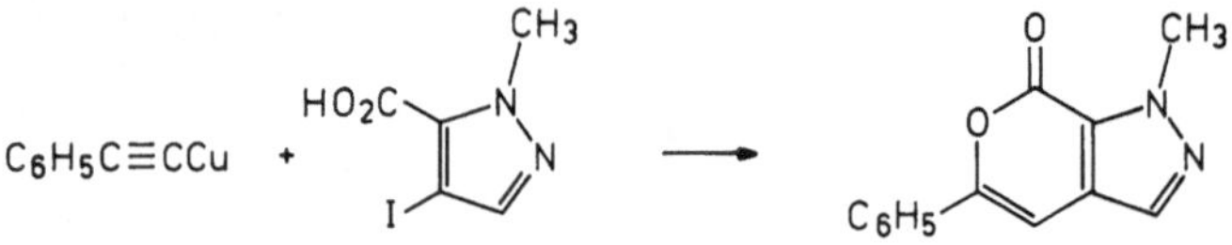

XIII   XIV

Additional ring systems are byproducts of substitution reactions and are described in Section 1.1.2.1.4.6 (Table 6, reaction of $C_6H_5C≡CCu$ and $2-(2-O_2NC_6H_4)C_6H_4C≡CC_6H_5$ [6]) and Section 1.1.2.1.4.7 (reaction of $C_6H_5C≡CCu$ and $2,4,6-(O_2N)_3C_6H_3$, see also Section 1.1.2.1.1, Table 1, No. 90) [7].

References:

[1] C. E. Castro, R. D. Stevens (J. Org. Chem. **28** [1963] 2163).
[2] R. D. Stevens, C. E. Castro (J. Org. Chem. **28** [1963] 3313/5).
[3] C. E. Castro, E. J. Gaughan, D. C. Owsley (J. Org. Chem. **31** [1966] 4071/8).
[4] S. A. Mladenović, C. E. Castro (J. Heterocycl. Chem. **5** [1968] 227/30).

[5] C. E. Castro, R. Havlin, U. K. Honvad, A. Malte, S. Moje (J. Am. Chem. Soc. **91** [1969] 6464/70).

[6] C. C. Leznoff, R. J. Hayward (Can. J. Chem. **49** [1971] 3596/601).

[7] O. Wennerström (Acta Chem. Scand. **25** [1971] 789/94).

[8] G. Batu, R. Stevenson (J. Org. Chem. **45** [1980] 1532/4).

[9] S. F. Vasilevskii, V. A. Gerasimov, M. S. Shvartsberg (Izv. Akad. Nauk SSSR Ser. Khim. **1981** 902/4; Bull. Acad. Sci. USSR Div. Chem. Sci. **1981** 683/5).

[10] M. S. Shvartsberg, S. F. Vasilevskii, T. V. Anisimova, V. A. Gerasimov (Izv. Akad. Nauk SSSR Ser. Khim. **1981** 1342/8; Bull. Acad. Sci. USSR Div. Chem. Sci. **1981** 1071/6).

1.1.2.1.5.4 Ring Closure Giving Carbocycles

This reaction is referred to as "Reaction Type 8 d" in Table 1 on pp. 17/41.

Inter- and Intramolecular Cyclization of Iodine-containing RC≡CCu. In most cases 2 to 6 molecules of RC≡CCu form the carbocycle in a "head-to-tail"-type intermolecular substitution reaction. Intramolecular substitutions are also known. The following examples are arranged by the number of the starting compound RC≡CCu as in Section 1.1.2.1.1, Table 1.

If $2\text{-IC}_6\text{H}_4\text{C}{\equiv}\text{CCu}$ is heated in pyridine (no temperature given) a 26% yield of I and a little II was isolated [1]. On the other hand, only some "crystallizable cyclic trimer" (presumably I) was obtained by heating $2\text{-IC}_6\text{H}_4\text{C}{\equiv}\text{CCu}$ 6 h at 110 °C in dimethylformamide or in pyridine. The main product is a linear polymer of the structure III containing some iodine. The yield of this compound was 90% in dimethylformamide and 83% in pyridine [8].

$3\text{-IC}_6\text{H}_4\text{C}{\equiv}\text{CCu}$ gives a 4.6% yield of IV after boiling 24 h in pyridine [13].

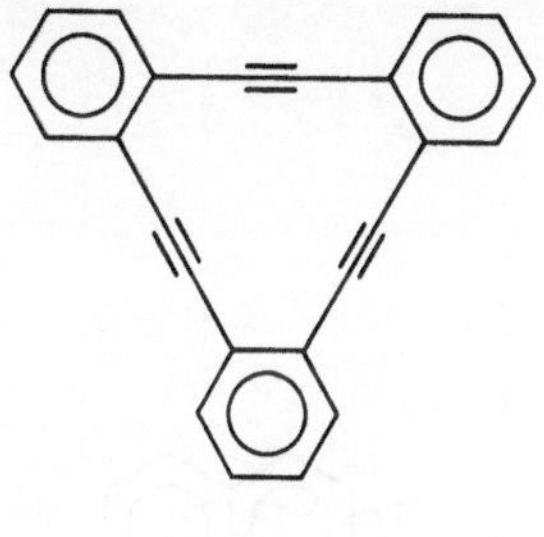

I

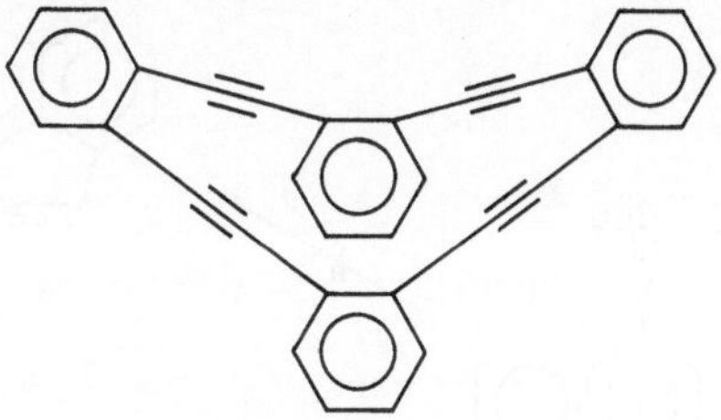

II

III

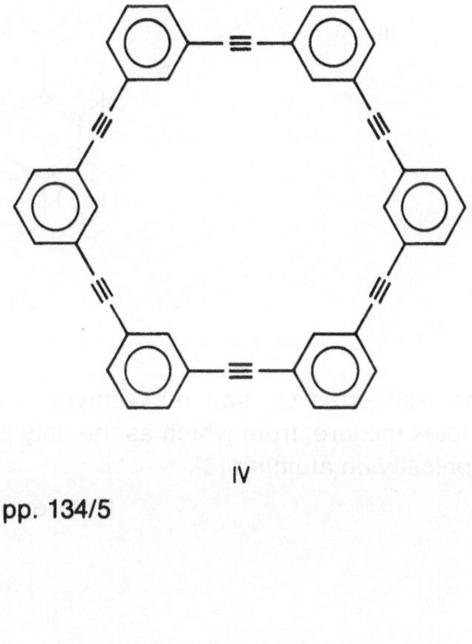

IV

Well-dried (E)-2-IC$_6$H$_4$CH=CHC≡CCu in boiling pyridine yields a mixture of V (14% yield) and VI (2.4% yield), which can be separated chromatographically on alumina [6].

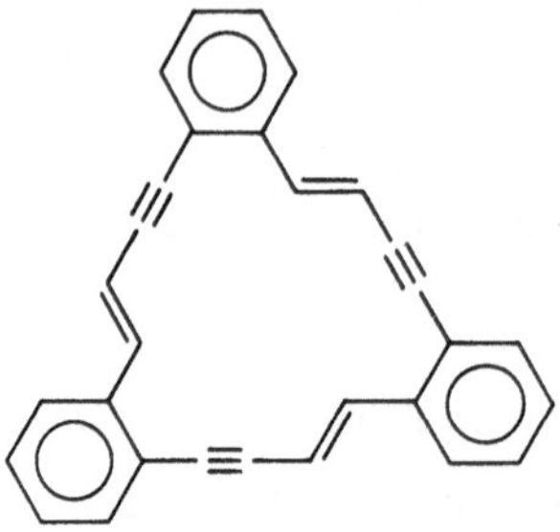

V 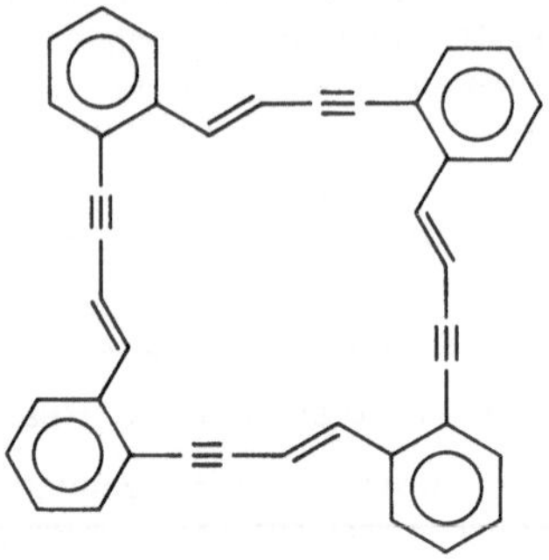   VI

The Cu salt of 1-ethynyl-8-iodonaphthalene VII in boiling pyridine is reported to yield 50% zethrene (VIII) [4, 7] or a mixture of VIII (52% yield) and IX (5% yield) [10]. The origin of the H atoms marked in formula VIII has not been studied. The way of the formation of compound IX is not clear. A Diels–Alder type reaction of X is improbable. An attempted trapping of X by boiling VII in the presence of 1,3-diphenylbenzo[c]furane in dimethylformamide gave no Diels–Alder adduct [10].

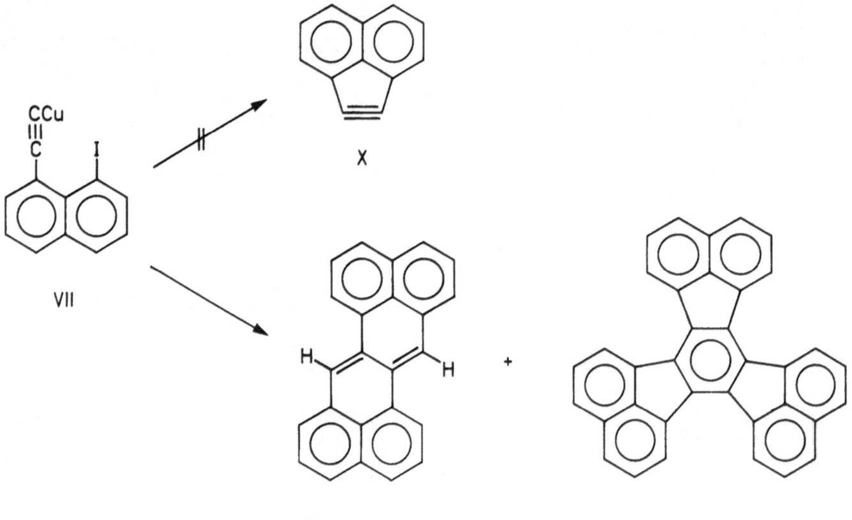

The well-dried Cu salt of 1-ethynyl-7-iodonaphthalene (XI) in boiling pyridine gives a complex mixture, from which as the only product 8% yield of XII could be isolated chromatographically on alumina [6].

References on pp. 134/5

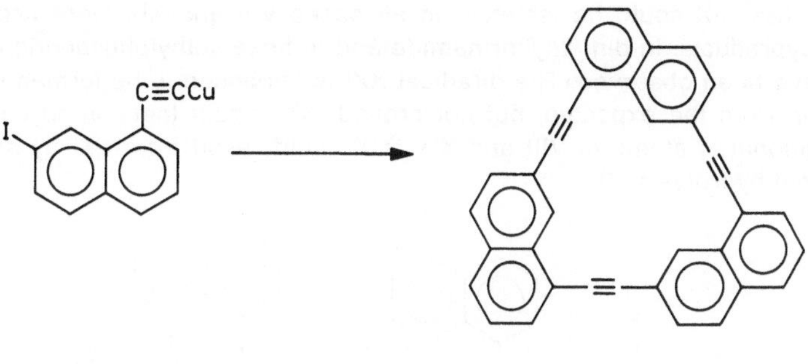

XI   XII

The previously described "head-to-tail reaction" of three or more molecules of a compound containing both the groups RC≡C and I is not the only possibility for producing carbocyclic ring systems from these bifunctional starting materials. $2\text{-}(2\text{-}IC_6H_4)C_6H_4C\equiv CCu$ (XIII) undergoes an intramolecular reaction in boiling pyridine. The first product is possibly the unstable carbene XIV, which dimerizes to the cumulene XV [5, 11].

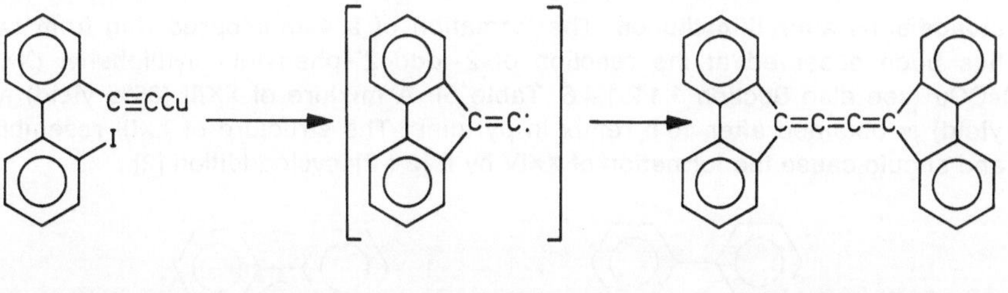

XIII   XIV   XV

If the distance between I and Cu is wide enough, two molecules can undergo the above mentioned "head-to-tail reaction". The Cu salts of the dienynes XVI in boiling pyridine with $R^1=CH_3$ (XVIa) yield after 48 h 8.4% XVIIa, with $R^1=CH_3OCH_2$ (XVIb) after 14 h 14.6% XVIIb [12].

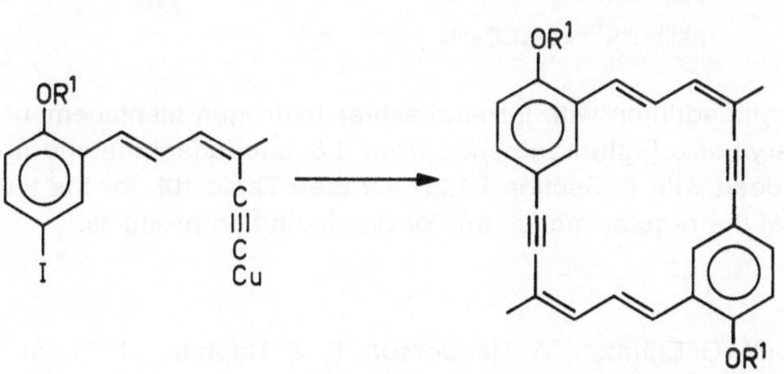

XVI   XVII

a : $R^1 = CH_3$
b : $R^1 = CH_3OCH_2$

The reaction products of XVIII in pyridine depend on the reaction conditions. After 90 min boiling a 22% yield of zethrene VIII (see p. 132) is formed. After 30 min at 80 °C a mixture of VIII (10% yield) and 7-iodozethrene XIX (11% yield) is formed. At 35 °C no VIII was obtained, but 6% XIX could be isolated. In all cases VIII and XIX were accompanied by unidentified byproducts. In dimethylformamide and in hexamethylphosphoric triamide similar results have been observed. The diradical XXI is supposed to be formed by an electrocyclic reaction from the expected, but not proved, XX. Again there is no evidence of the origin of additional H atoms in VIII and XIX. XIX could result from XXI by addition of CuI and subsequent hydrolysis [7].

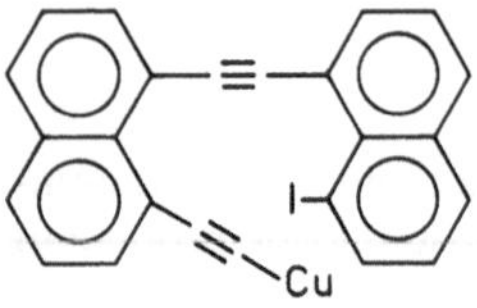

XVIII 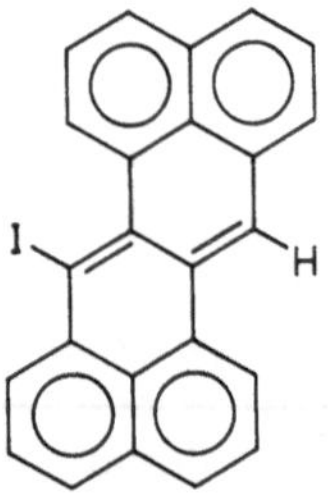　　　　　　　　XIX 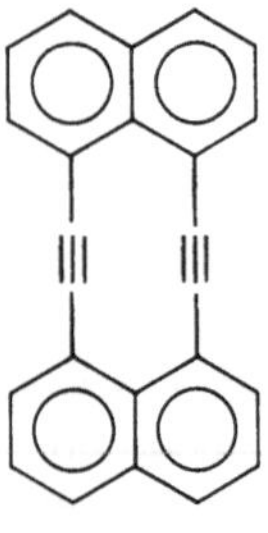　　　　　　　　XX 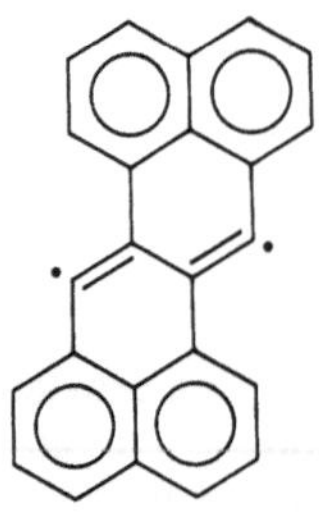　　　　　　　　XXI

Cycloadditions after Substitution. The formation of a 4-membered ring from two C≡C units has been observed at the reaction of 2-iodo-2′-phenylethynylbiphenyl (XXII) and $C_6H_5C\equiv CCu$ (see also Section 1.1.2.1.4.6, Table 6). A mixture of XXIII (65% yield) and XIV (10% yield) is obtained after 10 h reflux in pyridine. The structure of XXIII resembles that of XX and should cause the formation of XXIV by a [2+2]-cycloaddition [2].

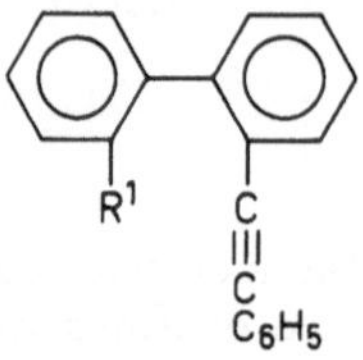

XXII : R^1 = I
XXIII : R^1 = C≡CC$_6$H$_5$

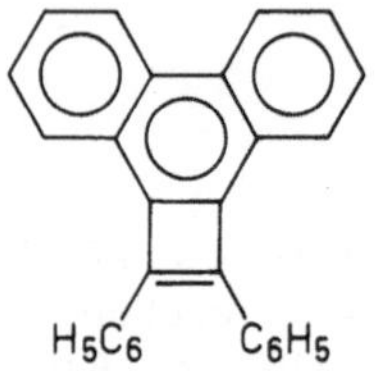

XXIV

A [2+2+2]-cycloaddition with intramolecular hydrogen displacement is involved in the formation of 7-arylbenzo[k]fluoranthenes from 1,8-diiodonaphthalene and RC≡CCu [3, 9]. This reaction is dealt with in Section 1.1.2.1.4.9 (see Table 10), for the typical reaction path is the formation of the regular mono- and/or disubstitution products.

References:

[1] J. D. Campbell, G. Eglinton, W. Henderson, R. A. Raphael (J. Chem. Soc. Chem. Commun. **1966** 87/9).
[2] S. A. Kandil, R. E. Dessy (J. Am. Chem. Soc. **88** [1966] 3027/34).
[3] J. Ipaktschi, H. A. Staab (Tetrahedron Letters **1967** 4403/8).
[4] H. A. Staab, A. Nissen, J. Ipaktschi (Angew. Chem. **80** [1968] 241; Angew. Chem. Intern. Ed. Engl. **7** [1968] 226).
[5] H. A. Staab, H. Mack, E. Wehinger (Tetrahedron Letters **1968** 1465/9).
[6] K. Endo, Y. Sakata, S. Misumi (Tetrahedron Letters **1970** 2557/60).
[7] R. H. Mitchell, F. Sondheimer (Tetrahedron **26** [1970] 2141/50).

[8] R. D. Stephens, W. C. Kray, Shell Oil Co. (U.S. 3749700 [1970/73]; C.A. **79** [1973] No. 146990).

[9] H. A. Staab, J. Ipaktschi (Chem. Ber. **104** [1971] 1170/81).

[10] H. A. Staab, J. Ipaktschi, A. Nissen (Chem. Ber. **104** [1971] 1182/90).

[11] H. A. Staab, E. Wehinger, W. Thorwart (Chem. Ber. **105** [1972] 2290/309).

[12] A. Tozuka, T. Otsubo, Y. Sakata, S. Misumi (Mem. Inst. Sci. Ind. Res. Osaka Univ. **30** [1973] 83/9; C.A. **79** [1973] No. 66092).

[13] H. A. Staab, K. Neunhoeffer (Synthesis **1974** 424).

1.1.2.1.6 Reactions with R′X to Form the Hydrocarbons R′H

This reaction is referred to as "Reaction Type 9" in Table 1 on pp. 17/41.

In certain cases the reaction of RC≡CCu and R′I or R′Br yields R′H, either as the sole product or in addition to the substitution product RC≡CR′ (Reaction Type 7). The origin of the hydrogen in the course of this "reductive elimination of halogen" was not investigated. The yields of R′H come up to over 50%. The H therefore could not originate from traces of water or from impurities, but could come from the solvents pyridine [1, 3 to 5], dimethylformamide, or glycol [2].

The conversion of R′X (X = I or Br) to R′H is especially favored in the bithienyl series. The reaction of $CH_3C≡CCu$ and Ia in pyridine at 105 °C gives Ib as the only product in a 52% yield. The expected product Ic is not formed [4].

$$(CH_3O)_2CHC≡C\underset{S}{\underbrace{}}\underset{S}{\underbrace{}}R^1 \qquad \begin{array}{l} Ia: R^1 = I \\ Ib: R^1 = H \\ Ic: R^1 = CH_3C≡C \end{array}$$

In many cases a mixture of the "normal" substitution product RC≡CR′ (Reaction Type 7) and of the "reduction product" R′H (Reaction Type 9) is obtained. Most examples belong to the reactions of RC≡CCu with 2-iodothiophenes and with 2-iodofurans [1, 3]. They are listed in Section 1.1.2.1.4.11, Tables 11 and 12.

In the course of the reaction of $C_6H_5C≡CCu$ and diiodinated aromatic compounds one iodine can be selectively substituted by hydrogen and the second one by $C_6H_5C≡C$. For the reaction of $1,2-I_2C_6F_4$ and $C_6H_5C≡CCu$ see Section 1.1.2.1.4.7 [5].

Proceeding from $C_6H_5C≡CCu$ and $C_6H_5COCH_2Br$ in glycol at 140 °C, the only product besides tars is $C_6H_5COCH_3$ (47%). $C_6H_5C≡CCu$ and 1-iodonaphth-2-ol react in dimethylformamide at 152 °C to form naphth-2-ol, but only in a yield of 2% [2].

References:

[1] R. E. Atkinson, R. F. Curtis, G. T. Phillips (Chem. Ind. [London] **1964** 2101/2).

[2] C. E. Castro, E. J. Gaughan, D. C. Owsley (J. Org. Chem. **31** [1966] 4071/8).

[3] R. E. Atkinson, R. F. Curtis, G. T. Phillips (J. Chem. Soc. C **1967** 2011/5).

[4] F. Bohlmann, J. Kocur (Chem. Ber. **108** [1975] 2149/52).

[5] R. S. Dickson, L. J. Michel (Australian J. Chem. **28** [1975] 1943/55).

1.1.2.1.7 Mannich Type Reactions

This reaction is referred to as "Reaction Type 10" in Table 1 on pp. 17/41.

The Mannich reaction of alkynes in the presence of copper ions can be understood as a substitution of Cu in RC≡CCu by the Mannich cation I to form the alk-2-ynylamine II.

$$HCHO + HNR_2^1 + H^+ \xrightarrow[-H_2O]{} CH_2 \overset{\oplus}{\cdots\cdots} NR_2^1 \xrightarrow{RC\equiv CCu} RC\equiv CCH_2NR_2^1$$

$$\quad\quad\quad\quad\quad\quad\quad\quad\quad\quad\quad\quad\quad\quad\quad I \quad\quad\quad\quad\quad\quad\quad\quad\quad\quad\quad II$$

To run Mannich reactions $RC\equiv CCu$ is not brought into the reaction batch as a pure substance, but formed there from alkynes and copper salts. Its presence is said to be observed in the reactions of $HOCH_2C\equiv CCu$ with $HN(C_2H_5)_2$ [7] and of $HOCR^2R^3C\equiv CCu$ or $R^4OCR^2R^3C\equiv CCu$ compounds with $HN(CH_2CH_2Cl)_2$ [6, 13]. It is first precipitated because of its insolubility and then consumed in the course of the reaction.

The intermediate formation of $RC\equiv CCu$ is also supported by the influence of the Cu on the reaction path of compounds, which contain not only the acetylenic H, but also a second active hydrogen able to undergo Mannich reactions. Compounds of the type $C_6H_5CO(CH_2)_nC\equiv CH$ react in the course of the Mannich reaction in the presence of Cu^I at the alkynyl carbon, but in the absence of copper at the $COCH_2$ position [10]. At any rate Cu^I salts increase the nucleophility of the acetylenic substrate towards the Mannich reagents [14]. The addition of 1% CuCl can enhance the reaction rate up to 30-fold and the yield is much higher [5]. Alkynes with donors in the 3-position undergo the Mannich reaction only in the presence of Cu^I [12].

A high number of publications deal with Cu-catalyzed Mannich reactions. In many cases the intermediate formation of $RC\equiv CCu$ and its reaction with the Mannich cation I to form the amine II can be strongly supposed [1 to 5, 8 to 12, 15]. If the Mannich reaction is catalyzed by Cu^{II}, the reduction of Cu^{II} to Cu^I by CH_2O must be assumed [6, 13].

The reaction of 1-aminobenzimidazolium salts with preformed $C_6H_5C\equiv CCu$ can be understood as a Mannich reaction, if the salt is written as a cyclic Mannich cation III: $R^1 = CH_3$, $C_4H_9O_2CCH_2$, $(CH_3)_2N(CH_2)_2$. The expected 2-alkynylbenzimidazoline IV corresponds to II at simple acyclic systems. It rearranges to form the pyrazole V, which crystallizes from the reaction batch. The corresponding chloride (III, Cl as the anion) does not react with $C_6H_5C\equiv CCu$ or with $C_6H_5C\equiv CH$ [16].

III   IV   V

References:

[1] L. B. Fisher (Usp. Khim. **27** [1958] 589/621).

[2] J.-A. Gautier, I. Marszak, M. Miocque (Bull. Soc. Chim. France **1958** 415/8).

[3] N. M. Libman, S. G. Kusnetzov (Zh. Obshch. Khim. **30** [1960] 1197/202; J. Gen. Chem. [USSR] **30** [1960] 1218/22).

[4] J.-A. Gautier, M. Miocque, N. M. Húng (Bull. Soc. Chim. France **1961** 2098/109).

[5] L. B. Fisher, I. L. Kotlyarevskii, E. K. Andrievskaya (Izv. Akad. Nauk SSSR Ser. Khim. **1964** 1543/5; Bull. Acad. Sci. USSR Div. Chem. Sci. **1964** 1455/7).

[6] R. I. Krugilikova, V. E. Pikalov (Izv. Vysshikh Uchebn. Zavedenii Khim. Khim. Tekhnol. **8** [1965] 349/51; C.A. **63** [1965] 11333).

[7] R. L. Salvador, D. Simon (Can. J. Chem. **44** [1966] 2570/5).

[8] M. I. Bardamova, R. N. Myasnikova, I. L. Kotlyarevskii (Izv. Akad. Nauk SSSR Ser. Khim. Nauk **1967** 443/5; Bull. Acad. Sci. USSR Div. Chem. Sci. **1967** 429/31).

[9] J.-A. Gautier, M. Miocque, L. Mascrier-Demagny (Compt. Rend. C **264** [1967] 778/81).

[10] J.-A. Gautier, M. Miocque, L. Mascrier-Demagny (Bull. Soc. Chim. France **1967** 1560/8).

[11] I. L. Kotlyarevskii, E. K. Andrievskaya, L. B. Fisher (Izv. Akad. Nauk SSSR Ser. Khim. Nauk **1967** 397/404; Bull. Acad. Sci. USSR Div. Chem. Sci. **1967** 373/7).

[12] G. R. Pettit, B. J. Danley (Can. J. Chem. **46** [1968] 792/5).

[13] I. N. Azerbaev, I. A. Poblavskaya, G. N. Kondaurov (Dokl. 4th Konf. Khim. Atsetilena, Alma-Ata 1972, Vol. 1, pp. 353/8; C.A. **79** [1973] No. 77998).

[14] M. Tramontini (Synthesis **1973** 703/75).

[15] M. M. Kwatra, D. Z. Simon, R. L. Salvador, P. D. Cooper (J. Med. Chem. **21** [1978] 253/7).

[16] L. Yu. Ukhin, V. N. Komissarov, Zh. I. Orlova, N. A. Dolgopolova (Zh. Obshch. Khim. **54** [1984] 1676/8); J. Gen. Chem. [USSR] **54** [1984] 1493/4).

1.1.2.1.8 Reactions with R'Cu

This reaction is referred to as "Reaction Type 11" in Table 1 on pp. 17/41.

The main product formed from $RC{\equiv}CCu$ and $R'Cu$ generally is $RC{\equiv}CR'$. If R and R' are aryl groups, heating is necessary and the reaction proceeds in several discrete steps involving the formation of Cu_6 and Cu_4 clusters.

Reactions of $RC{\equiv}CCu$ with CH_3Cu give $RC{\equiv}CCH_3$ and metallic Cu, if it is run at room temperature in ether. The $RC{\equiv}CCu$ compounds are prepared in situ from $RC{\equiv}CCO_2Cu$ (see Section 1.1.2.1.1, Preparation Method Va, p. 4). An addition of CH_3Cu to the $C{\equiv}C$ bond of $RC{\equiv}CCu$, as it is known from $RC{\equiv}CCH_3$, was not observed. The following $RC{\equiv}CCH_3$ have been prepared (yields in parentheses): $R = 3\text{-}ClC_6H_4$ (50%), $4\text{-}ClC_6H_4$ (55%), $3\text{-}FC_6H_4$ (45%), $3\text{-}CH_3C_6H_4$ (55%). With $R = C_6H_5$ (75%) and with $R = C_{12}H_{25}$ it is not clear whether $RC{\equiv}CCu$ or its LiI adduct (see Section 1.1.2.3) is reacted [1].

Reactions of $RC{\equiv}CCu$ with $2\text{-}(CH_3)_2NC_6H_4Cu$ in dimethylformamide at 65 °C give the yellow clusters $(RC{\equiv}C)_2Cu_6(C_6H_4N(CH_3)_2\text{-}2)_4$ (decomposition temperatures in parentheses): $R = C_6H_5$ (128 °C), $4\text{-}CH_3C_6H_4$ (138 °C), $4\text{-}CH_3OC_6H_4$ (165 °C), $2,4,6\text{-}(CH_3)_3C_6H_2$ (189 °C). If the reaction is carried out in dimethylformamide at 125 °C (heating for 2.5 h and workup with aqueous NH_3 is typical for these reactions), a mixture of $RC{\equiv}CC_6H_4N(CH_3)_2\text{-}2$, $2\text{-}(CH_3)_2NC_6H_4C_6H_4N(CH_3)_2\text{-}2$, $C_6H_5N(CH_3)_2$, and metallic Cu is obtained; see Table 18. The decomposition products of $(4\text{-}CH_3C_6H_4C{\equiv}C)_2Cu_6(C_6H_4N(CH_3)_2\text{-}2)_4$ in benzene at 80 °C correspond to this pattern; only the symmetric coupling product $2\text{-}(CH_3)_2NC_6H_4C_6H_4N(CH_3)_2\text{-}2$ could not be detected [2]. So the reactions at both temperatures are treated with the implicit understanding that in benzene and in dimethylformamide the same reactions occur and that the compounds listed in Table 18 are decomposition products of $(RC{\equiv}C)_2Cu_6(C_6H_4N(CH_3)_2\text{-}2)_4$ [2, 3].

Table 18
Reaction of RC≡CCu and 2-$(CH_3)_2NC_6H_4Cu$ according to

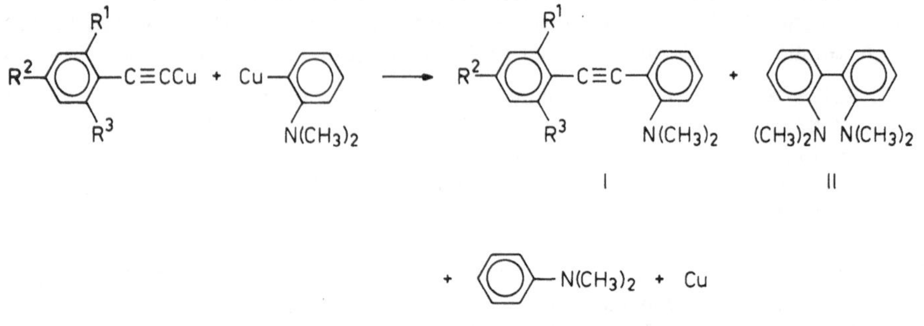

(belongs to Reaction Type 11, explanation see pp. 12/4) in $(CH_3)_2NCHO$ at 125 °C/2.5 h (last reaction 125 °C/4 h, then 150 °C/1 h).

R^1	R^2	R^3	yield of I in %	yield of II in %	yield of III in %
H	Cl	H	92	–	5
H	NO_2	H	76	10	12
H	H	H	90	0.5	6
H	CH_3	H	90	0.5	7
H	CH_3O	H	89	0.5	4
CH_3	CH_3	H	79	–	17
CH_3	CH_3	CH_3	29	1.2	65
CH_3	CH_3	CH_3	84	–	16

If the RC≡CCu reacted with 2-$(CH_3)_2NC_6H_4Cu$ at 125 °C is prepared in situ from RC≡CH and Cu^I, the yields drop considerably. Without solvent the yields are almost the same. The formation of the Cu can be well understood from the opposite polarization of the Cu–C bonds in the starting materials [2].

In order to understand the high temperature reaction of RC≡CCu and 2-$(CH_3)_2NC_6H_4Cu$, the decomposition of $(4-CH_3C_6H_4C≡C)_2Cu_6(C_6H_4N(CH_3)_2-2)_4$ in benzene at 80 °C has been studied in detail. First a dark red substance, which is soluble in benzene, insoluble in pentane, and highly oxygen-sensitive, is formed [2]. From NMR data and from the reaction products the structure $4-CH_3C_6H_4C≡CCu_2^ICu_2^0C_6H_4N(CH_3)_2-2$ was claimed. $4-CH_3C_6H_4C≡$ $CCu_4^ICu_2^0(C_6H_4N(CH_3)_2-2)_3$ is supposed to be its precursor [2, 3]. Prolonged heating of the Cu_6-compound at 80 °C causes a slow decomposition to give $4-CH_3C_6H_4C≡CC_6H_4N(CH_3)_2-2$, 2-$(CH_3)_2NC_6H_4Cu$, and dimethylaniline. As mentioned before, there is no distinct proof that with $4-CH_3C_6H_4C≡CCu$ and 2-$(CH_3)_2NC_6H_4Cu$ instead of $(4-CH_3C_6H_4C≡C)_2Cu_6(C_6H_4N(CH_3)_2-2)_4$ the reaction path is the same [2].

The yield pattern of the decomposition reaction can be partly explained by comparison of the reaction temperature of the mixture (125 °C) and the decomposition temperature of the cluster. If the latter is relatively high (e.g., 189 °C at R=2,4,6-$(CH_3)_3C_6H_2$), the side reaction product dimethylaniline is favored against the alkyne (65%); with R=C_6H_5 (decomposition temperature = 128 °C) only 6% dimethylaniline are formed [2].

An intra–aggregate valence disproportionation of $(RC\equiv C)_2Cu_6(C_6H_4N(CH_3)_2-2)_4$ with the formation of asymmetric $R'(RC\equiv C)Cu^{II}$ centers to give $RC\equiv CCu_4^ICu_2^0(C_6H_4N(CH_3)_2-2)_3$ and one mole of $RC\equiv CC_6H_4N(CH_3)_2-2$ was claimed in a later publication. Cluster reorganization and a reductive coupling step lead to metallic Cu and to a second mole of $RC\equiv CC_6H_4N(CH_3)_2-2$ [3].

Reactions of RC≡CCu with 4-(CH₃)₂NC₆H₄Cu at 120 °C in dimethylformamide follow a pattern similar to that of the 2–substituted compounds, but are only described with the aryl copper prepared in situ. With $R=4-ClC_6H_4$ the yield of $4-ClC_6H_4C\equiv CC_6H_4N(CH_3)_2-4$ is 81%, and 12% dimethylaniline are admixed. With $R=4-O_2NC_6H_4$ a mixture of 17% $4-O_2NC_6H_4C\equiv CC_6H_4N(CH_3)_2-4$, 42% $4-(CH_3)_2NC_6H_4C_6H_4N(CH_3)_2-4$, 20% dimethylaniline, and 7% $4-O_2NC_6H_4C\equiv CH$ is obtained. The high amount of the biphenyl is probably due to a oxidative coupling process by the presence of nitrosubstituted compounds. With $R=4-CH_3C_6H_4$, 48% $4-CH_3C_6H_4C\equiv CC_6H_4N(CH_3)_2-4$, 26% $4-(CH_3)_2NC_6H_4C_6H_4N(CH_3)_2-4$, and 11% dimethylaniline are isolated. With $R=4-CH_3OC_6H_4$ a mixture of 83% $4-CH_3OC_6H_4C\equiv CC_6H_4N(CH_3)_2-4$ and 12% dimethylaniline is formed [2].

References:

[1] R. Levene, J. Y. Becker, J. Klein (J. Organometal. Chem. **67** [1974] 467/71).
[2] G. Van Koten, R. M. W. Ten Hoedt, J. G. Noltes (J. Org. Chem. **42** [1977] 2705/11).
[3] J. G. Noltes (Phil. Trans. Roy. Soc. [London] A **308** [1982] 35/45).

1.1.2.1.9 Reactions with Nitrones to Form β-Lactames

This reaction is referred to as "Reaction Type 12" in Table 1 on pp. 17/41.

The reaction of RC≡CCu with nitrones $R^2CH=NR^1O$ to yield β–lactames is a useful method to prepare these heterocycles; see Table 19. There is a discrepancy in the literature concerning the geometry of the β–lactames formed. At room temperature in pyridine (0.5 to 4 h) cis–β–lactames [1] or a mixture of cis– and trans–β–lactames [2] have been found. At any rate, the primary product seems to be the cis compound. The rate of the rearrangement reaction obviously depends much on the substitution type. Generally the yields do not differ considerably whether pure RC≡CCu is reacted or whether it is formed in situ from $RC\equiv CH/CuCl/NH_3/H_2O/pyridine$ in the reaction batch [2].

Table 19
Reaction of RC≡CCu with Nitrones according to

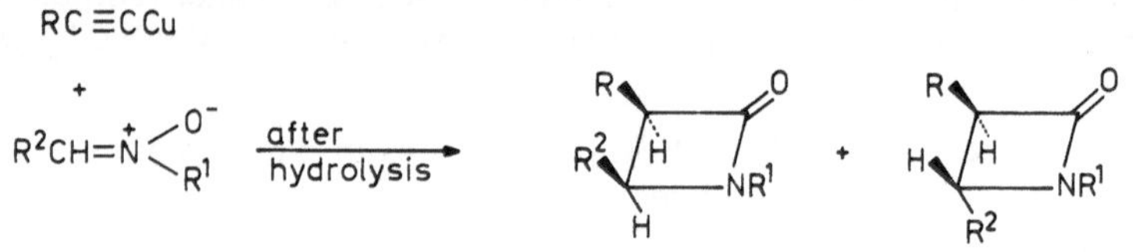

(Reaction Type 12, explanation see pp. 12/4).

R	R¹	R²	yield of II in %	yield of III in %	only yield of II+III given in %	remarks, product	Ref.
$C_2H_5O_2C$	C_6H_5	$4\text{-}ClC_6H_4$	—	—	50	acetylide prepared in situ (pyridine/H₂O, 1 h)	[2]
	C_6H_5	$4\text{-}CH_3C_6H_4$	39	13	—	acetylide prepared in situ (CH_3OH/H_2O, 4 h)	[2]
			—	—	42	—	[2]
		$C(CH_3)_2CH(CH_3)CH_2$	—	40	—	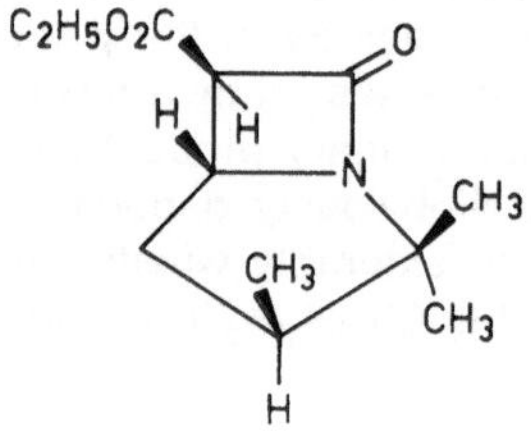	[2]
$n\text{-}C_4H_9$	C_6H_5	C_6H_5	61	5	—	—	[2]
	C_6H_5	$4\text{-}CH_3OC_6H_4$	—	—	55	—	[2]
	$4\text{-}CH_3C_6H_4$	$4\text{-}ClC_6H_4$	—	—	9	—	[2]
		$C(CH_3)_2CH(CH_3)CH_2$	—	40	—	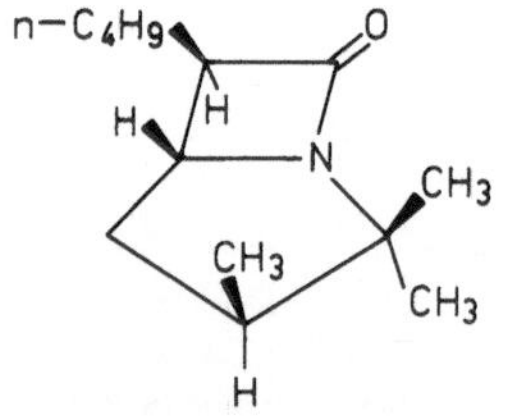	[2]
$(C_2H_5)_2NCH_2$	C_6H_5	$4\text{-}ClC_6H_4$	—	—	44	acetylide prepared in situ (CH_3OH/H_2O, 4 h)	[2]

References on p. 142

Table 19 [continued]

R	R¹	R²	yield of II in %	yield of III in %	only yield of II+III given in %	remarks, product	Ref.
C_6H_5	$4\text{-}ClC_6H_4$	C_6H_5	60	–	–	–	[1]
	C_6H_5	$2\text{-}ClC_6H_4$	51	–	–	–	[1]
	C_6H_5	C_6H_5	55	–	–	–	[1]
			20	12	–	–	[2]
	C_6H_5	$2\text{-}CH_3C_6H_4$	51	–	–	–	[1]
			23	2	–	–	[2]
	C_6H_5	$3\text{-}CH_3C_6H_4$	14	11	–	–	[2]
	C_6H_5	$4\text{-}CH_3C_6H_4$	32	8	–	–	[2]
	C_6H_5	$4\text{-}CH_3OC_6H_4$	–	–	28	–	[2]
	$4\text{-}CH_3C_6H_4$	$4\text{-}ClC_6H_4$	8	7	–	–	[2]
	$4\text{-}CH_3C_6H_4$	C_6H_5	11	14	–	–	[2]
	$C_6H_5C_6H_4$	$4\text{-}ClC_6H_4$	–	–	45	the position of C_6H_5 in R¹ is not given	[2]
	$(CH_2)_2OCH_2$		54	–	–	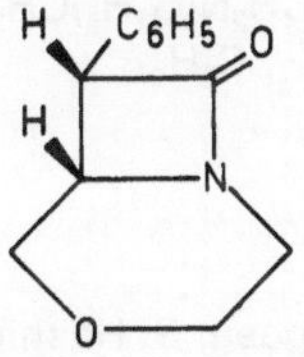	[2]
	$(CH_2)_4$		40	–	–	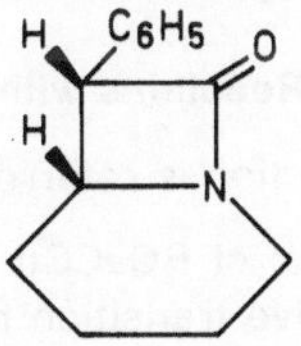	[2]
	$C(CH_3)_2CH(CH_3)CH_2$		–	40	–	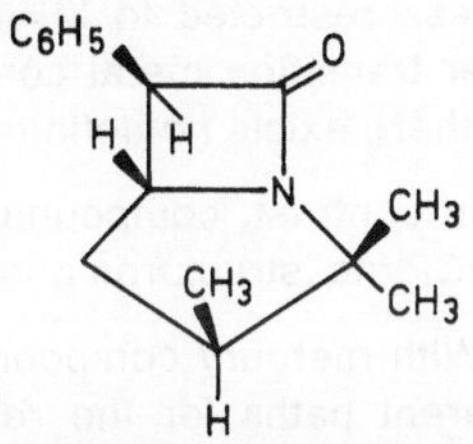	[2]

References on p. 142

Table 19 [continued]

R	R¹	R²	yield of II in %	yield of III in %	only yield of II+III given in %	remarks, product	Ref.
4-CH$_3$C$_6$H$_4$	C$_6$H$_5$	C$_6$H$_5$	35	15	—	—	[2]
HO–piperidine–CH$_2$– (N–CH$_3$)	C$_6$H$_5$	4-ClC$_6$H$_4$	25	25	—	acetylide prepared in situ (CH$_3$OH/H$_2$O, 4 h)	[2]
C$_6$H$_5$N(CH$_3$)CH$_2$	C$_6$H$_5$	4-ClC$_6$H$_4$	26	26	—		[2]
phthalimido–NCH$_2$–	C$_6$H$_5$	4-ClC$_6$H$_4$	—	—	12		[2]
	C$_6$H$_5$	4-CH$_3$C$_6$H$_4$	27	4	—		[2]
C$_6$H$_5$N(C$_2$H$_5$)CH$_2$	C$_6$H$_5$	4-ClC$_6$H$_4$	7	45	—	acetylide prepared in situ (CH$_3$OH/H$_2$O, 4 h)	[2]
	C$_6$H$_5$	4-CH$_3$OC$_6$H$_4$	—	—	39	—	[2]
4-(C$_2$H$_5$O$_2$C)C$_6$H$_4$NHCH$_2$	C$_6$H$_5$	4-ClC$_6$H$_4$	—	—	75	—	[2]
4-(CH$_3$)$_2$NC$_6$H$_4$COCH$_2$N(C$_2$H$_5$)CH$_2$	C$_6$H$_5$	4-ClC$_6$H$_4$	—	—	67	acetylide prepared in situ (CH$_3$OH/H$_2$O, 4 h)	[2]

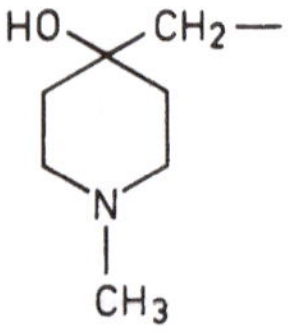

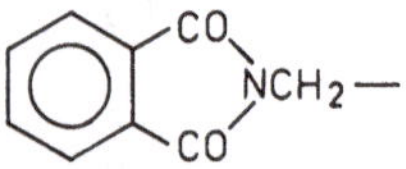

References:

[1] M. Kinugasa, S. Hashimoto (J. Chem. Soc. Chem. Commun. **1972** 466/7).
[2] L. K. Ding, W. J. Irwin (J. Chem. Soc. Perkin Trans. I **1976** 2382/6).

1.1.2.1.10   Reactions with Other Transition Metal Compounds

This reaction is referred to as "Reaction Type 18" in Table 1 on pp. 17/41.

Reactions of RC≡CCu with compounds containing groups ···M–X (M=transition metal) generally give transition metal complexes, and the reaction path usually seems to be very complicated. The simple reaction pattern RC≡CCu + ···M–X → RC≡CM··· +CuX is more or less restricted to M=Hg and Pt; in one case it also refers to a Ru compound. With other transition metal compounds the formation of RC≡CM··· as a first step is supposed, but there exists no definite proof. Examples are listed in Table 20.

In contrast, compounds containing SX, PX, SiX, or SnX groups generally give RC≡CS, RC≡CP etc. structures and are therefore treated in Section 1.1.2.1.4.14.

With mercury compounds the products depend strongly on the reaction conditions and different paths for the reactions have been suggested. Depending on the reaction time

RC≡CCu and HgCl$_2$ give (RC≡C)$_2$Hg or RC≡CHgCl (see Table 20). It is not clear whether the C$_6$H$_5$C≡CHgCl formed on long refluxing of C$_6$H$_5$C≡CCu and HgCl$_2$ in dioxane is produced by a direct substitution or from (C$_6$H$_5$C≡C)$_2$Hg and HgCl$_2$. The compounds (RC≡C)$_2$Hg are possibly produced via 2RC≡CCu · HgCl$_2$, which is known to be formed from the components in tetrahydrofuran (see Section 1.1.2.5). Reactions starting with RC≡CCu are the most important way to synthesize RC≡CHg structures [1].

With M = Re, Fe, Ru, and Rh, mixtures are generally produced and both the solvent and the molar ratio of the reactants are of importance. The reaction times are up to 65 h in refluxing toluene. In most cases the mixtures require separation by chromatographic methods. If RC≡CCu is reacted with phosphine complexes, the oxidation of a part of the phosphine unit to give OPR$_3'$ is typical and a copper mirror at the vessel wall is then observed.

Table 20

Reactions of RC≡CCu with Transition Metal Compounds (Reaction Type 18, explanation see pp. 12/4).

R	metal compound	products (yield in %), figures see pp. 148/9	molar ratio RC≡CCu:metal compound; solvent; remarks	Ref.
mercury compounds				
CH=CH	$HgCl_2$	$(CH_2=CHC\equiv C)_2Hg$ (40)	2:1; 15 min reflux in dioxan and workup with aqueous NH_3	[1]
$C_2B_{10}H_{11}$ (1,2-dicarba-closo-dodeca-boran(12)-1-yl)	CH_3HgBr	$C_2B_{10}H_{11}C\equiv CHgCH_3$ (47)	1:1; heptane/THF	[2]
$n-C_4H_9$	$HgCl_2$	$(n-C_4H_9C\equiv C)_2Hg$ (47)	2:1; 15 min reflux in dioxan and workup with aqueous NH_3	[1]
$4-BrC_6H_4$	$HgCl_2$	$(4-BrC_6H_4C\equiv C)_2Hg$ (78)	2:1; 15 min reflux in dioxan and workup with aqueous NH_3	[1]
$4-ClC_6H_4$	$HgCl_2$	$(4-ClC_6H_4C\equiv C)_2Hg$ (64)	2:1; 15 min reflux in dioxan and workup with aqueous NH_3	[1]
C_6H_5	$HgBr_2$	$(C_6H_5C\equiv C)_2Hg$	2:1; at room temperature in THF (40%) or C_2H_5OH (66%)	[1]
	$HgCl_2$	$C_6H_5C\equiv CHgCl$	"long time refluxing" with excess $HgCl_2$	[1]
		$(C_6H_5C\equiv C)_2Hg$	2:1; 15 min reflux in dioxan and workup with aqueous NH_3	[1]
	(E)–ClCH=CHHgCl	$(C_6H_5C\equiv C)_2Hg \cdot CuCl$	2 h at 40° in CH_3CN or THF	[13]
	$Hg(O_2CCH_3)_2$	$(C_6H_5C\equiv C)_2Hg$ (6)	2:1; 2 h at 25° in THF, workup with aqueous NH_3	[1]

$4-CH_3C_6H_4$	$HgCl_2$	$(4-CH_3C_6H_4C{\equiv}C)_2Hg$ (58)	2:1; 15 min reflux in dioxan and workup with aqueous NH_3	[1]
rhenium compounds				
C_6F_5	cis-$(CO)_3ReCl[P(C_6H_5)_3]_2$	X (21) + XI (6) + $[(C_6H_5)_3PCuCl]_4$	2.4:1; 20 h in THF	[6, 8]
C_6H_5	cis-$(CO)_3ReCl[P(C_6H_5)_3]_2$	I (65)	5:1; 6.5 h reflux in benzene	[8]
Iron compounds				
C_6F_5	$C_5H_5Fe(CO)_2Cl$	III, R=C_6F_5, X=Cl (20)	1:1; 2 h reflux in THF	[5]
$4-FC_6H_4$	$C_5H_5Fe(CO)_2Cl$	III, R=$4-FC_6H_4$, X=Cl (16) + ferrocene (34)	1:1; 4 h reflux in THF	[5]
C_6H_5	$C_5H_5Fe(CO)_2Br$	III, R=C_6H_5, X=Br (10)	1:1; 24 h reflux in THF	[5]
C_6H_5	$C_5H_5Fe(CO)_2Cl$	III, R=C_6H_5, X=Cl (42) + ferrocene (10)	1:1; 6 h reflux in acetone	[4, 5]
$4-CH_3C_6H_4$	$C_5H_5Fe(CO)_2Cl$	III, R=$4-CH_3C_6H_4$, X=Cl (27) + ferrocene (9)	1:1; 6 h reflux in THF	[5]
ruthenium compounds				
CH_3	$C_5H_5RuCl[P(C_6H_5)_3]_2$	IV (55)	2:1; 20 h reflux in benzene	[8]
C_6F_5	$C_5H_5RuCl[P(C_6H_5)_3]_2$	VI (33) + $[(C_6H_5)_3PCuCl]_4$	2:1; 7 h reflux in benzene	[8]
		VII	1:1.75; 13 h reflux in benzene	[8]
$4-FC_6H_4$	$C_5H_5RuCl[P(C_6H_5)_3]_2$	VII (5) + XIII, R=$4-FC_6H_4$ (56)	2:1; 6 h reflux in benzene	[7, 8]
C_6H_5	$C_5H_5RuCl[P(C_6H_5)_3]_2$	IV+V (together 82%)	2:1; 20 h reflux in benzene	[7, 8, 11]
$4-CH_3C_6H_4$	$C_5H_5RuCl[P(C_6H_5)_3]_2$	IV (7) + XIII, R=$4-CH_3C_6H_4$ (62)	2:1; 18 h reflux in benzene	[7, 8]
rhodium compounds				
C_6F_5	$RhCl[P(C_6H_5)_3]_3$	IX, R=C_6F_5, M=Rh (26) + $OP(C_6H_5)_3$	4:1; 22 h reflux in benzene	[10]
	trans-$CORhCl[P(C_6H_5)_2CH_3]_2$	VIII, R=C_6F_5, M=Rh (48)	1:1; 23 h reflux in benzene	[10]

Table 20 [continued]

R	metal compound	products (yield in %), figures see pp. 148/9	molar ratio RC≡CCu : metal compound; solvent; remarks	Ref.
	trans-CORhCl[P(C₆H₅)₃]₂	IX, R = C₆F₅, M = Rh (40) + [(C₆H₅)₃P]₃Cu₂Cl₂ (12)	1:1; 23 h reflux in toluene	[10]
4-FC₆H₄	trans-CORhCl[P(C₆H₅)₂CH₃]₂	VIII, R = 4-FC₆H₄, M = Rh (16)	4:1; 36 h reflux in benzene	[10]
	trans-CORhCl[P(C₆H₅)₃]₂	IX, R = 4-FC₆H₄, M = Rh (17) + [(C₆H₅)₃P]₃Cu₂Cl₂ (18) + OP(C₆H₅)₃	1:1; 21 h reflux in dimethoxyethane	[10]
C₆H₅	trans-CORhCl[P(C₆H₅)₂CH₃]₂	VIII, R = C₆H₅, M = Rh (29)	4:1; 4 h reflux in dimethoxyethane	[10]
	trans-CORhCl[P(C₆H₅)₃]₂	IX, R = C₆H₅, M = Rh (7) + [(C₆H₅)₃P]₃Cu₂Cl₂ + OP(C₆H₅)₃	4:1; 6 h reflux in THF	[10]
		IX, R = C₆H₅, M = Rh (3) + OP(C₆H₅)₃	4:1; 24 h reflux in benzene	[10]
4-CH₃C₆H₄	trans-CORhCl[P(C₆H₅)₂CH₃]₂	VIII, R = 4-CH₃C₆H₄, M = Rh (25)	4:1; 4 h reflux in dimethoxyethane	[10]
	trans-CORhCl[P(C₆H₅)₃]₂	IX, R = 4-CH₃C₆H₄, M = Rh (4) + [(C₆H₅)₃P]₃Cu₂Cl₂ (11)	4:1; 11 h reflux in THF	[10]
		IX, R = 4-CH₃C₆H₄, M = Rh (traces)	4:1; 21 h reflux in benzene	[10]

iridium compounds

R	metal compound	products (yield in %), figures see pp. 148/9	molar ratio RC≡CCu : metal compound; solvent; remarks	Ref.
C₆F₅	trans-COIrCl[P(C₆H₅)₂CH₃]₂	VIII, R = C₆F₅, M = Ir (9)	4:1; 3 h reflux in dimethoxyethane	[10]
	trans-COIrCl[P(C₆H₅)₃]₂	IX, R = C₆F₅, M = Ir (63) + XII (9) + [(C₆H₅)₃P]₃Cu₂Cl₂ (2) + OP(C₆H₅)₃ + Cu mirror	4:1; 65 h reflux in toluene	[10]
4-FC₆H₄	trans-COIrCl[P(C₆H₅)₃]₂	IX, R = 4-FC₆H₄, M = Ir (65) + [(C₆H₅)₃P]₃Cu₂Cl₂ (5) + OP(C₆H₅)₃ + Cu mirror	4:1; 50 h reflux in benzene	[10]

Gmelin Handbook
Cu-Org. Comp. 3

R	starting compound	product	conditions	Ref.
C_6H_5	trans-COIrCl[P(C_6H_5)$_2$CH$_3$]$_2$	VIII, R=C_6H_5, M=Ir (90)	4:1; 16 h reflux in benzene	[10]
	trans-COIrCl[P(C_6H_5)$_3$]$_2$	II (4)+IX, M=Ir (61) +[(C_6H_5)$_3$P]$_3$Cu$_2$Cl$_2$ (9) +OP(C_6H_5)$_3$+Cu mirror	4:1; 36 h reflux in benzene	[3, 10]
	COIrCl[(C_6H_5)$_2$PCH$_2$P(C_6H_5)$_2$]$_2$	C_6H_5C≡CIr(CO)[(C_6H_5)$_2$PCH$_2$P(C_6H_5)$_2$]$_2$CuCl	1:1; 30 min reflux in acetone	[14]
4-CH$_3$C$_6$H$_4$	trans-COIrCl[P(C_6H_5)$_3$]$_2$	IX, R=4-CH$_3$C$_6$H$_4$, M=Ir (73) +[(C_6H_5)$_3$P]Cu$_2$Cl$_2$ (12) +OP(C_6H_5)$_3$+Cu mirror	4:1; 36 h reflux in benzene	[10]

platinum compounds

R	starting compound	product	conditions	Ref.
CH$_2$=CH	cis-[(C$_2$H$_5$)$_3$P]$_2$PtCl$_2$	cis-[(C$_2$H$_5$)$_3$P]$_2$Pt(C≡CCH=CH$_2$)$_2$ (78)	2:1; 1 h in tetramethylethylene diamine/acetone at 25°	[12]
	trans-[(C$_2$H$_5$)$_3$P]$_2$PtCl$_2$	trans-[(C$_2$H$_5$)$_3$P]$_2$Pt(C≡CCH=CH$_2$)$_2$ (78)	2:1; 1 h in tetramethylethylene diamine/acetone at 25°	[12]
C_6F_5	cis-[(C_6H_5)$_3$P]$_2$PtCl$_2$	cis-[(C_6H_5)$_3$P]$_2$Pt(C≡CC$_6$F$_5$)$_2$ (56)	5:1; 4 h reflux in benzene	[9]
	[(C_6H_5)$_3$P]$_2$Pt(CH$_2$=CH$_2$)	cis-[(C_6H_5)$_3$P]$_2$Pt(C≡CC$_6$F$_5$)$_2$ (25)	2:1; 12 h reflux in benzene	[9]
4-FC$_6$H$_4$	[(C_6H_5)$_3$P]$_2$Pt(CH$_2$=CH$_2$)	trans(?)-[(C_6H_5)$_3$P]$_2$Pt(C≡CC$_6$H$_4$F-4)$_2$ (12)	2.1:1; 11 h reflux in THF	[9]
C_6H_5	[(C_6H_5)$_3$P]$_2$Pt(CH$_2$=CH$_2$)	trans-[(C_6H_5)$_3$P]$_2$Pt(C≡CC$_6$H$_5$)$_2$ (15)	2.4:1; 6 h reflux in benzene	[9]
	cis-[(C$_2$H$_5$)$_3$P]$_2$PtCl$_2$	cis-[(C$_2$H$_5$)$_3$P]$_2$Pt(C≡CC$_6$H$_5$)$_2$	2:1; 1 h in tetramethylethylene diamine/acetone at 25°	[12]
	trans-[(C$_2$H$_5$)$_3$P]$_2$PtCl$_2$	trans-[(C$_2$H$_5$)$_3$P]$_2$Pt(C≡CC$_6$H$_5$)$_2$	2:1; 1 h in tetramethylethylene diamine/acetone at 25°	[12]
4-CH$_3$C$_6$H$_4$	[(C_6H_5)$_3$P]$_2$Pt(CH$_2$=CH$_2$)	trans(?)-[(C_6H_5)$_3$P]$_2$Pt(C≡CC$_6$H$_4$CH$_3$-4)$_2$ (11)	2.1:1; 4.5 h reflux in THF	[9]

References on pp. 149/50

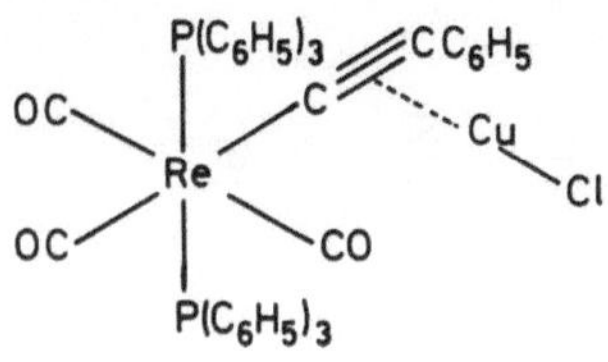

P(C₆H₅)₃, OC, Re, CO, I

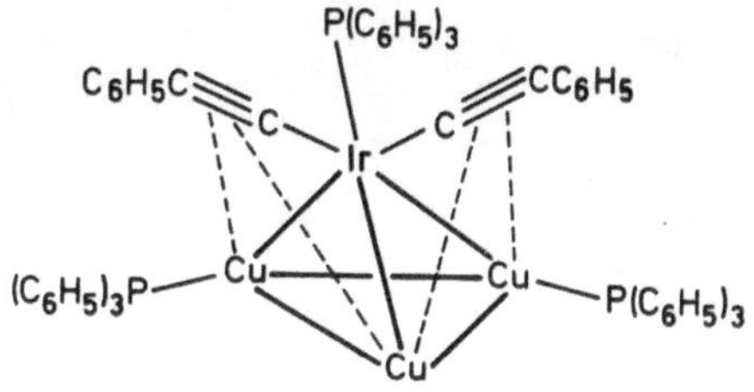

II

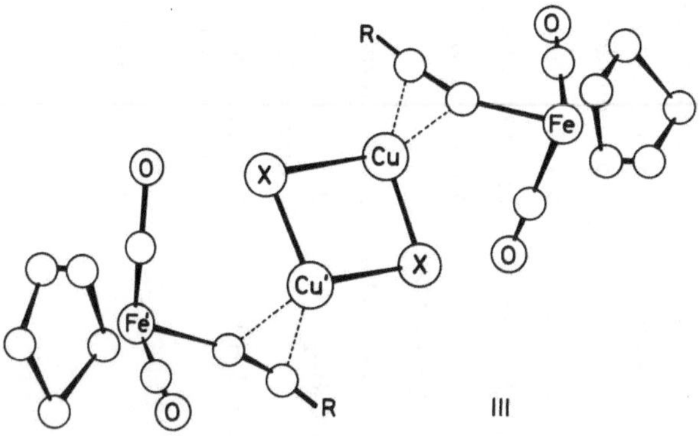

III

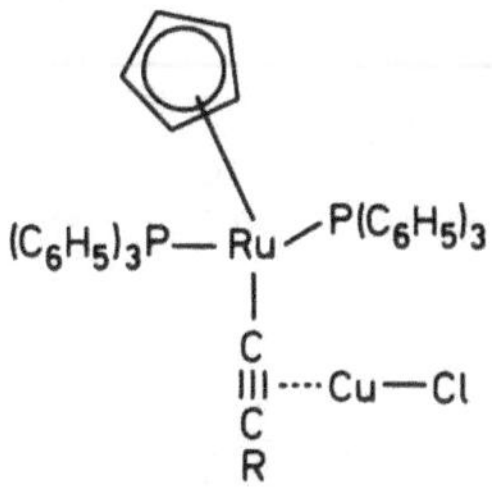

IV

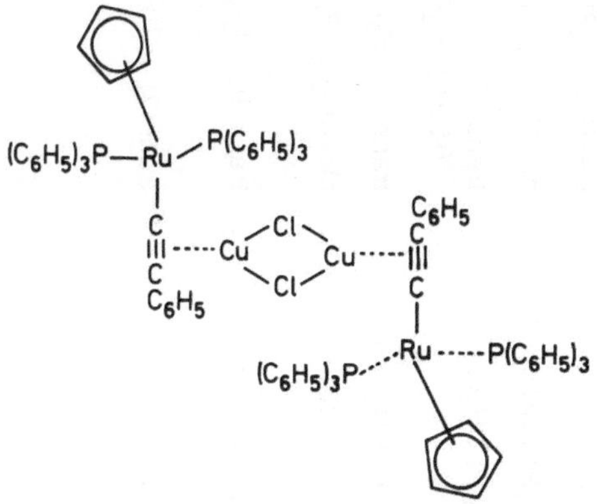

V

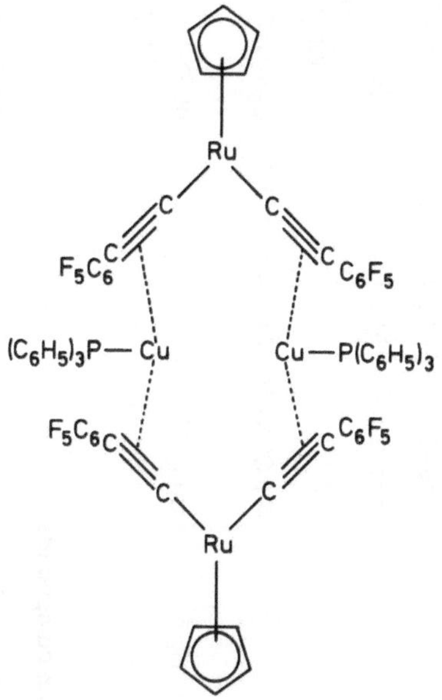

VI

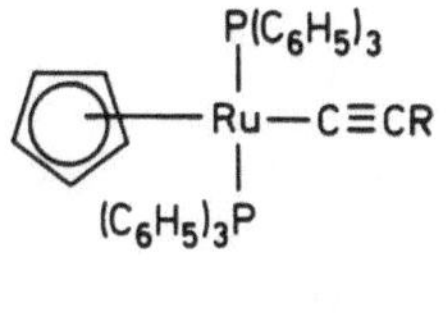

VII

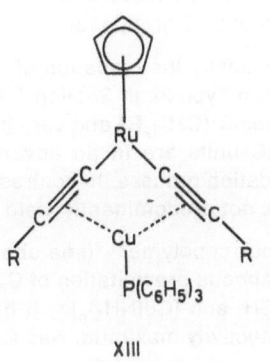

VIII : R^1 = CH$_3$

IX : R^1 = C$_6$H$_5$

X

XI

XII

XIII

References:

[1] A. M. Sladkov, L. Yu. Ukhin, Zh. I. Orlova (Izv. Akad. Nauk SSSR Ser. Khim **1968** 2586/90; Bull. Acad. Sci. USSR Div. Chem. Sci. **1968** 2446/9).

[2] L. I. Zakharkin, V. N. Kalinin, I. R. Gol'ding, A. M. Sladkov, A. V. Grebennikov (Zh. Obshch. Khim. **41** [1971] 823/8; J. Gen. Chem. [USSR] **41** [1971] 830/3).

[3] O. M. Abu Salah, M. I. Bruce, M. R. Churchill, S. A. Bezman (J. Chem. Soc. Chem. Commun. **1972** 858/9).

[4] M. I. Bruce, R. Clarke, J. Howard, P. Woodward (J. Organometal. Chem. **42** [1972] C107/ C109).

[5] O. M. Abu Salah, M. I. Bruce (J. Chem. Soc. Dalton Trans. **1974** 2302/4).

[6] O. M. Abu Salah, M. I. Bruce, A. D. Redhouse (J. Chem. Soc. Chem. Commun. **1974** 855).

[7] M. I. Bruce, O. M. Abu Salah, R. E. Davis, N. V. Raghavan (J. Organometal. Chem. **64** [1974] C48/C50).

[8] O. M. Abu Salah, I. M. Bruce (J. Chem. Soc. Dalton Trans. **1975** 2311/5).

[9] O. M. Abu Salah, M. I. Bruce (Australian J. Chem. **29** [1975] 73/7).

[10] O. M. Abu Salah, M. I. Bruce (Australian J. Chem. **29** [1975] 531/41).

[11] V. N. Narasimhachari (Diss. Univ. Texas 1975, pp. 1/215; Diss. Abstr. Intern. B **36** [1975]
       745; C.A. **83** [1975] No. 171198).
[12] K. Sonogashira, Y. Fujikura, T. Yatake, N. Toyoshima, S. Takahashi, N. Hagihara (J. Or-
       ganometal. Chem. **145** [1978] 101/8).
[13] I. A. Garbuzova, I. R. Gol'ding, N. N. Travkin, V. T. Aleksanyan, A. M. Sladkov (Koord.
       Khim. **5** [1979] 345/50; Soviet J. Coord. Chem. **5** [1979] 261/5; C.A. **90** [1979] No. 187079).
[14] A. T. Hutton, P. G. Pringle, B. L. Shaw (Organometallics **2** [1983] 1889/91).

1.1.2.2 "Mixed Acetylides" of the Type $R_n^1 R_{1-n}^2 C{\equiv}CCu$

The simultaneous reaction of two different alkynes (usually 1:1) with Cu^I compounds
in aqueous-ammoniacal solutions (Preparation Method I a in Section 1.1.2.1.1, see pp. 2/3)
yields substances which are called "mixed copper acetylides". Usually the alkyne mixture
is stirred into the copper solution. Though it is ostensibly an $R^1C{\equiv}CCu/R^2C{\equiv}CCu$ mixture,
their UV maxima may differ considerably from those of the individual acetylides. This is
the reason why they are usually treated like compounds. They are useful because of their
photoelectric properties. From joint analysis of the photoconductivity spectra and of the
$C{\equiv}C$ stretching frequencies a "block copolymer structure" can be attributed to these "mixed
acetylides". This structure involves intermolecular coordinations of Cu to $C{\equiv}C$, not a polymerization of $C{\equiv}C$ groups [3].

Consequently the oxidation of a "mixed acetylide" $R^1C{\equiv}CCu/R^2C{\equiv}CCu$ with $K_3[Fe(CN)_6]$
(cf. Reaction Type 2c in Section 1.1.2.1.1, p. 12) yields predominantly asymmetrically substi-
tuted diynes $R^1(C{\equiv}C)_2R^2$ and very little $R^1(C{\equiv}C)_2R^1$ and $R^2(C{\equiv}C)_2R^2$ [1]. Apparently the $R^1C{\equiv}C$
and $R^2C{\equiv}C$ units are in an advantageous position in this "block copolymer" in respect
to the oxidation process. In contrast, even a very thorough mechanical mixture of two acetylides does not predominantly yield $R^1(C{\equiv}C)_2R^2$.

A "block copolymer" (see above) resembling the "mixed acetylides" is also obtained
by simultaneous precipitation of $C_6H_5C{\equiv}CCu$ and C_6H_5SCu in aqueous solution from $C_6H_5C{\equiv}$
CH, C_6H_5SH, and $[Cu(NH_3)_n]^+$. It has a photoconductivity like the "mixed acetylides". The
photoconductivity maximum was found at 465 nm; the forbidden band width is 2.42 eV [6].

$NCC{\equiv}CCu/HOCH_2CH{=}CHC{\equiv}CCu$ is oxidized by $K_3[Fe(CN)_6]$ to give $HOCH_2CH{=}CH(C{\equiv}$
$C)_2CN$ [2].

$C_6H_5C{\equiv}CCu/2{-}O_2NC_6H_4C{\equiv}CCu$, oxidation in dilute aqueous $KOH/K_3[Fe(CN)_6]$ to give
$2{-}O_2NC_6H_4(C{\equiv}C)_2C_6H_5$ [1].

$CH_2{=}CHC{\equiv}CCu/C_6H_5C{\equiv}CCu$ [3]. The $C{\equiv}C$ band of this "mixed acetylide" was found at
1935 cm^{-1} (KBr) [4], and the photoconductivity maximum at $\sim$520 nm [4, 6]. A discrepancy
between the forbidden band widths from the photoconductivity data ($1.92{\pm}0.04$ eV [4] vs.
2.24 eV [6]) may possibly be attributed to a confusion of the data of the "mixed acetylides"
$CH_2{=}CHC{\equiv}CCu/C_6H_5C{\equiv}CCu$ and $Cu_2C_2/C_6H_5C{\equiv}CCu$ in [4].

$Cu_2C_2/C_6H_5C{\equiv}CCu$ [3]. The (molar?) ratio of Cu_2C_2 and $C_6H_5C{\equiv}CCu$ was given as 1:2.16.
In KBr, $C{\equiv}C$ bands at 1938 cm^{-1}, 1966 cm^{-1}, and 2037 cm^{-1} were measured [4]. The photoconductivity maximum was found at $\sim$620 nm [4, 6], and a UV maximum at 540 nm [6].
From the UV data a forbidden band width of 1.88 eV results. Again, the values from photoconductivity data differ (see above): 1.9 [5], 1.92 [6], $2.24{\pm}0.04$ eV [4]. The laser-induced photo-
decomposition threshold is three orders of magnitude higher than the photoconductivity
threshold; it is 16 kW · cm^{-2} under free-oscillation conditions and 0.5 MW · cm^{-2} under Q-
switching conditions [5].

References:

[1] A. Baeyer, L. Landsberg (Ber. Deut. Chem. Ges. **15** [1882] 57/61).
[2] P. J. Ashworth, E. R. H. Jones, G. H. Mansfield, K. Schlögl, J. M. Thompson, M. C. Whiting (J. Chem. Soc. **1958** 950/4).
[3] V. S. Myl'nikov, A. N. Dune, I. R. Gol'ding, A. M. Sladkov (Tr. 5th Mezhdunar. Konf. Organometallov. Soedin., Moscow 1972, Vol. 1, pp. 468/9; Summaries 5th Organometal. Compounds, Moscow 1972, pp. 470/1, Ref. No. 180).
[4] V. S. Myl'nikov, A. N. Dun'e, I. R. Gol'ding, A. M. Sladkov (Zh. Obshch. Khim. **42** [1972] 2543/6; J. Gen. Chem. [USSR] **42** [1972] 2532/5).
[5] V. S. Myl'nikov, A. A. Kharchenko, M. M. Sobolev (Kvantovaya Elektronika [Moscow] **3** [1976] 288/92; Soviet J. Quant. Electron. **6** [1976] 151/3).
[6] V. S. Myl'nikov (Zh. Nauchn. Prikl. Fotogr. Kinematogr. **24** [1979] 25/8; C.A. **91** [1979] No. 66221).

1.1.2.3 Complexes of RC≡CCu with Alkali and Mg Halides

A 1:1 complex of $CH_3OC(CH_3)_2C{\equiv}CCu$ and $LiSn(CH_3)_3$ is described in Section 1.1.2.4. "Mixed homocuprates" of the type [RC≡CCuR']M (M = Li, MgCl, MgBr) have been described in "Organocopper Compounds" 2, 1983, pp. 174/211 (M = Li) and 227/35 (M = MgCl or MgBr), as R' groups such as alkyl, alkenyl, or aryl generally are more reactive than the alkynyl groups RC≡C.

The generally highly insoluble, mostly deep red or brown RC≡CCu compounds can be dissolved in solvents such as ether, CH_3CN, tetrahydrofuran, or hexamethylphosphoric triamide in the presence of Li, Na, or Mg halides, and this even at temperatures of $-40\,°C$. The resulting yellow or colorless solutions obviously contain complexes of RC≡CCu with the halides and are much more reactive than solutions of RC≡CCu, which can be obtained only for certain R and certain solvents; see Section 1.1.2.1.1. It is not clear whether all RC≡CCu compounds form reagents of this type with LiI. $(CH_3)_3CC{\equiv}CCu$ could be an exception, see No. 12 in Table 21 [8]. In the case of NaCl, NaI, and LiF the complex reagent formation is very doubtful.

Obviously the type of the solvents is of high importance for the formation of the reagents. As their structure and, in some cases, even their stoichiometry are not known, they are formulated in this section as RC≡CCu,MX (cf. the remarks in "Organocopper Compounds" 1, pp. 4 and 6/7).

The reagents given in Table 21 are prepared by the following methods:

Method I: RC≡CCu + MX → RC≡CCu,MX.
> The reaction is carried out at $-40\,°C$ to $+20\,°C$. The end point of the addition can easily be observed, for the insoluble red acetylide is then consumed.

Method II: RC≡CM + CuX → RC≡CCu,MX.

References on p. 156

Table 21
Complexes RC≡CCu,MX (M = Li, Na, MgX; X = halogen).
For abbreviations and dimensions see p. X.

No.	reagent	method of preparation (solvent) properties, reactions, and further remarks	Ref.

lithium reagents

No.	reagent	method of preparation (solvent) / properties, reactions, and further remarks	Ref.
1	ClC≡CCu,LiI	II (CH$_3$CN/ether, OP(N(CH$_3$)$_2$)$_3$) no information about a possible halogen inter-change given yields 84% ClC≡CCH$_2$CH=CH$_2$ with CH$_2$=CHCH$_2$Br for reactions with R^{1}COCl see Table 22, p. 155	[5, 6]
2	CH$_3$C≡CCu,LiBr	I (CH$_3$CN) gives 7-(prop-1-inyl)cyclohepta-1,3,5-triene with tropylium salts in CH$_3$CN at 25°, isolated after precipitation of Cu$^+$ by Na$_2$S	[3]
3	C$_2$H$_5$OC≡CCu,LiI	II (CH$_3$CN/ether, ether) gives 55% C$_2$H$_5$OC≡CCH$_2$CH=CH$_2$ with CH$_2$=CHCH$_2$Br for reaction with n-C$_4$H$_9$COCl see Table 22, p. 155	[5, 6]
4	CH$_2$=C(CH$_3$)C≡CCu,LiBr	II (C$_6$H$_{14}$/(CH$_3$)$_2$SO/OP(N(CH$_3$)$_2$)$_3$) gives CH$_2$=C(CH$_3$)C≡CCH$_2$C≡COCH$_3$ with CH$_2$=C=C(I)OCH$_3$	[19]
5	cyclo-C$_3$H$_5$C≡CCu,LiBr	I (CH$_3$CN) reacts analogously to No. 2 (55% yield)	[3]
6	n-C$_3$H$_7$C≡CCu,LiI	II (ether) gives about 65% [n-C$_3$H$_7$C≡CCuCR^1=CH$_2$]Li with CH$_2$=CR^{1}Li in ether at 15 to 20°; R^1 = CH$_3$, CH(OC$_2$H$_5$)$_2$, Si(CH$_3$)$_3$	[17]
7	(CH$_3$)$_2$NCH$_2$C≡CCu,LiI	II (ether) gives [(CH$_3$)$_2$NCH$_2$C≡CCuR^1]Li with R^{1}Li in ether at 0°; R^1 = CH$_3$, CH$_2$=CH	[18]
8	(CH$_3$)$_3$SiC≡CCu,LiBr	II (C$_6$H$_{14}$/THF/OP(N(CH$_3$)$_2$)$_3$) gives (CH$_3$)$_3$SiC≡CCH$_2$C≡COCH$_3$ with CH$_2$=C=C(I)OCH$_3$	[19]
9	(CH$_3$)$_3$SiC≡CCu,LiI	II (THF) orange-red gives [(CH$_3$)$_3$SiC≡CCuR^1]MgCl with R^{1}MgCl in ether or ether/THF	[14, 20]
10	n-C$_4$H$_9$C≡CCu,LiBr	II (C$_6$H$_{14}$/(CH$_3$)$_2$SO/OP(N(CH$_3$)$_2$)$_3$) gives n-C$_4$H$_9$C≡CCH$_2$C≡COCH$_3$ with CH$_2$=C=C(I)OCH$_3$	[19]

References on p. 156

Table 21 [continued]

No.	reagent	method of preparation (solvent) properties, reagents, and further remarks	Ref.
11	n-$C_4H_9C\equiv CCu$,LiI	I (ether, $OP(N(CH_3)_2)_3$), II (ether) also from [n-$C_4H_9C\equiv CCuR^1$]Li · LiI and $CH_2=CHCH_2Br$; $R^1=CH_3$, n-C_4H_9, n-C_7H_{15} canary yellow solution in ether in part unclear whether No. 11 or n-$C_4H_9C\equiv CCu$ is reacted in ether at 0° equilibrium No. 11 + R^1Li $\rightleftharpoons$ [n-$C_4H_9C\equiv CCuR^1$]Li (see "Organocopper Compounds" 2, Section 1.1.1.2.4); $R^1=CH_3$, n-C_4H_9, n-C_7H_{15} for reactions with R^1COCl see Table 22, p. 155 gives 70% n-$C_4H_9C\equiv CCOC_6H_5$ with C_6H_5COCN in ether	[1, 6, 7, 11, 16]
12	$(CH_3)_3CC\equiv CCu$,LiI	II (THF, ether) orange solution in ether, at −49° crystallizes $(CH_3)_3CC\equiv CCu$ gives 65% $(CH_3)_3C(C\equiv C)_2C(CH_3)_3$ with BrCN the reaction with R^1Li in ether or THF gives $[(CH_3)_3CC\equiv CCuR^1]$Li; $R^1=CH_3$, $CH_2=CH$, $CH_2=C(CH_3)$, $(CH_3)_3SiC(=CH_2)$, $(C_2H_5O)_2CHC(=CH_2)$	[8, 12, 13, 15, 17]
13	$C_5H_5NC\equiv CCu$,LiI ($C_5H_5N=$pyrid-2-yl)	II (CH_3CN/ether) gives an unidentifiable product with BrCN	[15]
14	n-$C_5H_{11}C\equiv CCu$,LiBr	I ($OP(N(CH_3)_2)_3$) gives 83% n-$C_5H_{11}C\equiv CCH_2CH=CH_2$ with $CH_2=CHCH_2Br$	[2]
15	n-$C_5H_{11}C\equiv CCu$,LiCl	I (dimethylformamide, ether, $OP(N(CH_3)_2)_3$) gives 89% n-$C_5H_{11}C\equiv CCH_2CH=CH_2$ with $CH_2=CHCH_2Br$ for reaction with CH_3COCl see Table 22, p. 155	[1, 2]
16	n-$C_5H_{11}C\equiv CCu$,LiF	I ($OP(N(CH_3)_2)_3$) gives 45% n-$C_5H_{11}C\equiv CCH_2CH=CH_2$ with $CH_2=CHCH_2Br$	[2]
17	n-$C_5H_{11}C\equiv CCu$,LiI	I (ether, $OP(N(CH_3)_2)_3$) gives 83% n-$C_5H_{11}C\equiv CCH_2CH=CH_2$ with $CH_2=CHCH_2Br$ for reaction with CH_3COCl see Table 22, p. 155	[1, 2]
18	4-$O_2NC_6H_4C\equiv CCu$,LiCl	I (THF) gives $P(C\equiv CC_6H_4NO_2-4)_3$ with PCl_3 in THF and 4-$XC_6H_4P(C\equiv CC_6H_4NO_2-4)_2$ with 4-$XC_6H_4PCl_2$; X=Cl, H, CH_3O, $(CH_3)_2N$	[9]

References on p. 156

Table 21 [continued]

No.	reagent	method of preparation (solvent) properties, reagents, and further remarks	Ref.
19	$4-O_2NC_6H_4C{\equiv}CCu,LiBr$	I (THF) reacts with $4-XC_6H_4PCl_2$ as No. 18; the 15% yield with $X=H$ is enhanced to 35%, if one more mole LiBr is added	[9]
20	$C_6H_5C{\equiv}CCu,LiBr$	I (THF) gives 64% $C_6H_5P(C{\equiv}CC_6H_5)_2$ with $C_6H_5PCl_2$ in THF	[9]
21	$C_6H_5C{\equiv}CCu,LiI$	I (ether, ether/$OP(N(CH_3)_2)_3$), II (ether) in part unclear whether No. 21 or $C_6H_5C{\equiv}CCu$ is reacted gives 60% $C_6H_5C{\equiv}CCN$ and 26% $C_6H_5(C{\equiv}C)_2C_6H_5$ with BrCN, and $C_6H_5C{\equiv}CCH_3$ with $CuCH_3$ at 0° in ether gives 50% $C_6H_5C{\equiv}CC(C_6H_5){=}NC_6H_5$ with $C_6H_5C(Cl){=}NC_6H_5$ in ether at 25° reactions with R^1COCl see Table 22, p. 155	[1, 4, 6, 10, 15, 16]
22	$(C_6H_5)_2NC{\equiv}CCu,LiI$	II (CH_3CN, ether) gives 20% $(C_6H_5)_2NC{\equiv}CCH_2CH{=}CH_2$ with $CH_2{=}CHCH_2Br$	[5]
23	$n-C_{12}H_{25}C{\equiv}CCu,LiI$	II (ether) gives $n-C_{12}H_{25}C{\equiv}CCH_3$ with $CuCH_3$ in ether at 0°	[10]

sodium reagents

No.	reagent	method of preparation (solvent) properties, reagents, and further remarks	Ref.
24	$C_2H_5OCH_2C{\equiv}CCu,NaI$	II (THF, $OP(N(CH_3)_2)_3$) gives 25% $C_2H_5OCH_2C{\equiv}CCH_2CH{=}CH_2$ with $CH_2{=}CHCH_2Br$	[5]
25	$n-C_4H_9C{\equiv}CCu,NaCl$	II (THF, $OP(N(CH_3)_2)_3$) different halogen anions from added salts present, therefore mixtures possible gives 54% $n-C_4H_9C{\equiv}CCH_2CH{=}CH_2$ with $CH_2{=}CHCH_2Br$	[5]
26	$n-C_4H_9C{\equiv}CCu,NaI$	II (THF) different halogen anions from added salts present, therefore mixtures possible gives 75% $n-C_4H_9C{\equiv}CCH_2CH{=}CH_2$ with $CH_2{=}CHCH_2Br$	[5]

magnesium reagents

No.	reagent	method of preparation (solvent) properties, reagents, and further remarks	Ref.
27	$n-C_4H_9C{\equiv}CCu,MgBr_2$	II (ether, CH_3CN/ether) for reaction with $n-C_4H_9COCl$ see Table 22, p. 155	[6]
28	$n-C_5H_{11}C{\equiv}CCu,MgCl_2$	I ($OP(N(CH_3)_2)_3$) gives 41% $n-C_5H_{11}C{\equiv}CCH_2CH{=}CH_2$ with $CH_2{=}CHCH_2Br$	[2]
29	$n-C_5H_{11}C{\equiv}CCu,MgBr_2$	I ($OP(N(CH_3)_2)_3$) gives 73% $n-C_5H_{11}C{\equiv}CCH_2CH{=}CH_2$ with $CH_2{=}CHCH_2Br$	[2]

References on p. 156

The reaction $RC \equiv CCu,MX + R^1COCl \rightarrow RC \equiv CCOR^1 + CuCl + MX$ is a very useful method to prepare ynones; see Table 22. With $RC \equiv CCu,LiI$ in ether the yield depends on the order of the addition of reactant and solvent. Adding an acid halide first and hexamethylphosphoric triamide (tetramethylethylenediamine or triethylamine) afterwards gives higher yields than the reverse order [1].

Table 22
Preparation of Ynones according to $RC \equiv CCu,MX + R^1COCl \rightarrow RC \equiv CCOR^1 + MX + CuCl$ (M = Li or MgX; X = halogen).
Typically the reaction is carried out in ether or ether/hexamethylphosphoric triamide at 20 °C for 1 to 15 h.

$RC \equiv CCu,MX$	R^1	remarks (yield of $RC \equiv CCOR^1$ in %)	Ref.
$ClC \equiv CCu,LiI$	CH_3	(<5)	[6]
	$n-C_4H_9$	high yield, the product polymerizes easily	[6]
$C_2H_5OC \equiv CCu,LiI$	$n-C_4H_9$	(50)	[6]
$n-C_4H_9C \equiv CCu,LiI$	CCl_3	(36)	[6]
	$ClCH_2$	(9)	[6]
	CH_3	in ether (25 to 53), diglyme (10), hexamethyl-phosphoric triamide (2), or in ether followed by hexamethylphosphoric triamide (73) no reaction in $N(C_2H_5)_3$	[1, 6]
	$CH_2=CH$	(80)	[6]
	$ClCO(CH_2)_2$	gives $n-C_4H_9C \equiv CCO(CH_2)_2COC \equiv CC_4H_9-n$ (10)	[6]
	$CH_3CH=CH$	(80)	[6]
	$C_2H_5OCH_2$	(73)	[6]
	$n-C_4H_9$	(53 to 96)	[1, 6]
	$ClCO(CH_2)_4$	gives $n-C_4H_9C \equiv CCO(CH_2)_4COC \equiv CC_4H_9-n$ (26)	[6]
	C_6H_5	(86)	[6]
	$C_6H_5C \equiv C$	(40)	[6]
$n-C_4H_9C \equiv CCu,MgBr_2$	$n-C_4H_9$	(36)	[6]
$n-C_5H_{11}C \equiv CCu,LiCl$	CH_3	(65)	[1]
$n-C_5H_{11}C \equiv CCu,LiI$	CH_3	(82)	[1]
$C_6H_5C \equiv CCu,LiI$	CH_3	(82)	[1, 6]
	C_6H_5	(90)	[1, 6]
	$C_6H_5C \equiv C$	(50)	[6]

References on p. 156

References:

[1] J. F. Normant, M. Bourgain (Tetrahedron Letters **1970** 2659/62).
[2] J. F. Normant, M. Bourgain, A.-M. Rone (Compt. Rend. C **270** [1970] 354/7).
[3] C. E. Hudson, N. L. Bauld (J. Am. Chem. Soc. **94** [1972] 1158/63).
[4] J. F. Normant (Synthesis **1972** 63/80, 70/80).
[5] M. Bourgain, J. F. Normant (Bull. Soc. Chim. France **1973** 1777/9).
[6] M. Bourgain, J. F. Normant (Bull. Soc. Chim. France **1973** 2137/42).
[7] M. Bourgain, J. Villieras, J. F. Normant (Compt. Rend. C **276** [1973] 1477/80).
[8] H. O. House, M. Umen (J. Org. Chem. **38** [1973] 3893/901).
[9] B. I. Stepanov, L. I. Chekunina, A. I. Bokanov (Zh. Org. Khim. **43** [1973] 2648/54; J. Org. Chem. [USSR] **43** [1973] 2627/31).
[10] R. Levene, J. Y. Becker, J. Klein (J. Organometal. Chem. **67** [1974] 467/71).

[11] J. P. Marino, D. M. Floyd (J. Am. Chem. Soc. **96** [1974] 7138/40).
[12] H. O. House, C.-Y. Chu, J. M. Wilkins, M. J. Umen (J. Org. Chem. **40** [1975] 1460/9).
[13] J. P. Marino, J. S. Farino (Tetrahedron Letters **1975** 3901/4).
[14] M. W. Logue, G. L. Moore (J. Org. Chem. **40** [1975] 131/2).
[15] P.-L. Compagnon, B. Grosjean (Synthesis **1976** 448/9).
[16] R. Robin (Compt. Rend. C **282** [1976] 281/2).
[17] R. K. Boeckman, M. Ramaiah (J. Org. Chem. **42** [1977] 1581/6).
[18] D. B. Ledlie, G. Miller (J. Org. Chem. **44** [1979] 1006/7).
[19] J. M. Oostveen, H. Westmijze, P. Vermeer (J. Org. Chem. **45** [1980] 1158/60).
[20] J. Drouin, G. Rousseau (J. Organometal. Chem. **289** [1985] 223/30).

1.1.2.4   Complexes of RC≡CCu with Compounds of Further Main Group Elements

The polymeric structure of RC≡CCu is broken down in suitable solvents by S, N, P, As, and Sb donor compounds by complex formation. Most of the complexes have been isolated as crystalline solids, others are not as well characterized or even exist only in solution. With stibines, no complex has been isolated. In most cases only the stoichiometry of the complexes is known, but not their aggregation. Only in the cases of $C_6H_5C≡CCu \cdot P(CH_3)_3$ and $C_6H_5C≡CCu \cdot P(C_6H_5)_3$ (Nos. 23 and 32) have tetrameric structures been proved by X-ray analysis. For No. 32, however, insoluble, probably polymeric isomers exist too (structures unknown). For other 1:1 complexes, tetrameric solid state structures have also been assumed. Molecular weight determinations generally show lower aggregation in solution dependent on the type of acetylide, ligand, solvent, and sometimes the concentration. For complexes with a ligand:Cu ratio >1, generally lower aggregation is assumed than for 1:1 complexes.

All cases where complexes are proved or can be expected are listed in Table 23. Isolated compounds with known stoichiometry are formulated with a point (like $C_6H_5C≡CCu \cdot NH_3$); a probable complex formation with unknown stoichiometry is symbolized by a comma (like $n\text{-}C_4H_9C≡CCu,PCl_3$) as explained in "Organocopper Compounds" 1, 1985, p. 6. Supposed complex formation with S-, N-, and P-donor compounds is discussed in the following.

Sulfur Complexes. The reaction of $(CH_3)_3SiC≡CLi$ with $CuBr \cdot S(CH_3)_2$ in ether at $-20\ °C$ has been said to give $(CH_3)_3SiC≡CCu$ (see Section 1.1.2.1.1) [51]. The same authors formulate the structure $(CH_3)_3SiC≡CCuS(CH_3)_2$ in a further publication without any explanation [50].

Amine Complexes. The compounds $C_6H_5C≡CCu$, $4\text{-}CH_3C_6H_4C≡CCu$, and $4\text{-}ClC_6H_4C≡CCu$ are soluble in cyclic amines like pyrrolidine, piperidine, azepane, and morpholine. The IR frequency of the C≡C bond is found in the region of 2030 cm^{-1}, which led these authors to claim the formation of dissolved amine complexes of RC≡CCu although an isolation was not possible [21]. On the other hand, no complex has been obtained from C_6H_5C≡CCu and 4,4'-bipyridine in toluene or ether [13]. C_6H_5C≡CCu, electrochemically formed in CH_3CN or CH_3COCH_3 containing $[N(C_2H_5)_4]ClO_4$, gives a dark red solution with 2,2'-bipyridine [49], whereas 2,2'-bipyridine and C_6H_5C≡CCu prepared as usual give no product in toluene or ether [13]. Solutions of RC≡CCu in pyridine are supposed to contain at least traces of a RC≡CCu-pyridine complex [19]. From solubility phenomena the formation of similar complexes with dimethylformamide and hexamethylphosphoric triamide has been supposed, but no reliable work is done about this question. The in situ preparation of RC≡CCu from RC≡CH, activated Cu, and K_2CO_3 in pyridine is believed to be favored by a pyridine complex formation [15]. C_6H_5C≡CCu can be extracted by $HN(C_2H_5)_2$, to give a coarse crystalline product, no complex formation with the amine occurs [8]. For better characterized complexes see Table 23.

Phosphorus Complexes are in general well characterized and crystallized compounds. Often the ratio RC≡CCu to PR_3^1 varies and sometimes it depends on the P:Cu ratio in the reaction batch.

Complexes with P:Cu ratios of 4:1 seem to exist too, although no solid compounds have been isolated from the corresponding solutions. The IR spectra of solutions of C_6H_5C≡CCu or CH_2=$C(CH_3)C$≡CCu and $P(CH_3)_3$ in benzene at P to Cu ratios of 1, 2, 3 or 4 show similar bands. The complexes C_6H_5C≡CCu · $nP(CH_3)_3$ (n = 1 to 3) are known as pure substances; see Table 23 [5]. From C_6H_5C≡CCu and $CuCl$ · $P(OC_2H_5)_3$ no products could be obtained in organic solvents, even after heating [10]. For better characterized complexes see Table 23.

[$CH_3OC(CH_3)_2C$≡CCuSn($CH_3)_3$]Li. This 1:1 complex of $CH_3OC(CH_3)_2C$≡CCu and $LiSn(CH_3)_3$ is formulated in its ionic form, but no structural proof is given. It is prepared by addition of solid $CH_3OC(CH_3)_2C$≡CCu to a solution of $LiSn(CH_3)_3$ in tetrahydrofuran at −48 °C [35]. The reaction of this solution (4 h at −48 °C) with R^1C≡$CCO_2C_2H_5$ yields (E)- and (Z)-$(CH_3)_3SnCR^1$=$CHCO_2C_2H_5$. With R^1 = CH_3 (yield 82%), only the (E)-compound is formed [35, 38], with R^1 = $(CH_3)_3C$, the (E):(Z) ratio is 8:62 (yield 76%), and with R^1 = $(CH_3)_3C$-$Si(CH_3)_2OCH_2$ this ratio is about 96:4 (total yield 82%) [35]. The reaction with R^1C≡CCO_2R^2 in tetrahydrofuran (30 min at −48 °C and 3 h at 0 °C) yields (E)-$(CH_3)_3SnCR^1$=$C(CO_2R^2)$-$Sn(CH_3)_3$. The following compounds were reacted (R^1, R^2, yield): CH_3, C_2H_5, 74%; C_2H_5, C_2H_5, 76%; $(CH_3)_2CH$, CH_3, 74%; n-C_6H_{13}, CH_3, 86%; cyclo-C_3H_5, CH_3, 73%; 2-(cyclopent-1-enyl)ethyl, CH_3, 71%; cyclohex-2-enylmethyl, CH_3, 82%; $(CH_3)_3CSi(CH_3)_2OCH_2$, C_2H_5, 70% [38].

The reaction of [$CH_3OC(CH_3)_2C$≡CCuSn($CH_3)_3$]Li and R^1OCH_2CH=$CHCH_2Br$ at −30 °C in tetrahydrofuran gives mixtures of $R^1OCH_2CH(CH$=$CH_2)Sn(CH_3)_3$ and R^1OCH_2CH=$CHCH_2Sn(CH_3)_3$. The configuration at the double bond is not changed during reaction (isomeric purity >99%). The following R^1OCH_2CH=$CHCH_2Br$ were reacted (R^1, E or Z, yield of $R^1OCH_2CH(CH$=$CH_2)Sn(CH_3)_3$/yield of R^1OCH_2CH=$CHCH_2Sn(CH_3)_3$): C_2H_5, Z, 8%/39%; C_2H_5, E, 47%/25%; i-C_3H_7, Z, 31%/39%; $C_6H_5CH_2$, Z, 4%/38%; $C_6H_5CH_2$, E, 16%/8%; and $C_6H_5CH(CH_3)$, Z, 8%/68% [42].

Explanation for Table 23. The compounds are in most cases prepared by the following methods:

Method I: Formation of the complexes mRC≡CCu · nD from RC≡CCu and the ligand D. The type of the solvent or the absence of any solvent is indicated in parentheses because it is very important in this formation reaction.

References on pp. 174/5

Method II: Formation of the complexes by ligand exchange, for instance

$$RC \equiv CCu \cdot D^1 \xrightarrow{D^2} RC \equiv CCu \cdot D^2 \text{ or}$$

$$RC \equiv CCu \cdot D^1 \cdot D^2 \xrightarrow{D^3} RC \equiv CCu \cdot D^1 \cdot D^3$$

or by addition of a second ligand, for instance

$$RC \equiv CCu \cdot D^1 \xrightarrow{D^2} RC \equiv CCu \cdot D^1 \cdot D^2$$

Method III: Formation of the complexes $mRC \equiv CCu \cdot nD$ from $RC \equiv CM$ and mixtures or complexes of CuX and D, where $M = H$, Li, K and $X = Cl$, I, N_3, $C_5(CH_3)_5$. If azide complexes like $P(C_6H_5)_3 \cdot C_{12}H_8N_2 \cdot CuN_3$ are reacted with very acid acetylenes, N_2 is liberated.

Table 23
Complexes of RC≡CCu with N, P, and As Compounds

($C_{12}H_8N_2$ = 1,10-phenanthroline, $C_{16}H_{16}N_2$ = 3,4,7,8-tetramethyl-1,10-phenanthroline).
Further information on compound numbers preceded by an asterisk is given at the end of the table on pp. 170/4.
For abbreviations and dimensions see p. X.

No.	complex	method of preparation (solvent, yield), properties, remarks and reactions	Ref.
complexes with N compounds			
1	1,7-$C_2B_{10}H_{11}$C≡CCu · NH_3 (1,7-$C_2B_{10}H_{11}$ = 1,7-dicarba-closo-dodecaboran(12)-1-yl)	I (CH_3OH/H_2O) colorless, insoluble in hexane and in CH_3OH/25% aqueous NH_3 (1:1) gives 1,7-$C_2B_{10}H_{11}$C≡CCu after 16 h at 1 Torr and 20°	[39]
2	2$C_2H_5O_2$CC≡CCu · $C_{12}H_8N_2$	III (benzene/ethanol): from excess $C_2H_5O_2$CC≡CH and CuN_3 · $P(C_6H_5)_3$ · $C_{12}H_8N_2$ at 25°/15 min pale yellow, m.p. 158° (dec.) sparingly soluble in benzene and hexane IR (Nujol): 1685 (d, C=O), 1935 + 2100 (C≡C)	[44]
3	2$C_2H_5O_2$CC≡CCu · $C_{16}H_{16}N_2$	III (benzene/ethanol): from excess $C_2H_5O_2$CC≡CH and CuN_3 · $P(C_6H_5)_3$ · $C_{16}H_{16}N_2$ at 25°/15 min white, m.p. 160° (dec.) sparingly soluble in benzene and hexane IR (Nujol): 1675 (C=O), 1920 + 2090 (C≡C)	[44]
4	$C_2H_5O_2$CC≡CCu · $C_{16}H_{16}N_2$ · 0.25H_2O	II: from No. 9, $C_{16}H_{16}N_2$ and $C_2H_5O_2$CC≡CH in wet benzene, more slowly in dry benzene; or from No. 54 and $C_2H_5O_2$CC≡CH in benzene salmon pink after washing with ether the H_2O content inferred only from the thermogravimetric analysis reacts easily with H_2, O_2, and CO_2 (no products given) gives with cyclo-C_6H_{11}NC in benzene cyclo-C_6H_{11}NCCuC≡CCO$_2C_2H_5$($C_{16}H_{16}N_2$) (see Section 1.1.4.1.3) gives with $P(C_6H_5)_3$ in benzene No. 54 IR (Nujol): 1660 (C=O), 2060 (C≡C)	[44]

References on pp. 174/5

Table 23 [continued]

No.	complex	method of preparation (solvent, yield), properties, remarks and reactions	Ref.
5	$2(CH_3)_2C(CN)CH_2C{\equiv}CCu \cdot NH_3$	III (M=H; liquid NH_3) dark brown, diamagnetic stable to air, sensitive against acids IR (solid): 248 (CuC), 320 (CC≡), 475 (CN), 664 (NH_3), 704 (CH_2), 857+918 (CC≡), 1093 (CH_3), 1255 (NH_3), 1414 (CH_2), 1600 (NH_3), 1934 (C≡C), 2205 (CN), 2845 (CH_2), 2920 (CH_3), 2957 (CH_2), 3155+3240+3318 (NH) differential thermal analysis: at 105° loss of NH_3 (endothermic), various exothermic peaks, decomposes at 200 to 250° to give a brown powder, at 500° in high vacuum formation of Cu (X-ray analysis)	[20]
6	$C_6H_5C{\equiv}CCu \cdot NH_3$	I, III (M=H or K), both in liquid NH_3; or from [($C_6H_5C{\equiv}C)_2Cu]Na \cdot 2NH_3$ and CuI in liquid NH_3 (80%) colorless flakes, sparingly soluble in liquid NH_3 gives $C_6H_5C{\equiv}CCu$ under N_2 at room temperature with CO in liquid NH_3 a soluble CO complex is supposed	[2 to 4]
*7	$2C_6H_5C{\equiv}CCu \cdot C_{12}H_8N_2$	I (in ether at 0°, in toluene at 60°) red-brown, becomes orange at 145°, decomposes above 173° IR (Nujol): 2040 (C≡C)	[13, 49]

complexes with P compounds

No.	complex	method of preparation (solvent, yield), properties, remarks and reactions	Ref.
8	$CH_3C{\equiv}CCu \cdot P(C_2H_5)_3$	I (toluene; unsuccessful without solvent) yellow needles from hexane very air-sensitive, dissociates readily, e.g., in contact with benzene	[3, 5, 6]
9	$2C_2H_5O_2CC{\equiv}CCu \cdot 3P(C_6H_5)_3$	III (benzene/ethanol): from $C_2H_5O_2CC{\equiv}CH$ and $\{CuN_3[P(C_6H_5)_3]_2\}_2$ in C_2H_5OH/benzene (5:1) under formation of N_2 and $P(C_6H_5)_3$ white, m.p. 155° sparingly soluble in C_2H_5OH/benzene (5:1) molecular weight (CH_2Cl_2) 843 (theoretical 1107), slightly dependent on concentration	[44]

References on pp. 174/5

No.	Compound	Description	Ref.
10	$CH_2=C(CH_3)C\equiv CCu \cdot P(CH_3)_3$	IR (Nujol): 1690 (d, C=O), 2050 (C≡C) gives in benzene with cyclo-C_6H_{11}NC [cyclo-C_6H_{11}NCCuC≡C$CO_2C_2H_5$P(C_6H_5)$_3$]$_2$ gives with $C_2H_5O_2$CC≡CH and $C_{16}H_{16}N_2$ No. 4 (faster in presence of a little H_2O), with $C_2H_5O_2$CC≡CH in benzene slowly, in CH_2Cl_2 rapidly No. 53, with $C_{16}H_{16}N_2$ in cyclohexane No. 54 gives with anhydrous HCl in benzene $C_2H_5O_2$CC≡CH and 2CuCl · 3P(C_6H_5)$_3$	[5, 6]
*11	$n\text{-}C_3H_7C\equiv CCu$, $P(C_4H_9\text{-}n)_3$	I (benzene) yellow, m.p. 82 to 82.5° degree of association cryoscopically measured in benzene is 3.3 to 3.4, in nitrobenzene 2.0 to 2.2 IR (KBr): 2031, 2065 (C≡C)	[27, 32]
*12	$n\text{-}C_3H_7C\equiv CCu$, $P(N(CH_3)_2)_3$	I (hexamethylphosphoric triamide) I (ether, ether/THF, THF) in [31] for a "hexamethyl phosphorous triamide complex of $n\text{-}C_3H_7$C≡CCu ... satisfactory elemental analyses were obtained", but no stoichiometry of the compound is given	[17, 25, 26, 29 to 31, 37, 40]
*13	$n\text{-}C_3H_7C\equiv CCu$, $OP(N(CH_3)_2)_3$	I (ether), obviously only obtained in solution formulated in [36] as a 1:2 compound without any proof	[27, 36]
14	$CH_2=CHCO_2CH_2C\equiv CCu$, $P(C_6H_5)_3$	I (dimethylformamide) IR: no C≡C band	[18]
15	$n\text{-}C_4H_9C\equiv CCu$, PCl_3	I (no solvent) IR: compared with $n\text{-}C_4H_9$C≡CCu, the C≡C frequency is increased by 40 cm^{-1}	[14]
16	$(CH_3)_3CC\equiv CCu \cdot P(CH_3)_3$	I (benzene) yellow-green, m.p. 105 to 106° degree of association cryoscopically measured in benzene is 2.6 to 2.8, in nitrobenzene 2.0 IR (KBr): 2050 (C≡C)	[5, 6]

Table 23 [continued]

No.	complex	method of preparation (solvent, yield), properties, remarks and reactions	Ref.
17	$(CH_3)_3CC{\equiv}CCu \cdot 2P(CH_3)_3$	I (benzene) colorless, m.p. 95 to 98° degree of association cryoscopically measured in benzene is 1.0, in nitrobenzene 0.4 to 0.9 IR (KBr, Nujol): no $C{\equiv}C$ band	[5, 6]
18	$(CH_3)_3CC{\equiv}CCu \cdot P(C_6H_5)_3$	III (methanol): from $(CH_3)_3CC{\equiv}CH$ and $CuCl \cdot 4P(C_6H_5)_3$ (molar ratio = 30:7), catalyzed by CH_3ONa (62%) yellow solid IR (Nujol): 2055 ($C{\equiv}C$)	[43]
19	$CH_2{=}C(CH_3)CO_2CH_2C{\equiv}CCu$, $P(C_6H_5)_3$	I (dimethylformamide) IR: no $C{\equiv}C$ band	[18]
20	$C_6F_5C{\equiv}CCu \cdot P(C_6H_5)_3$	III (methanol): from $C_6F_5C{\equiv}CH$ and $CuCl \cdot 4P(C_6H_5)_3$ (molar ratio = 70:17), catalyzed by CH_3ONa (87%) yellow solid IR (Nujol): 2064 ($C{\equiv}C$) gives with $H_2Os_3(CO)_{10}$ at $-23°$ in toluene after 2 h 98% 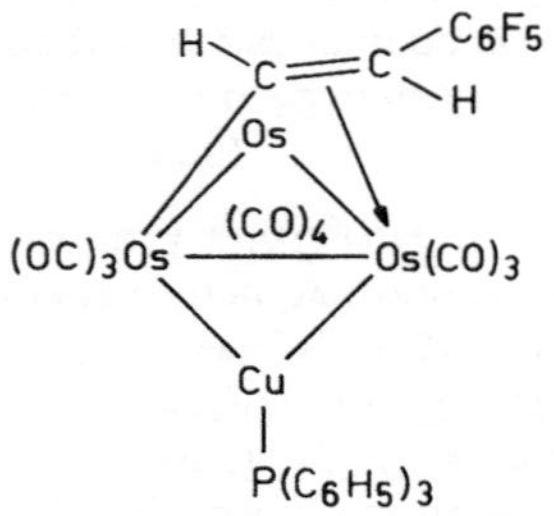	[43, 48]

No.	Compound	Description	Ref.
21	$4\text{-}O_2NC_6H_4C{\equiv}CCu \cdot P(C_6H_5)_3$	I (benzene) orange, m.p. 222 to 224° (dec.) degree of association cryoscopically measured in benzene or in nitro-benzene is ~2 IR (KBr): 2015 (C≡C) gives on contact with C_2H_5OH a bright red powder, m.p. and analysis unchanged	[5, 6]
22	$4\text{-}O_2NC_6H_4C{\equiv}CCu \cdot 2P(C_2H_5)_2C_6H_5$	I (benzene) unstable, deep red, m.p. 75° IR (KBr): 2025 (C≡C)	[5, 6]
23	$C_6H_5C{\equiv}CCu \cdot P(CH_3)_3$	I (benzene) proved to be a tetramer by X-ray analysis, see Section "Tetranuclear Compounds"	[5, 6, 9]
24	$C_6H_5C{\equiv}CCu \cdot 2P(CH_3)_3$	I (benzene) yellow–green, m.p. 94 to 96° in benzene and in nitrobenzene monomeric IR (KBr): 2035 + 2048 (C≡C)	[5, 6]
25	$C_6H_5C{\equiv}CCu \cdot 3P(CH_3)_3$	I (benzene) unstable, pale cream, m.p. 88 to 93° IR (KBr): 2034 + 2051 (C≡C)	[5, 6]
26	$C_6H_5C{\equiv}CCu, P(OCH_3)_3$	I (THF) not isolated, stoichiometry not estimated in dimethylformamide at 35° under CO_2 equilibrium with $C_6H_5C{\equiv}CCO_2Cu \cdot P(OCH_3)_3$ (at least 11% carboxylated, in THF at 80° only 4%)	[19, 22]

Table 23 [continued]

No.	complex	method of preparation (solvent, yield), properties, remarks and reactions	Ref.
27	$C_6H_5C{\equiv}CCu \cdot P(C_2H_5)_3$	I (benzene) yellow, m.p. 139.5° degree of association cryoscopically measured in benzene is 2.5 to 2.9, in nitrobenzene 2; earlier published data (3.3 to 3.8 and 3, respectively) have been corrected IR (KBr): $2018 + 2048$ (C≡C) gives $C_6H_5C{\equiv}CCu$ after standing in air for 1 to 2 d	[3, 5, 6]
28	$C_6H_5C{\equiv}CCu \cdot P(C_3H_7-i)_3$	III (benzene): from CuCl · $2P(C_3H_7-i)_3$ and $C_6H_5C{\equiv}CLi$ (1:1), 3 h/25° (33%) III (toluene): from $C_5(CH_3)_5Cu \cdot P(C_3H_7-i)_3$ and $C_6H_5C{\equiv}CH$ (1:1) at −78° yellow–green solid, m.p. 76° can be precipitated with pentane from benzene and toluene solutions monomeric in benzene according to osmometric measurements ^1H NMR (C_6D_6): 1.29 (dd, CH_3, $J_{PH} = 12.8$, $J_{HH} = 6.4$), 2.0 (m, CH), 7.6 (m, C_6H_5) ^{13}C NMR (C_6D_6): 20.32, 22.84 (both d, PR'_3, $J_{PC} = 5.2$, 10.3), 125.46, 128.19, 128.84, 131.02 (all s, C_6H_5) ^{31}P NMR (C_6D_6): 18.0 (br) IR (KBr): 2030 (C≡C)	[46]
29	$C_6H_5C{\equiv}CCu, P(C_2H_5)_2C_6H_5$	no preparation method given gives with $4-ClCOC_6H_4COCl$ in THF $4-C_6H_5C{\equiv}CCOC_6H_4COC{\equiv}CC_6H_5$ in high yield	[11]
30	$C_6H_5C{\equiv}CCu, P(C_4H_9-n)_3$	I (THF) not isolated, stoichiometry not estimated 1:3 acetylide–phosphine mixtures are sometimes formulated as $C_6H_5C{\equiv}CCu \cdot 3P(C_4H_9-n)_3$ in dimethylformamide at 35° under CO_2 equilibrium with $C_6H_5C{\equiv}CCO_2Cu \cdot P(C_4H_9-n)_3$ (at least 33% carboxylated; in THF at 80°, 50%) can be used for reversible CO_2 fixation, diagrams given reacts with CH_3I to give $C_6H_5C{\equiv}CCH_3$	[19, 22]

31   $C_6H_5C{\equiv}CCu \cdot P(C_6H_5)_2CH_3$

no preparation method given
the reaction with trans–IrCl(CO)(P(C₆H₅)₃)₂ in THF gives 11% [24]

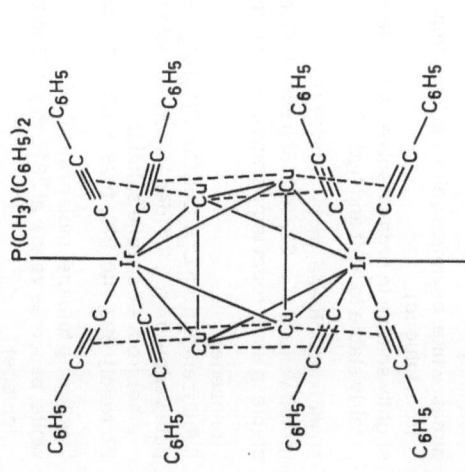

*32 $C_6H_5C{\equiv}CCu \cdot P(C_6H_5)_3$

—

I (benzene, 60°, molar ratio 0.77:1)
also from the mother liquor of the preparation of No. 32 ("Form D")
according to [33]
bright yellow, moderately soluble in benzene
IR (KBr): 1890, 1960, 2020, 2050
Raman: 998, 1173, 1200, 1592, 2050
UV (mineral oil): 380
with CuK$_\alpha$ radiation the reflections of highest intensity are d = 4.46, 4.91, 10.6, and 10.8 Å
equilibrium reaction with CO₂ and conversion to special forms of
$C_6H_5C{\equiv}CCu \cdot P(C_6H_5)_3$ see under No. 32 [33]

33 $2C_6H_5C{\equiv}CCu \cdot 3P(C_6H_5)_3$

Table 23 [continued]

No.	complex	method of preparation (solvent, yield), properties, remarks and reactions	Ref.
34	$C_6H_5C{\equiv}CCu$, PCl_3	I (no solvent) IR: compared with $C_6H_5C{\equiv}CCu$ the $C{\equiv}C$ frequency is increased by 40 cm^{-1} conversion to $P(C{\equiv}CC_6H_5)_3$ possible, no details given	[7, 14]
35	$4\,C_6H_5C{\equiv}CCu \cdot 3(C_6H_5)_2PCH_2P(C_6H_5)_2$	I (toluene, at 60°, molar ratio 1.5:1) pale yellow, becomes deeper yellow at 154°, m.p. 161 to 163° (recrystallized from toluene) stable in air at room temperature, fairly stable to heat and hydrolysis, slightly soluble in organic solvents	[13]
36	$C_6H_5C{\equiv}CCu \cdot (C_6H_5)_2P(CH_2)_2P(C_6H_5)_2$	I (toluene) almost white, becomes yellow at 215°, m.p. 219 to 221° (recrystallized from toluene) slightly soluble in organic solvents, stable in air at room temperature, fairly stable to heat and hydrolysis	[13]
37	$4\,C_6H_5C{\equiv}CCu \cdot 3(C_6H_5)_2P(CH_2)_2P(C_6H_5)_2$	I (refluxing toluene, molar ratio 2:1) lemon yellow, m.p. 229 to 233° (recrystallized from toluene) stable in air at room temperature, slightly soluble in organic solvents	[13]
38	$2\,C_6H_5C{\equiv}CCu \cdot (C_6H_5)_2P(CH_2)_2P(C_6H_5)_2$	II (benzene): from $C_6H_5C{\equiv}CCu \cdot P(C_6H_5)_3$, 18 h reflux (70%) pale yellow, m.p. 212° (dec.) after reprecipitation with light petroleum ether from a benzene solution IR (Nujol): 692, 740, 870, 1025, 1100, 1200, 1466, 1495, 1540, 2020	[43]
39	$2\,C_6H_5C{\equiv}CCu \cdot 3(C_6H_5)_2P(CH_2)_2P(C_6H_5)_2$	I (refluxing toluene, molar ratio 0.66:1) white, becomes yellow at 201°, m.p. 205 to 208° (recrystallized from toluene) stable in air at room temperature, slightly soluble in organic solvents	[13]

References on pp. 174/5

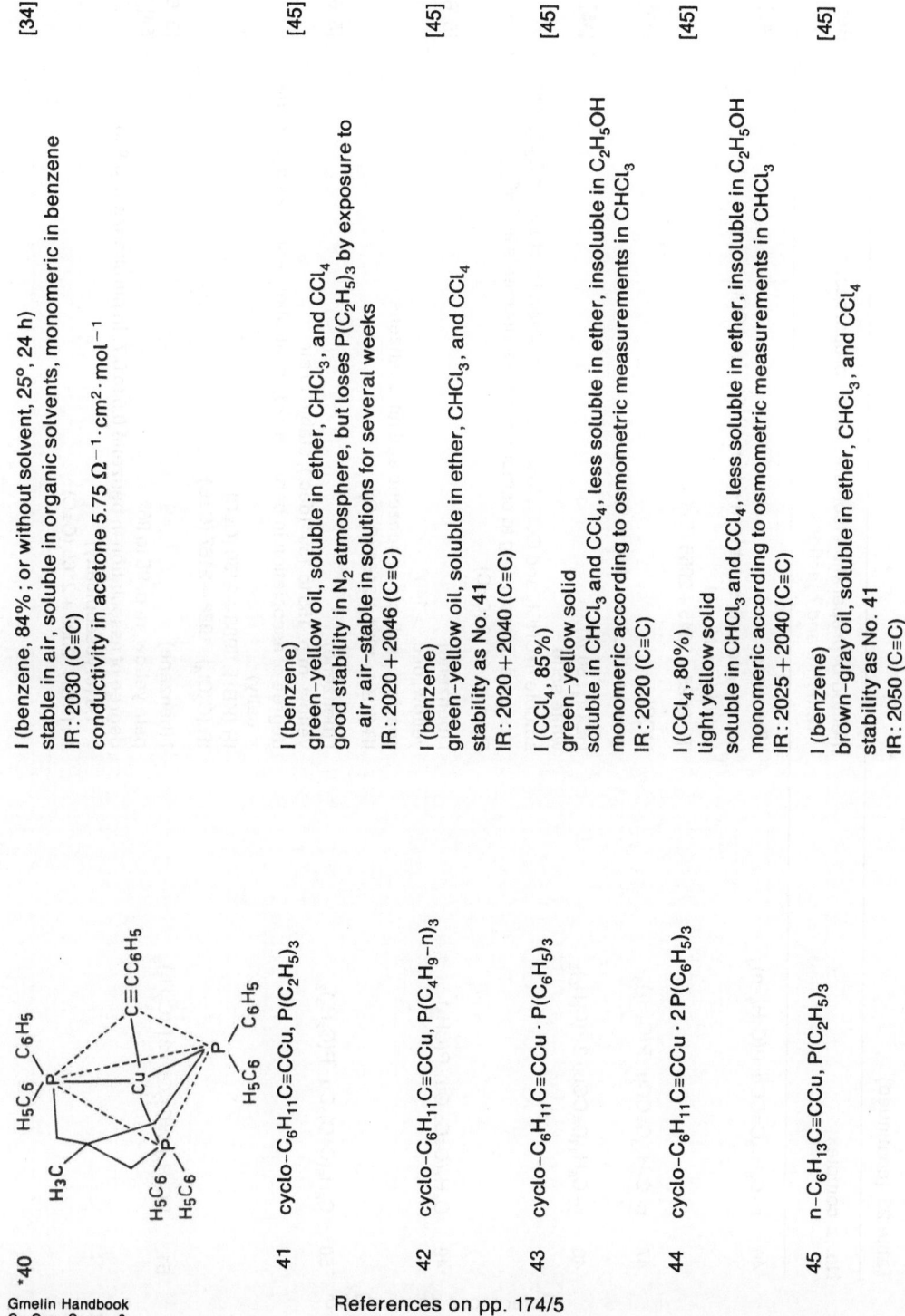

*40

I (benzene, 84%; or without solvent, 25°, 24 h)
stable in air, soluble in organic solvents, monomeric in benzene
IR: 2030 (C≡C)
conductivity in acetone 5.75 $\Omega^{-1} \cdot cm^2 \cdot mol^{-1}$

[34]

41   cyclo-$C_6H_{11}C$≡CCu, $P(C_2H_5)_3$

I (benzene)
green–yellow oil, soluble in ether, $CHCl_3$, and CCl_4, good stability in N_2 atmosphere, but loses $P(C_2H_5)_3$ by exposure to air, air–stable in solutions for several weeks
IR: 2020 + 2046 (C≡C)

[45]

42   cyclo-$C_6H_{11}C$≡CCu, $P(C_4H_9-n)_3$

I (benzene)
green–yellow oil, soluble in ether, $CHCl_3$, and CCl_4
stability as No. 41
IR: 2020 + 2040 (C≡C)

[45]

43   cyclo-$C_6H_{11}C$≡CCu · $P(C_6H_5)_3$

I (CCl_4, 85%)
green–yellow solid
soluble in $CHCl_3$ and CCl_4, less soluble in ether, insoluble in C_2H_5OH
monomeric according to osmometric measurements in $CHCl_3$
IR: 2020 (C≡C)

[45]

44   cyclo-$C_6H_{11}C$≡CCu · $2P(C_6H_5)_3$

I (CCl_4, 80%)
light yellow solid
soluble in $CHCl_3$ and CCl_4, less soluble in ether, insoluble in C_2H_5OH
monomeric according to osmometric measurements in $CHCl_3$
IR: 2025 + 2040 (C≡C)

[45]

45   n-$C_6H_{13}C$≡CCu, $P(C_2H_5)_3$

I (benzene)
brown–gray oil, soluble in ether, $CHCl_3$, and CCl_4
stability as No. 41
IR: 2050 (C≡C)

[45]

References on pp. 174/5

Table 23 [continued]

No.	complex	method of preparation (solvent, yield), properties, remarks and reactions	Ref.
46	$n\text{-}C_6H_{13}C\equiv CCu, P(C_4H_9\text{-}n)_3$	I (benzene) brown–gray oil, soluble in ether, $CHCl_3$, and CCl_4 stability as No. 41 IR: $2025 + 2045 + 2065$ ($C\equiv C$)	[45]
47	$n\text{-}C_6H_{13}C\equiv CCu \cdot P(C_6H_5)_3$	I (CCl_4) green solid, insoluble in $CHCl_3$	[45]
48	$n\text{-}C_6H_{13}C\equiv CCu \cdot 2P(C_6H_5)_3$	I (CCl_4, 80%) white solid soluble in $CHCl_3$ and CCl_4, less soluble in ether, insoluble in C_2H_5OH monomeric according to osmometric measurements in $CHCl_3$ Raman: 2116 ($C\equiv C$)	[45]
49	$C_6H_5(C\equiv C)_2Cu \cdot P(CH_3)_3$	I (benzene) yellow, dec. $> 140°$ sparingly soluble in benzene and nitrobenzene IR (KBr): $1981 + 2161$ ($C\equiv C$)	[5, 6]
50	$C_6H_5(C\equiv C)_2Cu \cdot P(C_2H_5)_3$	I (benzene) yellow, m.p. 155 to 156° (dec.), stable to air degree of association in benzene ~ 3, in nitrobenzene ~ 2 (cryoscopically) IR (KBr): $1988 + 2159$ ($C\equiv C$) IR (C_2Cl_4): $1988 + 2167$ ($C\equiv C$)	[5, 6]
51	$2C_6H_5(C\equiv C)_2Cu \cdot 3P(C_2H_5)_3$	I (benzene) pale yellow, m.p. 95 to 96° degree of association in benzene 0.5 to 0.7, in nitrobenzene 0.9 to 1.0 (cryoscopically) IR (KBr): $2008 + 2165$ ($C\equiv C$)	[5, 6, 54]

References on pp. 174/5

52 $C_6H_5(C\equiv C)_2Cu \cdot P(C_3H_7-n)_3$

IR (benzene): 2009 + 2171 (C≡C)
space group Pbca – D_{2h}^{15} (No. 61), a = 14.85, b = 20.8, c = 26.7 Å, d_m = 1.182 g · cm⁻³, Z = 8; tetrameric structure "similar to No. 23" (no further details)
on standing in air or on heating in propanol No. 50 is formed   [5, 6]

53 $C_2H_5O_2CC\equiv CCu \cdot C_{12}H_8N_2 \cdot P(C_6H_5)_3$

I (benzene)
yellow, m.p. 87°
degree of association in benzene 3.0 to 3.4, in nitrobenzene 2.0 (cryoscopically)
IR (KBr): 1984 + 2164 (C≡C)
IR (C_2Cl_4): 1986 + 2166 (C≡C)
II: rapidly formed from No. 9 and $C_{12}H_8N_2$ in CH_2Cl_2 solution or more slowly in suspension in benzene
yellow, m.p. 150° (dec.)
insoluble in toluene
molecular weight (CH_2Cl_2) = 523 (theoretical 603), slightly dependent on concentration
IR (Nujol): 1660 (close d, CO), 2050 (C≡C)
does not react with $C_2H_5O_2CC\equiv CH$ even at 50°   [44]

54 $C_2H_5O_2CC\equiv CCu \cdot C_{16}H_{16}N_2 \cdot P(C_6H_5)_3$

II (cyclohexane): from No. 9 and $C_{16}H_{16}N_2$ in suspension
II (benzene): from No. 4 and $P(C_6H_5)_3$
yellow, m.p. 160° (dec.)
insoluble in benzene
IR (Nujol): 1665 (C=O), 2045 (C≡C)
gives with $C_2H_5O_2CC\equiv CH$ in benzene No. 4   [44]

complexes with As compounds

55 $n-C_4H_9C\equiv CCu, AsCl_3$

I (no solvent)   [14]

56 $C_6H_5C\equiv CCu, AsCl_3$

I (no solvent)
IR: compared with $C_6H_5C\equiv CCu$ the C≡C frequency is increased by 40 cm⁻¹   [14]

*Further information:

2C$_6$H$_5$C≡CCu · C$_{12}$H$_8$N$_2$ (Table **23**, No. **7**; C$_{12}$H$_8$N$_2$ = 1,10-phenanthroline). It can also be prepared by electrochemical oxidation of a copper foil in dry CH$_3$CN or CH$_3$COCH$_3$ containing 4% C$_6$H$_5$C≡CH, 0.05% [N(C$_2$H$_5$)$_4$]ClO$_4$, and 1,10-phenanthroline [49].

Crystallizes from pyridine/ether and is stable in air at room temperature. It is fairly stable to heat and hydrolysis, readily soluble in organic solvents [13].

On standing for weeks, or at reflux for a longer period in Cl(CH$_2$)$_2$Cl, a reaction was observed to give CH$_2$=CHCl, C$_6$H$_5$C≡CH, and a complex of CuCl, C$_{12}$H$_8$N, and Cl(CH$_2$)$_2$Cl. Neither structural nor stoichiometric data for the complex are given [13].

n-C$_3$H$_7$C≡CCu,D (D = P(C$_4$H$_9$-n)$_3$, P(N(CH$_3$)$_2$)$_3$, OP(N(CH$_3$)$_2$)$_3$; Table **23**, Nos. **11**, **12**, **13**) react with R^{1}Li compounds to give reagents of the type [n-C$_3$H$_7$C≡CCuR^1]Li; see Section 1.1.1.2.4 in "Organocopper Compounds" 2, 1983, pp. 174/211.

No.	solvent	R^1 (yields of subsequent products of [n-C$_3$H$_7$C≡CCuR^1]Li in %)	Ref.
11	OP(N(CH$_3$)$_2$)$_3$	(E)- or (Z)-CH=CHCR^{2}R^{3}OSi(CH$_3$)$_3$, R^2 = alkenyl and R^3 = alkyl	[27, 32]
12	ether or ether/C$_6$H$_{14}$	CH$_3$ (37), C$_2$H$_5$, (E)-CH=CHOC$_2$H$_5$, n-C$_4$H$_9$ (52), t-C$_4$H$_9$ (70), C$_6$H$_5$ (44), (E)-CH=CHCH(C$_5$H$_{11}$-n)OSi(CH$_3$)$_2$C(CH$_3$)$_3$ (28)	[40]
12	ether	CH=CHCH$_2$CH(C$_4$H$_9$-n)OSi(CH$_3$)$_2$C(CH$_3$)$_3$, (E)-CH=CHCH(C$_5$H$_{11}$-n)OSi(CH$_3$)$_2$C(CH$_3$)$_3$ (>86), (E)-CH=C(C$_6$H$_5$)CH(C$_5$H$_{11}$-n)OC$_5$H$_9$O; C$_5$H$_9$O = tetrahydropyran-2-yl	[26, 29, 31]
12	THF	CH=CH$_2$, n-C$_4$H$_9$, t-C$_4$H$_9$, (Z)-CH=CHCH$_2$CH(CH$_3$)OC$_5$H$_9$O	[17, 25]
13	ether	C$_2$H$_5$	[36]
13	OP(N(CH$_3$)$_2$)$_3$	(E)- or (Z)-CH=CHCR^{2}R^{3}OSi(CH$_3$)$_3$, R^2 = alkenyl and R^3 = alkyl	[27, 32]

Analogously [n-C$_3$H$_7$C≡CCuR^1]Li can be prepared and reacted in OP(N(CH$_3$)$_2$)$_3$ in situ using n-C$_3$H$_7$C≡CH, Li, CuI, and R^{1}Br instead of n-C$_3$H$_7$C≡CCu under ultrasonic treatment [37]. No. 13 gives with C$_2$H$_5$MgBr in ether [n-C$_3$H$_7$C≡CCuC$_2$H$_5$]MgBr [36].

C$_6$H$_5$C≡CCu · P(C$_6$H$_5$)$_3$ (Table **23**, No. **32**). Different products are claimed to have this formula. The best characterized is [C$_6$H$_5$C≡CCuP(C$_6$H$_5$)$_3$]$_4$, in the following called Form A. The lack of data and the contradictory results do not permit a reliable estimation of further products. Tentatively, a classification into the "Forms" B to E is undertaken; two more products are too poorly characterized to be evaluated. The basic data used for this classification are compared in the following tables.

forms	IR (cm^{-1}; Nujol, KBr, or spread on NaCl disk)					Ref.
A				2020		[52]
B					2043	[5, 6, 43]
C	1895		1960		2052	[33]
D	1890	1917	1960		2052	[33]
E			1935			2060	[52]

forms	Raman (cm⁻¹)			Ref.

forms		Raman (cm^{-1})				Ref.
C		1583		2018 (vs)		[33]
D	538	1190	1594 (vs)	1914	2053	[33]

The Forms C and D also show identical or almost identical strong Raman bands at 252/3, 525, 780/2, 998 (vs), 1027, 1097/8, 1175/6, 1198/201, and 3056 cm^{-1} [33].

forms	interplanar spacing d (Å; CuK$_\alpha$)	UV (nm; mineral oil suspension)	Ref.
C	4.48, 4.82, 11.6, 11.8	380	[33]
D	4.32, 4.62, 10.4, 11.5, 12.4	380, 435	[33]

Form A, $[C_6H_5C{\equiv}CCuP(C_6H_5)_3]_4$, is obtained by treatment of $Cu[P(C_6H_5)_3]_2BH_4$ with $C_6H_5C{\equiv}$ CH and KOH in the molar ratio 1:1:1 in a benzene/benzyl alcohol (1:1) solution. Yellow crystals suitable for X-ray analysis were obtained by slow diffusion of hexane into the solution. It is the first known metal complex containing $\mu_3-\eta^1$ bridging alkynyls and can also be obtained by refluxing Form E in CH_3CN. Form A and Form E have the same melting point (value not given) [52].

The compound is monoclinic with the space group C2/c $-$ C$_{2h}^4$ (No. 13), Z = 4, a = 14.866 (4), b = 24.552 (4), c = 23.345 (9) Å, β = 95.89 (2)°, d$_m$ = 1.338 g/cm³. The final anisotropic refinement led to the conventional R value of 0.054. The structure (see **Fig. 5**, p. 172) consists of a tetrahedral skeleton of metal atoms bonded to four terminal phosphine molecules and to four μ_3-bridging phenylacetylide ligands (which behave essentially as 2e donors). Average distances (in Å) and bond angles are:

Cu-Cu	2.603	Cu-P	2.228	Cu-C	2.185	C≡C	1.174
P-Cu-C	113.7°	C-Cu-C	104.6°				

The copper-alkynyl triple bridges are markedly asymmetric, with Cu-C contacts in the range 2.072 (4) to 2.380 (4) Å. The observed very short C≡C distances are indicative of a triple bond (cf. 1.20 Å in C_2H_2) and for the absence of π-back-donation from the metal to the ligand π^* orbitals. The C(1)-C(2)-C(3) and C(9)-C(10)-C(11) fragments are strictly linear (average angle 178°), and the interactions of atoms C(2) and C(10) with metal triangles Cu(1)-Cu(1')-Cu(2') and Cu(1)-Cu(2)-Cu(2'), respectively, are negligible (average C $\cdots$ Cu distance 3.104 Å): both these features are inconsistent with a substantial ligand-to-metal π-donation [52].

Form B is prepared by the following methods:
a) By shaking the components (molar ratio 1:1) in benzene until all RC≡CCu is dissolved [5, 6, 12, 41].
b) From $[(C_6H_5)_3P]_2CuSC_6H_5$ and $C_6H_5C{\equiv}CH$, no conditions given [12].
c) From $[(C_6H_5C{\equiv}C)_2Cu][N(P(C_6H_5)_3)_2]$ and $C_6H_5C{\equiv}CAu \cdot P(C_6H_5)_3$ at room temperature in CH_2Cl_2 (yield 95%), besides $[(C_6H_5C{\equiv}C)_2Au]^-$ [41].
d) From $[(C_6H_5C{\equiv}C)_2Cu]^-$ and $AuCl \cdot P(C_6H_5)_3$ in CH_2Cl_2, besides $C_6H_5C{\equiv}CCu$ [41].
e) From $C_6H_5C{\equiv}CH$ and $[CuCl \cdot P(C_6H_5)_3]_4$ (molar ratio = 7:29) in CH_3OH with catalysis by a little CH_3ONa, yield 90% [43].
f) From $C_6H_5C{\equiv}CCu$ and $P(C_6H_5)_3$ in C_2H_5OH, yield 80% (no further data are given) [43].

References on pp. 174/5

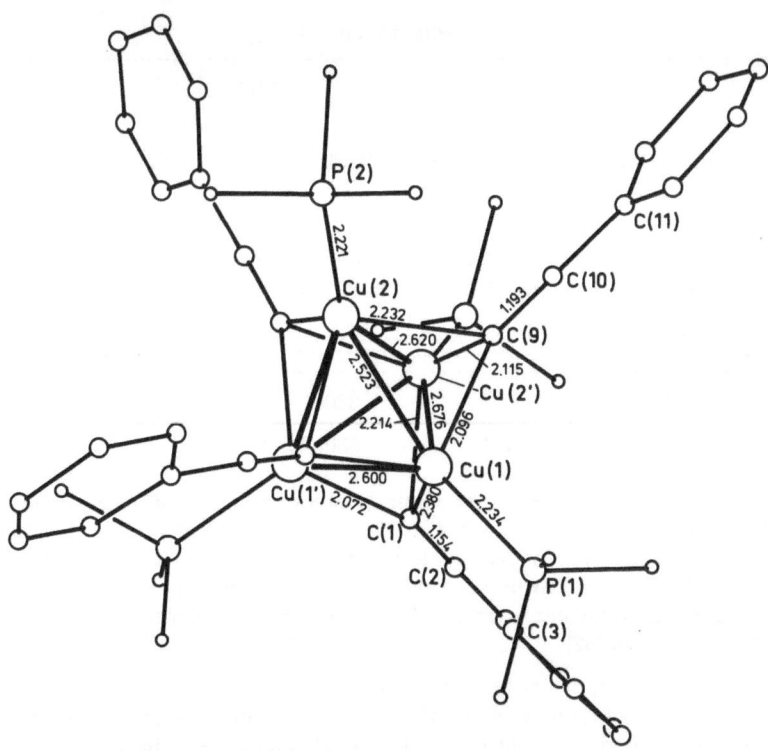

Fig. 5. Structure of $[C_6H_5C\equiv CCuP(C_6H_5)_3]_4$-(No. 32, Form A).

Form B is yellow-green after recrystallization from toluene [5, 6] and pale yellow after recrystallization from benzene [43]. Its melting point is at 212 to 214 °C (dec.). It is too sparingly soluble in benzene or nitrobenzene to estimate the degree of association by osmometric methods [5, 6]. It shows an IR band at 2043 cm^{-1} in Nujol [43] and in KBr [5, 6].

Form C. The yellow form has been prepared by boiling Form D in benzene or from concentrated solutions of the 2:3 complex No. 33 in benzene simply upon standing or by shaking the components in benzene [33]. However, this last method is similar to that used in [5] to prepare Form B. From the only IR band given in [5] no conclusion about the identity of the product is possible. The products from the preparation methods used in [33] may not be identical, and it is not mentioned, from which product the IR data are recorded. This fact must be also taken into account in the further data given.

Form C is slightly soluble in benzene and decomposes upon standing in air with appearance of a new Raman band at 1926 cm^{-1} [33].

Form D. The yellowish green form [33] is prepared by the following methods:
a) By boiling $C_6H_5C\equiv CCu$ and $P(C_6H_5)_3$ (molar ratio = 10:23) in benzene:H_2O = 19:1 for 1 h. A residue is filtered off; at 15 °C, Form D crystallizes (44%). The mother liquor gives 59% of No. 33 after half of the solvent is evaporated [33].
b) On long standing of concentrated solutions of Form C or of the 2:3 complex No. 33 in CH_2Cl_2 or $CHCl_3$ [33].

Form D is very slightly soluble in benzene. It decomposes on standing in air with appearance of a new Raman band at 1926 cm^{-1} [33].

A substance, which is similarly prepared as under a), by addition of $P(C_6H_5)_3$ to a suspension of $C_6H_5C\equiv CCu$ in benzene (molar ratio not given) at 50 °C and then boiling for 1 min [48, 55], may be identical with Form D. It reacts with $H_2Os_3(CO)_{10}$ in toluene in 19 h at room temperature to give after chromatography on Kieselgel GF$_{254}$ (development with ether/cyclohexane = 1:4) a mixture of 41% yellow I, 3% orange-red II, 3% yellow III, 8% $HOs_3Cu(CH= CC_6H_5)(CO)_8$, and four other unidentified products. The reaction gives after 3.5 h at −30 °C 69% I and 13% III [48].

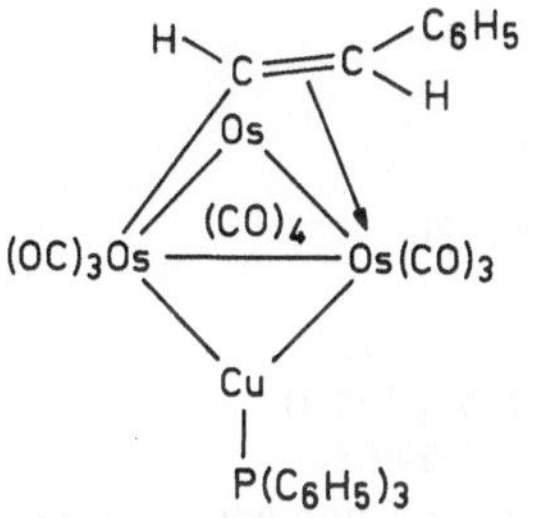

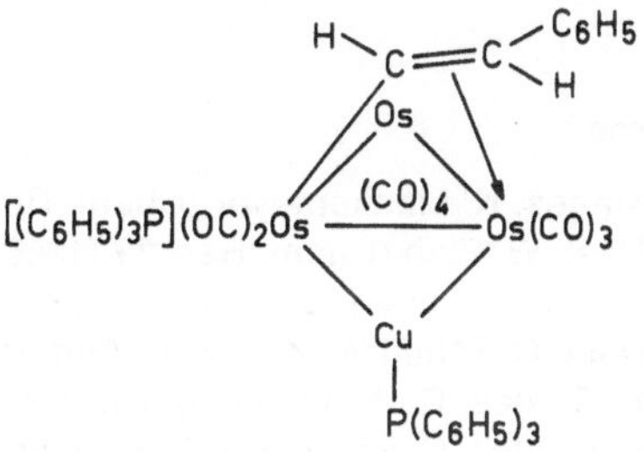

Form E is prepared when a "solution of Form A is stirred for several days". An "insoluble fluorescent yellow-green compound ... separates (melting point and analysis unchanged)". It is not clear whether it is a solution in benzene:benzyl alcohol = 1:1 or a benzene/benzyl alcohol/hexane solution. The melting point is not given for Form E or for Form A [52].

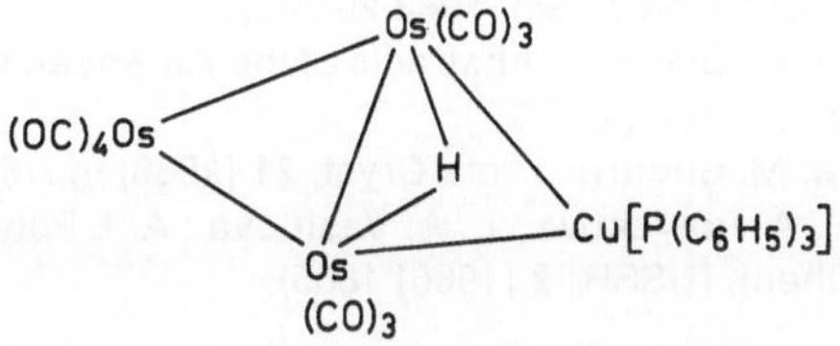

No structural data are said to be available, but it is believed to be polymeric. It is supposed that the compound may contain differently bonded phenylacetylide ligands (i.e., π- and σ-bonded) with substantial ligand-to-metal π-donation (which has been excluded for Form A), possibly similar to the situation found in [53] for $[(CH_3C\equiv C)_2BeN(CH_3)_2]_2$ [47, 52].

Form E was characterized by X-ray photoelectron spectroscopy (XPES) and by Auger spectroscopy. The core level photoelectron spectra ($Cu2p_{3/2} = 933.0$ eV; $P2p = 131.5$ eV), the $Cu(L_3VV)$ Auger line (a kinetic energy of 915.7 eV corresponds to that line), and the Auger parameters are discussed in terms of both the different coordination environments and of the polarizability of the ligands [47].

On refluxing in CH_3CN it gives Form A [52].

Two **further products** cannot be assigned to any of the forms mentioned before. From $C_6H_5C\equiv CCu$ and $Pt(P(C_6H_5)_3)_4$ a compound "$C_6H_5C\equiv CCu \cdot P(C_6H_5)_3$" and further tertiary phosphine–copper complexes are formed (no conditions given) [23]. "$C_6H_5C{\equiv}CCu \cdot P(C_6H_5)_3$" can also be formed by decarboxylation of $C_6H_5C{\equiv}CCO_2Cu \cdot P(C_6H_5)_3$, but only in equilibrium. At ordinary pressure and 35 °C in dimethylformamide under CO_2 at least 17% can be recarboxylated [22].

$C_{49}H_{44}CuP_3$ (Table 23, No. 40). The reaction with R^1H in tetrahydrofuran gives $R^1Cu\{(C_6H_5)_2PCH_2\}_3CCH_3$, where R^1H is $4\text{-}CH_3C_6H_4OH$ (40% at room temperature), $4\text{-}CH_3CO_2C_6H_4CN$ (46% at 0 °C), and $4\text{-}CH_3CO_2C_6H_4NO_2$ (33% at 0 °C) [34].

The reaction with $CH_2(CN)_2$ yields a yellow powder (structure not revealed). No reaction occurs with CH_3CN, $CH_2(CO_2C_2H_5)_2$, or CH_3NO_2 in boiling tetrahydrofuran [34].

References:

[1] F. Klages, K. Mönkemeyer (Chem. Ber. **85** [1952] 109/22).
[2] R. Nast, W. Pfab (Chem. Ber. **89** [1956] 415/21).
[3] D. Blake, G. Calvin, G. E. Coates (Proc. Chem. Soc. **1959** 396/7).
[4] R. Nast, C. Schultze (Z. Anorg. Allgem. Chem. **307** [1960] 15/21).
[5] G. E. Coates, C. Parkin (J. Inorg. Nucl. Chem. **22** [1961] 59/67).
[6] G. E. Coates, C. Parkin (in: S. Kirschner, Advances in the Chemistry of the Coordination Compounds, Plenum, New York 1961, pp. 173/9).
[7] A. M. Sladkov, L. Y. Ukhin, V. V. Korshak (Izv. Akad. Nauk SSSR Ser. Khim. **1963** 2213/5; Bull. Acad. Sci. USSR Div. Chem. Sci. **1963** 2043/5).
[8] P. W. R. Corfield, H. M. M. Shearer (Abstracts of the American Crystallographic Association, Bozeman, Montana 1964, pp. 96/7).
[9] P. W. R. Corfield, H. M. M. Shearer (Acta Cryst. **21** [1966] 957/65).
[10] M. F. Shostakovskii, I. A. Polyakova, L. A. Vasil'eva, A. I. Polyakov (Zh. Org. Khim. **2** [1966] 1899; J. Org. Chem. [USSR] **2** [1966] 1865).

[11] A. M. Sladkov, I. R. Gol'ding (Khim. Atsetilena Tr. 3rd Vses. Konf., Dushanbe, Tadzh. SSR, 1968 [1972], pp. 45/7; C.A. **79** [1973] No. 5431).
[12] W. T. Reichle (Inorg. Nucl. Chem. Letters **1969** 981/4).
[13] A. Camus, N. Marsich (J. Organometal. Chem. **21** [1970] 249/58).
[14] A. M. Sladkov, I. R. Gol'ding (Izv. Akad. Nauk SSSR Ser. Khim. **1970** 2644; Bull. Acad. Sci. USSR Div. Chem. Sci. **1970** 2495).
[15] M. S. Shvartsberg, A. A. Moroz (Izv. Akad. Nauk SSSR Ser. Khim. **1971** 1582/5; Bull. Acad. Sci. USSR Div. Chem. Sci. **1971** 1488/90).
[16] L. I. Zakharkin, V. N. Kalinin, I. R. Gol'ding, A. M. Sladkov, A. V. Grebennikov (Zh. Obshch. Khim. **41** [1971] 823/8; J. Gen. Chem. [USSR] **41** [1971] 830/3).
[17] E. J. Corey, D. J. Beames (J. Am. Chem. Soc. **94** [1972] 7210/1).
[18] S. Sh. Rashidova, V. A. Barabanov, D. D. Il'yasova (Uzb. Khim. Zh. **18** [1974] 63/6; C.A. **82** [1975] No. 17434).
[19] T. Tsuda, K. Ueda, T. Saegusa (J. Chem. Soc. Chem. Commun. **1974** 380/1).
[20] M. C. Barral, V. Moreno, A. Santos (Anales Quim. **71** [1975] 770/4; C.A. **84** [1976] No. 180367).

[21] I. R. Gol'ding, I. A. Garbusova, A. M. Sladkov, V. T. Aleksanyan (Khim. Atsetilena Tr. 5th Vses. Konf., Tbilisi 1975, p. 455; C.A. **88** [1978] No. 189798).
[22] T. Tsuda, Y. Chujo, T. Saegusa (J. Chem. Soc. Chem. Commun. **1975** 963/4).
[23] O. M. Abu Salah, M. I. Bruce (Australian J. Chem. **29** [1975] 73/7).
[24] O. M. Abu Salah, M. I. Bruce (Australian J. Chem. **29** [1975] 531/41).
[25] E. J. Corey, P. Ulrich, J. M. Fitzpatrick (J. Am. Chem. Soc. **98** [1976] 222/4).

[26] E. J. Corey, M. Shibasaki, K. C. Nicolaou, C. L. Malmsten, B. Samuelsson (Tetrahedron Letters **1976** 737/40).

[27] B. F. Middleton, M. J. Weiss, C. V. Grudzinskas, S.-M. L. Chen, American Cyanamid Co. (Brit. 2006186 [1979]; Ger. Offen. 2837530 [1979]; C.A. **91** [1979] No. 38996).

[28] A. Camus, N. Marisch, G. Nardin, L. Randaccio (Inorg. Chim. Acta **23** [1977] 131/44).

[29] P. W. Collins, E. Z. Dajani, D. R. Driskill, M. S. Bruhn, C. J. Jung, R. Pappo (J. Med. Chem. **20** [1977] 1152/9).

[30] H. C. Kluender, Miles Laboratories Inc. (U.S. 4065493 [1977/78]; U.S. 4100356 [1977/78]; U.S. 4100357 [1977/78]; U.S. 4127727 [1977/78]; C.A. **88** [1978] No. 190216).

[31] R. T. Buckler, D. L. Garling (Tetrahedron Letters **1978** 2257/60).

[32] M. B. Floyd, M. J. Weiss, C. V. Grudzinskas, S.-M. L. Chen, American Cyanamid Co. (Brit. 2009173 [1978/79]; C.A. **92** [1980] No. 214968).

[33] I. A. Garbuzova, I. R. Gol'ding, A. M. Sladkov, Ya. V. Genin, D. Ya. Tsvankin, V. T. Aleksanyan (Izv. Akad. Nauk SSSR Ser. Khim. **1978** 1328/31; Bull. Acad. Sci. USSR Div. Chem. Sci. **1978** 1154/6).

[34] K. Hiraki, Y. Fuchita, Y. Morita (Bull. Chem. Soc. Japan **51** [1978] 2012/4; C.A. **89** [1978] No. 129631).

[35] E. Piers, J. M. Chong, H. E. Morton (Tetrahedron Letters **22** [1981] 4905/8).

[36] M. Lopp, A. Pals, U. Lille (Eesti NSV Tead. Akad. Toimetised Keem. **29** [1980] 191/5).

[37] J. L. Luche, C. Pétrier, A. L. Gemal, N. Zikra (J. Org. Chem. **47** [1982] 3805/6).

[38] E. Piers, J. M. Chong (J. Org. Chem. **47** [1982] 1602/4).

[39] L. I. Zakharkin, A. I. Kovredov, V. A. Ol'shevskaya (Izv. Akad. Nauk SSSR Ser. Khim. **1982** 673/5; Bull. Acad. Sci. USSR Div. Chem. Sci. **1982** 599/602).

[40] R. J. Batten, J. D. Coyle, R. J. K. Taylor (Synthesis **1980** 910/1).

[41] O. M. Abu-Salah, A. R. Al-Ohaly (J. Organometal. Chem. **255** [1983] C39/C40).

[42] Y. Naruta, K. Maruyama (J. Chem. Soc. Chem. Commun. **1983** 1264/5).

[43] M. I. Bruce, M. G. Humphrey, J. G. Matisons, S. K. Roy, A. G. Swincer (Australian J. Chem. **37** [1984] 1955/61).

[44] G. La Monica, G. Ardizzoia, S. Cenini, F. Porta (J. Organometal. Chem. **273** [1984] 263/73).

[45] E. C. Royer, M. C. Barral (Inorg. Chim. Acta **90** [1984] L47/L49).

[46] H. Werner, H. Otto, T. Ngo-Khac, C. Burschka (J. Organometal. Chem. **262** [1984] 123/36).

[47] C. Battistoni, G. Mattogno, E. Paparazzo, L. Naldini (Inorg. Chim. Acta **102** [1985] 1/3).

[48] M. I. Bruce, E. Horn, J. G. Matisons, M. R. Snow (J. Organometal. Chem. **286** [1985] 271/87).

[49] R. Kumar, D. G. Tuck (J. Organometal. Chem. **281** [1985] C47/C48).

[50] J. Enda, T. Matsutani, I. Kuwajima (Tetrahedron Letters **25** [1984] 5307/10).

[51] J. Enda, I. Kuwajima (J. Am. Chem. Soc. **107** [1985] 5495/501).

[52] L. Naldini, F. Demartin, M. Manassero, M. Sansoni, G. Rassu, A. Zoroddu (J. Organometal. Chem. **279** [1985] C42/C44).

[53] N. A. Bell, I. W. Nowell, H. M. M. Shearer (J. Chem. Soc. Chem. Commun. **1982** 147/8).

[54] H. M. M. Shearer, P. W. R. Corfield (unpublished from [5]).

[55] M. I. Bruce, E. Horn, J. G. Matisons, M. R. Snow (Australian J. Chem. **37** [1984] 1163/70).

1.1.2.5 Complexes of RC≡CCu with Cu, Au, Zn, or Hg Compounds

The complexes are given in Table 24. Some of the products are very poorly characterized. Compounds with known stoichiometry are formulated with a point (like $C_6H_5C≡CCu \cdot CH_3CO_2Cu$); a probable complex formation with unknown stoichiometry is indicated by a comma (like n-$C_5H_{11}C\equiv CCu,ZnCl_2$) as explained in "Organocopper Compounds" 1, 1985, p. 6.

$C_2B_{10}H_{11}C\equiv CCu$ ($C_2B_{10}H_{11}$ = 1,2-dicarba-closo-dodecaboran(12)-1-yl) and HgX_2 (X = halogen) do not form complexes. This is possibly a question of a polarization of the $C\equiv C$ bond different from that of the typical $RC\equiv CCu$ for electronic and steric reasons [8].

Table 24
Complexes of $RC\equiv CCu$ with Cu, Au, Zn, or Hg Compounds.
For abbreviations and dimensions see p. X.

No.	compound	preparation, properties, and remarks	Ref.
complexes with Cu or Au compounds			
1	$CH\equiv CCu \cdot CuCl$	at certain acidities intermediate formation assumed from C_2H_2 in $CuCl/HCl/NH_4Cl/H_2O$ from emf measurements [12]; also without NH_4Cl [11], and similarly in $Cu^I/Cu^{II}/Cl_2$ (green solution) [13] dissociates in weakly acidic solutions to give "$[ClCu_2C\equiv C]^-$" gives Cu_2C_2 in neutral or alkaline solutions	[11 to 13]
2	$CH\equiv CCu \cdot CuOH$	from dilute aqueous solutions of C_2H_2 and Cu^I salt, neutralized by NH_3 or NaOH; precipitates from more conc. solutions or after drying in air contain more C than corresponds to the formula white or red-brownish, explosive after drying over P_2O_5 in vacuum soluble in cold dilute HCl or HNO_3 and in warm conc. H_2SO_4 gives C_2H_2 with NH_3, CH_3CO_2H, or H_2S	[2]
3	$C_4H_3SC\equiv CCu \cdot CH_3CO_2Cu$ (C_4H_3S = thien-2-yl)	structure derived only from analogy from solutions of $C_4H_3SC\equiv CCu$ in boiling CH_3CO_2H, precipitation with ice-water "amorphous", oxidation gives $C_4H_3S(C\equiv C)_2C_4H_3S$	[4]
4	n-$C_3H_7CH(OH)CH_2C\equiv CCu \cdot CuCl$	from n-$C_3H_7CH(OH)CH_2C\equiv CH$ and a $CuCl-[C_2H_5NH_3]Cl$ complex in $CH_3OH/C_2H_5NH_2$ can be converted to n-$C_3H_7CH(OH)CH_2C\equiv CH$ in a 97% yield (no details)	[5]
5	$C_6H_5C\equiv CCu \cdot CH_3CO_2Cu$	from $C_6H_5C\equiv CCu$ and CH_3CO_2H; crystallizes from concentrated solutions after boiling under CO_2, precipitation with ice-water also possible	[1, 3]

References on p. 178

Table 24 [continued]

No.	compound	preparation, properties, and remarks	Ref.
	$C_6H_5C{\equiv}CCu \cdot CH_3CO_2Cu$ [continued]	orange–yellow, soluble in organic solvents except petroleum ether gives on boiling in CH_3CO_2H in presence of air $C_6H_5C{\equiv}CCH{=}CHC_6H_5$ (cf. the so-called "Straus reaction", Section 1.1.2.1.2; the title compound is possibly an intermediate of this reaction)	
6	$C_6H_5C{\equiv}CCu \cdot CuBr$	from $(i\text{-}C_3H_7)_2NCH_2C{\equiv}CC_6H_5$ and 2 mol CuBr in CH_3CN at 25°, then 65°/30 min, $[CH_2{=}N(C_3H_7\text{-}i)_2]Br$ also obtained yellow after washing with dilute HCl and then H_2O m.p. 270° (dec.) IR (KBr): 1930 (C≡C) gives 25% $(C_6H_5C{\equiv}C)_2$ with excess 30% H_2O_2 after 20 h stirring at room temperature	[10]

7   [14]

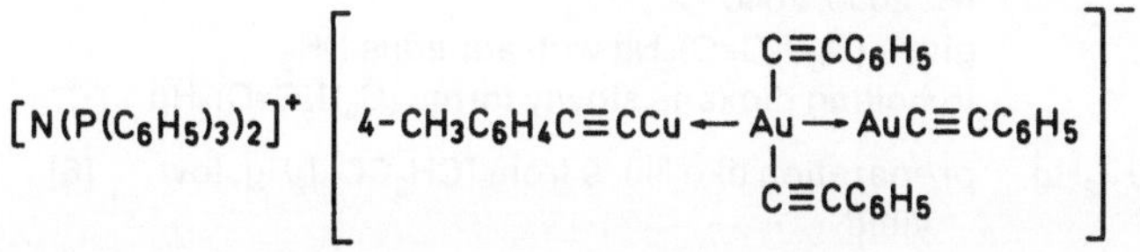

$$[N(P(C_6H_5)_3)_2]^+ \left[\begin{array}{c} C{\equiv}CC_6H_5 \\ | \\ 4\text{-}CH_3C_6H_4C{\equiv}CCu \longleftarrow Au \longrightarrow AuC{\equiv}CC_6H_5 \\ | \\ C{\equiv}CC_6H_5 \end{array} \right]^-$$

or

$$[N(P(C_6H_5)_3)_2]^+ \left[\begin{array}{c} C_6H_5C{\equiv}C\diagdown{}_{Au}\diagup{}^{C{\equiv}CC_6H_5} \\ \downarrow \\ 4\text{-}CH_3C_6H_4C{\equiv}CCu\diagup{}^{Au}\diagdown{}_{C{\equiv}CC_6H_5} \end{array} \right]^-$$

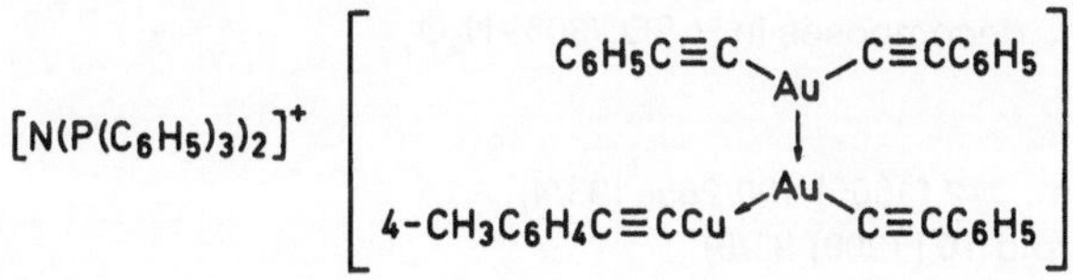

from excess $4\text{-}CH_3C_6H_4C{\equiv}CCu$ in acetone and

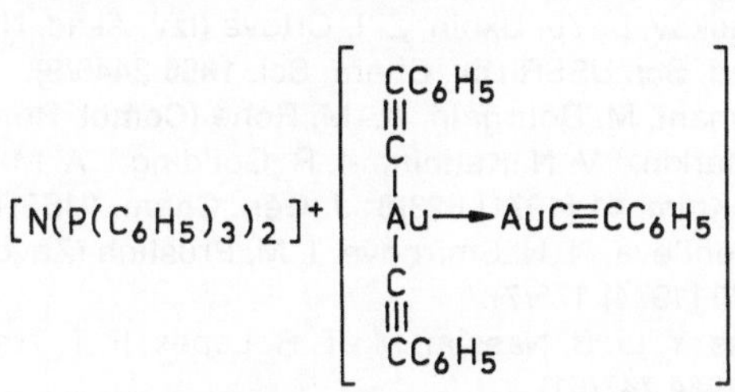

$$[N(P(C_6H_5)_3)_2]^+ \left[\begin{array}{c} CC_6H_5 \\ ||| \\ C \\ | \\ Au \longrightarrow AuC{\equiv}CC_6H_5 \\ | \\ C \\ ||| \\ CC_6H_5 \end{array} \right]^-$$

in moderate yields
stable yellow solid
IR: 2070 + 2110 (C≡C)
from the molecular weight of 854 (theor. 1413) the 1:1 electrolytic character is supported

References on p. 178

Table 24 [continued]

No.	compound	preparation, properties, and remarks	Ref.

complexes with Zn or Hg compounds

8 n–$C_5H_{11}C{\equiv}CCu,ZnCl_2$ — the formation of a complex from n–C_5H_{11}-$C{\equiv}CCu$ and $ZnCl_2$ in hexamethylphosphoric triamide is supposed from the enhanced reactivity towards $CH_2{=}CHCH_2Br$, compared with n–$C_5H_{11}C{\equiv}CCu$; n–C_5H_{11}-$C{\equiv}CCH_2CH{=}CH_2$ is formed in a 32% yield — [7]

9 2$C_6H_5C{\equiv}CCu \cdot HgBr_2$ — from $C_6H_5C{\equiv}CCu$ and $HgBr_2$ in THF, dioxane, or C_2H_5OH at 25°; lower yields in water yellow, insoluble in organic solvents and in H_2O, dissolves in pyridine with decomposition gives $C_6H_5C{\equiv}CCu$ with KI in C_2H_5OH/H_2O and $(C_6H_5C{\equiv}C)_2Hg$ with aqueous NH_3 — [6]

10 2$C_6H_5C{\equiv}CCu \cdot HgCl_2$ — preparation like No. 9 from $HgCl_2$ colorless IR: 2000, 2040 gives $(C_6H_5C{\equiv}C)_2Hg$ with aqueous NH_3 in boiling dioxane slowly forms $(C_6H_5C{\equiv}C)_2Hg$ — [6]

11 2$C_6H_5C{\equiv}CCu \cdot [CH_3CO_2]_2Hg$ — preparation like No. 9 from $[CH_3CO_2]_2Hg$, low yield — [6]

12 2$C_5H_5FeC_5H_4C{\equiv}CCu \cdot HgCl_2$ — preparation not given decomposes in $H_2SO_4/30\%\ H_2O_2$ — [9]

References:

[1] F. Straus (Liebigs Ann. Chem. **342** [1905] 190/265, 193/4).

[2] R. M. Schierl (Z. Calciumcarbid **10** [1909] 93/5).

[3] P. Piganiol (Acétylène: Homologues et Dérivés, Masson, Paris 1945, p. 263).

[4] A. Vaitiekunas, F. F. Nord (J. Org. Chem. **19** [1954] 902/6).

[5] M. Gaudemar (Compt. Rend. **258** [1964] 4803/4).

[6] A. M. Sladkov, L. Yu. Ukhin, Z. I. Orlova (Izv. Akad. Nauk SSSR Ser. Khim. **1968** 2586/90; Bull. Acad. Sci. USSR Div. Chem. Sci. **1968** 2446/9).

[7] J. F. Normant, M. Bourgain, A.-M. Rone (Compt. Rend. C **270** [1970] 354/7).

[8] L. I. Zakharkin, V. N. Kalinin, I. R. Gol'ding, A. M. Sladkov, A. V. Grebennikov (Zh. Obshch. Khim. **41** [1971] 823/8; J. Gen. Chem. [USSR] **41** [1971] 830/3).

[9] E. H. Terent'eva, N. N. Smirnova, I. M. Pruslina (Zavodsk. Lab. **40** [1974] 143/5; Ind. Lab. [USSR] **40** [1974] 175/7).

[10] S. Searles, Y. Li, B. Nassim, M.-T. R. Lopes, P. T. Tran, P. Crabbé (J. Chem. Soc. Perkin Trans. I **1984** 747/51).

[11] O. M. Temkin, R. M. Flid, E. D. German, T. A. Onishchenko (Kinetika Kataliz 2 [1961] 205/13; Kinet. Catal. [USSR] 2 1961] 190/8).

[12] O. A. Chaltykyan (Zh. Obshch. Khim. **18** [1948] 1626/38).

[13] R. Ohlson (Acta Chem. Scand. **20** [1966] 585/6).

[14] O. M. Abu-Salah (J. Organometal. Chem. **270** [1984] C26/C28).

1.1.2.6 Compounds of the Type [RC≡CCuH]Li

The structure of these reagents has not yet been determined exactly. In this section the formulation [RC≡CCuH]Li is preferred, but in the literature "RC≡CLi · CuH" and "HLiCuC≡CR" are also common. A similar product with a Cu:Li ratio of about 3 exists too [2, 3].

[RC≡CCuH]Li has only been prepared in solution from RC≡CLi and (CuH)$_n$ in ether or tetrahydrofuran at −40 to −78 °C. This red or brown solution is a useful means for reducing the structural unit C=CC=O to give CHCHC=O with high regiospecificity and high functional selectivity (comparable to that of the hydrogenation). The reaction is usually applied to cyclic enones. In cases where the β-carbon is highly substituted, added hexamethylphosphoric triamide appears to facilitate somewhat the reduction of the double bond. Other normally reducible functionalities are unaffected. Usually the molar ratio of substrate to complex is 1:6 [1 to 3].

[n-C$_3$H$_7$C≡CCuH]Li is formed from n-C$_3$H$_7$C≡CLi and (CuH)$_n$ at −40 °C in ether or in tetrahydrofuran [1, 2]. The ethereal solution and 2-bromononane react to form 40% (at 25 °C) or 5% (at 0 °C) n-nonane and 4% nonene with second-order kinetics. This reaction does not occur in tetrahydrofuran [2].

The following enones were selectively hydrogenated at the C=C double bond; generally an antiparallel entry of the hydride is observed.

enone	yield, temperature (in °C)	Ref.
reactions in THF		
CH$_2$=C(CH$_3$)COC$_2$H$_5$	85% at −20°	[1]
CH$_2$=CH(CH$_2$)$_3$CO$_2$CH$_3$	42% at −120°	[1]
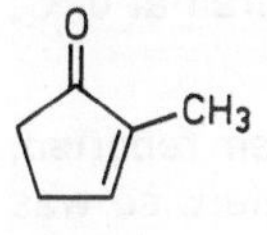 (cyclopentenone with CH$_3$)	70% at −20°	[1]
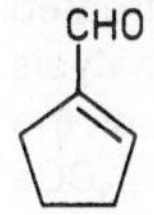 (cyclopentene with CHO)	86% at −78°	[1]
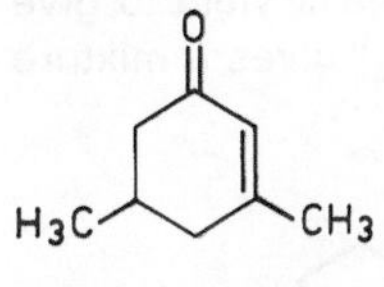 (cyclohexenone, H$_3$C, CH$_3$ substituents)	90% at −20° (cis:trans = 9:1)	[1]
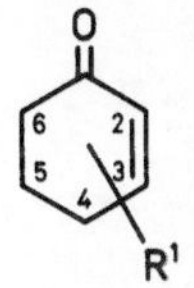 R^1 = H	in ether 30% at −40°, 90% at 0°; THF retards this reaction and causes side-reactions	[2]
R^1 = 2-CH$_3$	64% at −20°	[1]
R^1 = 6-CH$_3$	50% at −20°	[1]
R^1 = 6-CH(CH$_3$)$_2$	95% at −20°	[1]
R^1 = 2-CH$_2$CH$_2$O$_2$CCH$_3$	76% at −20°	[1]

References on p. 181

enone	yield, temperature (in °C)	Ref.
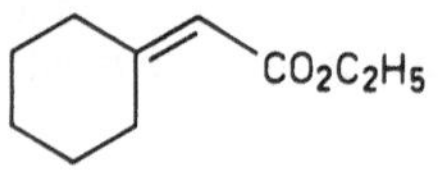 (cyclohexylidene) $CO_2C_2H_5$	47% at −20°	[1]

reactions in THF in the presence of 10% $OP(N(CH_3)_2)_3$

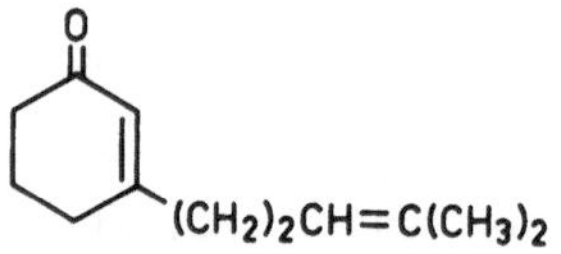 (cyclohexenone) $(CH_2)_2CH=C(CH_3)_2$	74% at −20°, no hydrogenation of the side chain	[1]
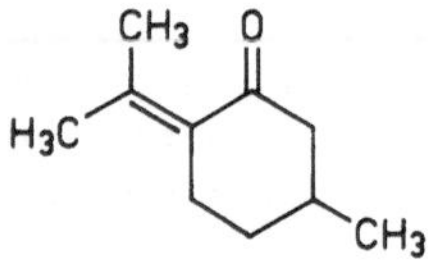 CH_3 H_3C (cyclohexanone with CH_3)	85% at −20°	[1]
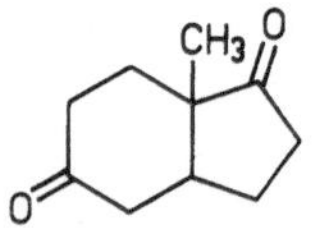 CH_3 (bicyclic dione)	22% at −20°	[1]
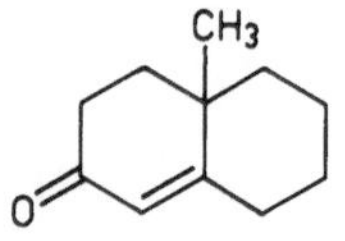 CH_3 (bicyclic enone)	67% at −20° (cis:trans = 7:3)	[1]

[n-C_4H_9C≡CCuH]Li is formed from n-C_4H_9C≡CLi and $(CuH)_n$ in tetrahydrofuran at 0 °C; no further conditions or subsequent reactions are reported [3].

[($CH_3)_3$CC≡CCuH]Li. The literature data are contradictory. There has been reported, $(CuH)_n$ "does not dissolve, when treated in ether with $(CH_3)_3$CC≡CLi" [3]; otherwise was stated, "diethylether is definitely superior to tetrahydrofuran" [2]. When $(CuH)_n$ is treated in tetrahydrofuran with $(CH_3)_3$CC≡CLi, it dissolves at a Cu:Li ratio of 3:1. The analysis of the brown solution gives 2.6:1, which is probably due to the pyridine content of the $(CuH)_n$ prepared according to [4]. With an excess of this solution (corresponding to $(CH_3)_3$CC≡ CLi · 3CuH) 3,5,5-trimethylcyclohex-2-enone is reduced at −27 °C in a 48% yield to give 3,5,5-trimethylcyclohexanone. Under the same conditions I (R^1=H or CH_3) gives a mixture of II and III, where the cis isomer is predominant (77:23 with R^1=H) [3].

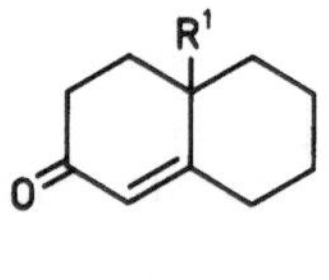

$$R^1 \qquad\qquad R^1 \qquad\qquad R^1$$

I 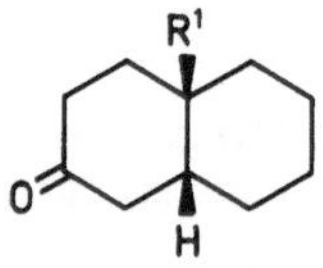   II 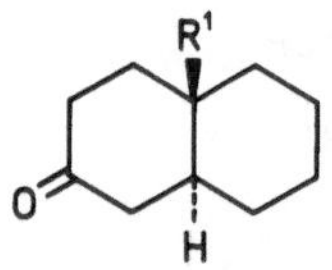   III

With a Cu:Li ratio of 1:1 (corresponding to [($CH_3)_3$CC≡CCuH]Li) in tetrahydrofuran the yield is lowered and higher boiling byproducts are formed [3]. From [($CH_3)_3$CC≡CCuH]Li or "LiCuHC≡CC($CH_3)_3$" and cyclohex-2-enone after 1 h in ether (reagent:substrate ratio=

6:1) at −40 °C 50% and at 0 °C 90% cyclohexanone is formed. At a reagent:substrate ratio of 4:1 the yields are somewhat lower. Using tetrahydrofuran instead of ether retards the reaction and causes more byproducts. Hexamethylphosphoric triamide does not influence the reaction. The reactivity of $[(CH_3)_3CC\equiv CCuH]Li$ is higher than that of $[n-C_3H_7C\equiv CCuH]Li$ [2].

References:

[1] R. K. Boeckman Jr., R. Michalak (J. Am. Chem. Soc. **96** [1974] 1623/5).

[2] S. Masamune, G. S. Bates, P. E. Georghiou (J. Am. Chem. Soc. **96** [1974] 3686/8).

[3] H. O. House, J. C. DuBose (J. Org. Chem. **40** [1975] 788/90).

[4] G. M. Whitesides, J. S. Filippo Jr., E. R. Stredronsky, C. P. Casey (J. Am. Chem. Soc. **91** [1969] 6542/4).

1.1.2.7   Salts of the Anions $[(RC\equiv C)_2Cu^I]^-$ and $[R^1C\equiv CCu^IC\equiv CR^2]^-$

The bonding schema resembles that of the complex anion $[(RC\equiv C)_3Cu]^{2-}$; see Section 1.1.2.8. The cations in these salts are often complexes, e.g., $[Ni(NH_3)_6]^{2+}$. Depending on the preparation method, these compounds sometimes contain stoichiometric amounts of NH_3. For examples see Table 25.

The asymmetric complexes $[(RC\equiv C)R'Cu]M$ (R' = alkyl, alkenyl, aryl) transfer the group R' more easily, whereas $RC\equiv C$ is strongly bonded. They are dealt with therefore in "Organo-copper Compounds" 2, 1983, in Sections 1.1.1.2.4 (M = Li) and 1.1.1.2.6 (M = MgX') with examples in Tables 12 on p. 175 ff. and 17 on p. 228 ff.

The symmetric complexes in Table 25 are prepared by the following methods. If not stated otherwise, they have not been isolated from the preparation mixture.

Method I: $RC\equiv CCu + RC\equiv CM \rightarrow [(RC\equiv C)_2Cu]M$, M = Li or K, in liquid NH_3 or in an ether.

Method II: $2RC\equiv CM + CuI \rightarrow [(RC\equiv C)_2Cu]M + MI$, M = Li or Na, in liquid NH_3 or in an ether. In liquid NH_3 complexes of NH_3 can be formed.

Method III: $[(RC\equiv C)_3Cu]K_2 + CuI \rightarrow [(RC\equiv C)_2Cu]K + RC\equiv CCu + KI$ in liquid NH_3.

Method IV: $4RC\equiv CM + 2Cu^{2+} \rightarrow [(RC\equiv C)_2Cu]^- + R(C\equiv C)_2R + Cu^+ + 4M^+$, M = alkali metal, in liquid NH_3. Cu^{2+} is probably first reduced by $RC\equiv CM$.

$[(RC\equiv C)_2Cu]Li$ and bicyclic enones yield bicyclic or tricyclic products, respectively, dependent on the substitution type of the enone. $[(n-C_3H_7C\equiv C)_2Cu]Li$ (No. 3 in Table 25) reacts with I (R¹ = CH_3SO_3 or $4-CH_3C_6H_4SO_3$) at −78 °C in ether/CH_2Cl_2 to give 83% II (1RS,5RS,6RS compound). Under the same conditions $[\{(CH_3)_3CC\equiv C\}_2Cu]Li$ (No. 6 in Table 25) gives with I (R¹ = $4-CH_3C_6H_4SO_3$) no II, but 50% III (R = $(CH_3)_3C$ and R¹ = $4-CH_3C_6H_4SO_3$). This structure is only "tentatively assigned" to the reaction product. In ether at −78 °C $[(n-C_3H_7C\equiv C)_2Cu]Li$ and IV give 38% III (R = $n-C_3H_7$ and R¹ = H). In both cases the stereochemistry of III is unclear. All reactions are complete after 1 h [9].

I   II   III   IV

                          References on p. 184

Table 25
Salts of the Anions [(RC≡C)$_2$Cu]$^-$.
For abbreviations and dimensions see p. X.

No.	salt	method of preparation (solvent, yield) properties, remarks, reactions	Ref.
1	[(CH≡C)$_2$Cu]K	III (liquid NH$_3$) yields metallic Cu with K, see also "Kupfer" B3, 1965, p. 1044	[2, 3]
2	[(CH$_3$C≡C)$_2$Cu]K	I, III (100%) yellow solid, very slightly soluble in liquid NH$_3$ reacts with K like No. 1, yields with CuI CH$_3$C≡CCu	[2, 3]
3	[(n-C$_3$H$_7$C≡C)$_2$Cu]Li	II (ether) gives n-C$_3$H$_7$(C≡C)$_2$C$_3$H$_7$-n with (E)-C$_6$H$_5$CH=CHCOC$_6$H$_5$ for the reaction with bicyclic enones see p. 181	[9]
4	[{(CH$_3$)$_3$SiC≡C}$_2$Cu]Li	II (THF/C$_6$H$_{14}$, with CuBr instead of CuI) gives 70% (CH$_3$)$_3$SiC≡CC(CH$_3$)=C=C(CH$_3$)-CH$_2$OH with 2-(prop-1-ynyl)-2-methyl-oxirane at 25° in THF, but only in presence of Pd[P(C$_6$H$_5$)$_3$]$_4$	[11]
5	[(n-C$_4$H$_9$C≡C)$_2$Cu]Li	II (ether/THF) reacts with (E)-C$_2$H$_5$CH=C(CH$_3$)CHO to give (E)-C$_2$H$_5$CH=C(CH$_3$)CH(OH)C≡CC$_4$H$_9$-n	[8]
6	[{(CH$_3$)$_3$CC≡C}$_2$Cu]Li	II (ether) for the reaction with bicyclic enones see p. 181	[9]
7	[(n-C$_5$H$_{11}$C≡C)$_2$Cu]Li	I (ether) reacts with CH$_3$COCl to give 33% n-C$_5$H$_{11}$C≡CCOCH$_3$	[5]
8	[(C$_6$H$_5$C≡C)$_2$Cu]Li	II (ether or THF with [ICuP(C$_4$H$_9$-n)$_3$]$_4$ instead of CuI) yields C$_6$H$_5$(C≡C)$_2$C$_6$H$_5$ with O$_2$ at −78° and CH$_2$=C(C≡CC$_6$H$_5$)CH$_2$SO$_2$C$_6$H$_4$CH$_3$-4 with CH$_2$=C=CHSO$_2$C$_6$H$_4$CH$_3$-4 yields C$_6$H$_5$CH$_2$CN, obviously via C$_6$H$_5$C≡CNH$_2$, with (C$_6$H$_5$)$_2$P(O)ONH$_2$ at −20° in THF	[4, 7, 10]
9	[(C$_6$H$_5$C≡C)$_2$Cu]Na · 2NH$_3$	II (25%) colorless crystals, stable under NH$_3$ atmosphere reacts with CuI in liquid NH$_3$ to give 80% C$_6$H$_5$C≡CCu · NH$_3$ (see Section 1.1.2.4) and with [Ni(NH$_3$)$_6$](SCN)$_2$ in liquid NH$_3$ to give No. 11	[1, 2]
10	[(C$_6$H$_5$C≡C)$_2$Cu]K	III (liquid NH$_3$) yields metallic Cu with K	[2, 3]

References on p. 184

Table 25 [continued]

No.	salt	method of preparation (solvent, yield) properties, remarks, reactions	Ref.
11	$[(C_6H_5C\equiv C)_2Cu]_2[Ni(NH_3)_6] \cdot 2NH_3$	for preparation see under No. 9 brownish violet solid, yields after 3 h at room temperature in high vacuum 60% of No. 12	[2]
12	$[(C_6H_5C\equiv C)_2Cu]_2Ni \cdot 4NH_3$	for preparation see under No. 11 formulated as "$NiCu_2(C\equiv CC_6H_5)_4 \cdot 4NH_3$" black solid, soluble in benzene	[2]
13	$[(C_6H_5C\equiv C)_2Cu^I]_2[Cu^{II}(NH_3)_4] \cdot xNH_3$	IV (liquid NH_3) x not estimated, upon evaporation of all NH_3 $C_6H_5C\equiv CCu$ and $C_6H_5(C\equiv C)_2C_6H_5$ are formed	[2]
14	$[(C_6H_5C\equiv C)_2Cu^I]_2[Cu^{II}(H_2N(CH_2)_2NH_2)] \cdot 2NH_3$	for preparation see under No. 15 green–yellow solid yields $C_6H_5C\equiv CCu$ and $C_6H_5(C\equiv C)_2C_6H_5$ on exposure to high vacuum	[2]
15	$[(C_6H_5C\equiv C)_2Cu^I]_2[Cu^{II}(H_2N(CH_2)_2NH_2)_2] \cdot 12NH_3$	IV (in presence of $H_2N(CH_2)_2NH_2$) violet crystals after drying in high vacuum at $-78°$ yields No. 14 at $-30°$ in high vacuum	[2]
16	$[(C_6H_5C\equiv C)_2Cu^I][N(P(C_6H_5)_3)_2]$	prepared from Nos. 9 or 10 (?) more stable than Nos. 9 and 10 gives $[(C_6H_5C\equiv C)_2Au]^-$ and $C_6H_5C\equiv CCu$ $\cdot P(C_6H_5)_3$ with $C_6H_5C\equiv CAu \cdot P(C_6H_5)_3$ in CH_2Cl_2 at room temperature (see Section 1.1.2.4) gives besides $C_6H_5C\equiv CCu$ and $C_6H_5C\equiv CCu$ $\cdot P(C_6H_5)_3$ (see Section 1.1.2.4) also $[(C_6H_5-C\equiv C)_2Au]^-$ with $AuCl \cdot P(C_6H_5)_3$ in CH_2Cl_2 gives $[(C_6H_5C\equiv C)_2Au]^-$ and $C_6H_5C\equiv CCu$ with $C_6H_5C\equiv CAu$ or $[C_6H_5C\equiv CAuCl]^-$ in CH_2Cl_2	[12]

$[4-C_6H_5C_6H_4NHCOCH_2C\equiv CCuC\equiv CCH=CHC_2H_2OC_5H_{11}]Li$ (see IV)

The asymmetric complex acetylide is obtained from $4-C_6H_5C_6H_4NHCOCH_2C\equiv CCu$ and V in tetrahydrofuran/hexamethylphosphoric triamide. Oxidation with iodine in the same solvent mixture gives VI in a 44% yield [6].

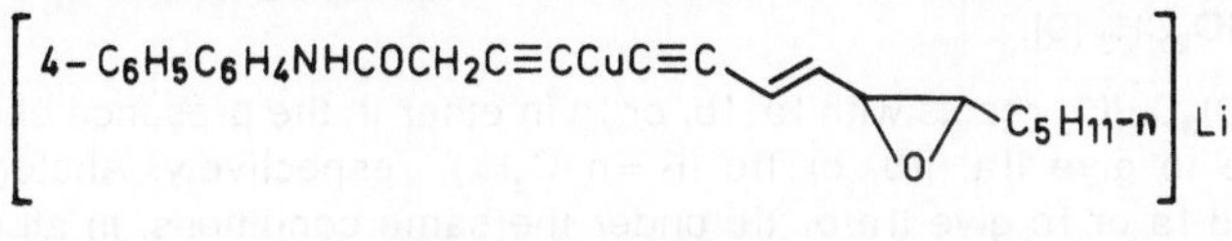

IV

References on p. 184

$LiC\equiv C$ — (structure) — C_5H_{11}-n

$4\text{-}C_6H_5C_6H_4NHCOCH_2(C\equiv C)_2$ — (structure) — C_5H_{11}-n

V   VI

References:

[1] R. Nast (Z. Naturforsch. **8b** [1953] 381/3).
[2] R. Nast, W. Pfab (Chem. Ber. **89** [1956] 415/21).
[3] R. Nast, P.-G. Kirst, G. Beck, J. Gremm (Chem. Ber. **96** [1963] 3302/5).
[4] M. L. H. Green (Organometallic Compounds II: The Transition Elements, 3rd Ed., Methuen, London 1968, pp. 271/6).
[5] M. Bourgain, J. F. Normant (Bull. Soc. Chim. France **1973** 2137/42).
[6] E. J. Corey, G. W. J. Fleet, M. Karo (Tetrahedron Letters **1973** 3963/6).
[7] K. Koosha, J. Berlan, M.-L. Capmau (Compt. Rend. C **276** [1973] 1633/6).
[8] C. Chuit, J. P. Foulon, J. F. Normant (Tetrahedron **37** [1980] 1385/9).
[9] D. J. Hannah, R. A. J. Smith, I. Teoh, R. T. Weavers (Australian J. Chem. **34** [1981] 181/8).
[10] G. Boche, M. Bernheim, W. Schrott (Tetrahedron Letters **1982** 5399/402).
[11] H. Kleijn, J. Meijer, G. C. Overbeek, P. Vermeer (Recueil J. Roy. Neth. Chem. Soc. **101** [1982] 97/101).
[12] O. M. Abu-Salah, A. R. Al-Ohaly (J. Organometal. Chem. **255** [1983] C39/C40).

1.1.2.8 Salts of the Anions $[(RC\equiv C)_3Cu^I]^{2-}$

The small volume taken up by the acetylide group in the region of the valence orbitals of copper makes anions $[(RC\equiv C)_nCu^I]^{(n-1)-}$ possible. Probably a considerable back-donation occurs from the filled d-orbitals on the metal into antibonding π^*-orbitals of the acetylide. The bonding scheme is like that for N_2 complexes, but the anionic acetylide ligands seem to be better σ-donors than N_2. In this respect acetylides resemble cyanide ion.

Salts of the anions $[(RC\equiv C)_3Cu]^{2-}$ can be prepared from copper acetylides and alkali metal acetylides, or from alkali metal acetylides and copper halides or complexes thereof. Ethers and liquid NH_3 are suitable solvents. With the latter, NH_3 complexes can be formed. In Table 26 the preparation procedure is indicated by the following methods:

Method I: $RC\equiv CCu + 2RC\equiv CM \rightarrow [(RC\equiv C)_3Cu]M_2$, $M = Li$ or K (cf. Method I in Section 1.1.2.7).

Method II: $3RC\equiv CM + CuX \rightarrow [(RC\equiv C)_3Cu]M_2 + MX$, $M = Li$ or K; $X = Br$, I, or CN (cf. Method II in Section 1.1.2.7).

Method III: $3RC\equiv CLi + CuBr \cdot S(CH_3)_2 \rightarrow [(RC\equiv C)_3Cu]Li_2 + LiBr + S(CH_3)_2$.

$[(RC\equiv C)_3Cu]Li_2$ have become interesting for the preparation of alkynylamines according to the equation $[(RC\equiv C)_3Cu]Li_2 + 2(CH_3)_2NX \rightarrow 2RC\equiv CN(CH_3)_2 + RC\equiv CCu + 2LiX$, where $X = PO_2(C_6H_5)_2$ or SO_2CH_3 [9].

$[(n\text{-}C_3H_7C\equiv C)_3Cu]Li_2$ reacts with Ia, Ib, or Ic in ether in the presence of hexamethylphosphoric triamide to give IIa, IIb, or IIc ($R = n\text{-}C_3H_7$), respectively. Analogously, $[(n\text{-}C_4H_9\text{-}C\equiv C)_3Cu]Li_2$ and Ia or Ic give IIa or IIc under the same conditions. In all cases, the regiospecificity is high and the yields are between 85 and 95% [7].

References on p. 187

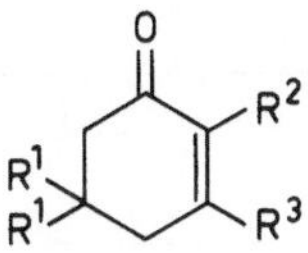

a : $R^1 = R^2 = R^3 = H$

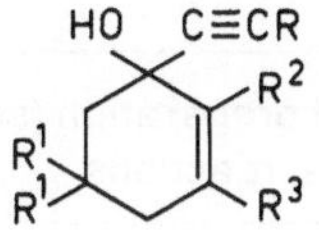

b : $R^1 = R^3 = H$; $R^2 = CH_3$

c : $R^1 = R^3 = CH_3$; $R^2 = H$

I   II

$[(n-C_3H_7C{\equiv}C)_3Cu]Li_2$ and III give under the conditions outlined before IV (yields 85% at $R^1 = H$, 95% at $R^1 = CH_2{=}CHCH_2$) [7].

$[(n-C_4H_9C{\equiv}C)_3Cu]Li_2$ and V in dioxane (only partly dissolved) at 30 °C give 62% VI. $[(C_6H_5C{\equiv}C)_3Cu]Li_2$ and V in ethereal solution at 30 °C also yield VI (no yield given), $C_6H_5C{\equiv}CH$, and two higher molecular products of unknown structure [5].

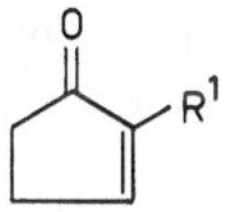

III 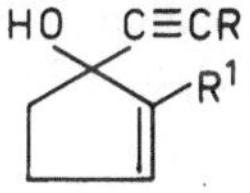   IV 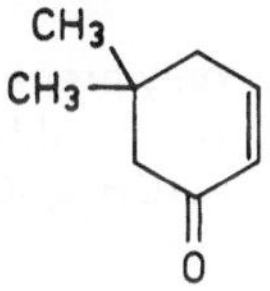   V 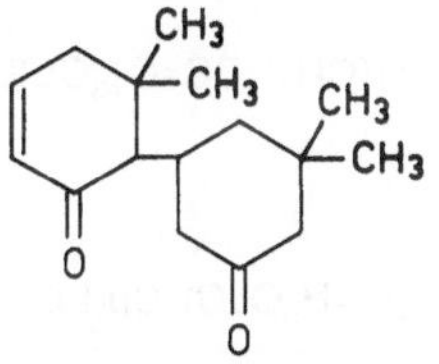   VI

Table 26
Salts of the Anions $[(RC{\equiv}C)_3Cu]^{2-}$.
For abbreviations and dimensions see p. X.

No.	compound	method of preparation (solvent) properties, reactions	Ref.
1	$[(CH{\equiv}C)_3Cu]K_2$	I, II (X=I, yield 75%; both in NH_3) white, diamagnetic, not shock-sensitive, sparingly soluble in liquid NH_3 susceptibility $\chi_g = -0.7 \times 10^{-6}$ cm$^3 \cdot$ g^{-1} (cylinder method) reacts with H_2O to give Cu_2C_2 and C_2H_2, with K in liquid NH_3 to give metallic Cu see also "Kupfer" B 3, 1965, p. 1044	[1 to 3]
2	$[(CH_3C{\equiv}C)_3Cu]K_2$	I, II (X=I; both in NH_3) white, diamagnetic, not shock-sensitive, sparingly soluble in liquid NH_3 reacts with H_2O to give $CH_3C{\equiv}CCu$, with K in liquid NH_3 to give metallic Cu gives $[(CH_3C{\equiv}C)_2Cu]K$ (see Section 1.1.2.7) with CuI in liquid NH_3	[1 to 3]
3	$[(n-C_3H_7C{\equiv}C)_3Cu]Li_2$	only prepared in solution: I, III (both in ether), II (X=I; ether:hexamethylphosphoric triamide=4:1) for reactions with cyclic enones see above reacts with $(CH_3)_2NX$ (X=$PO_2(C_6H_5)_2$ or SO_2CH_3) to give 69 to 87% $n-C_3H_7C{\equiv}CN(CH_3)_2$	[7, 9]

References on p. 187

Table 26 [continued]

No.	compound	method of preparation (solvent) properties, reactions	Ref.
4	$[\{(CH_3)_3SiC\equiv C\}_3Cu]Li_2$	only prepared in ethereal solution: II ($X = Br$), III gives 67% $(CH_3)_3SiC\equiv CN(CH_3)_2$ with $(CH_3)_2NPO_2(C_6H_5)_2$ at $-20°$	[9]
5	$[(n\text{-}C_4H_9C\equiv C)_3Cu]Li_2$	only prepared in solution: I (ether), II ($X = I$; dioxane), II ($X = CN$; ether), III (ether) for reactions with cyclic enones see p. 185 reacts with $(CH_3)_2NX$ ($X = PO_2(C_6H_5)_2$ or SO_2CH_3) to give 78% $n\text{-}C_4H_9C\equiv CN(CH_3)_2$	[5, 7, 9]
6	$[\{(CH_3)_3CC\equiv C\}_3Cu]Li_2$	only prepared in ethereal solution: II ($X = CN$), III reacts with $(CH_3)_2NPO_2(C_6H_5)_2$ at $-20°$ to give 71% $(CH_3)_3CC\equiv CN(CH_3)_2$	[9]
7	$[(C_6H_5C\equiv C)_3Cu]Li_2$	only prepared in solution: I (ether), II ($X = Br$; ether), II ($X = I$; ether or dioxane), II ($X = CN$; ether), III (ether) reacts with $n\text{-}C_5H_{11}I$ to give $<1\%$ $C_6H_5C\equiv CC_5H_{11}\text{-}n$ gives 79% (E)$\text{-}CH_3CH\text{=}CHCH(OSi(CH_3)_3)C\equiv CC_6H_5$ with (E)$\text{-}CH_3CH\text{=}CHCHO$ and $(CH_3)_3SiCl$ reacts with $(CH_3)_2NX$ ($X = PO_2(C_6H_5)_2$ or SO_2CH_3) to give up to 83% $C_6H_5C\equiv CN(CH_3)_2$	[5, 6, 8, 9]
8	$[(C_6H_5C\equiv C)_3Cu]K_2$	for preparation see under No. 9 yellowish powder reacts with $Ba(SCN)_2$ in liquid NH_3 to give No. 10 yields with K or Li in liquid NH_3 a red solution containing $[(C_6H_5C\equiv C)_3Cu^0]^{3-}$ with precipitation of $LiNH_2$; in presence of Ba^{2+} the $[(C_6H_5C\equiv C)_3Cu^0]_2Ba_3$ is precipitated (see Section 1.1.2.9)	[2 to 4]
9	$[(C_6H_5C\equiv C)_3Cu]K_2 \cdot 2NH_3$	II ($X = I$; liquid NH_3) yellowish needles, stable in an NH_3 atmosphere yields No. 8 after 3 h at room temperature in high vacuum	[2]
10	$[(C_6H_5C\equiv C)_3Cu]Ba$	from $[(C_6H_5C\equiv C)_3Cu^0]_2Ba_3$ (see Section 1.1.2.9) and O_2; see also No. 8 yellowish crystals IR (KBr): 690, 750 (both CH), 900 to 1200 (CH aromatic), 1440, 1480, 1575, 1592 (all C–C), 2045 (C$\equiv$C), 3030 (CH aromatic)	[3]
11	$[(C_6H_5SC\equiv C)_3Cu]Li_2$	only prepared in ethereal solution: II ($X = CN$), III gives 17 to 45% $C_6H_5SC\equiv CN(CH_3)_2$ with $(CH_3)_2NPO_2(C_6H_5)_2$ at $-20°$	[9]

Table 26 [continued]

No.	compound	method of preparation (solvent) properties, reactions	Ref.
12	[(cyclo-$C_6H_{11}C\equiv C)_3Cu$]Li_2	only prepared in ethereal solution: II (X = CN), III gives cyclo-$C_6H_{11}C\equiv CN(CH_3)_2$ with $(CH_3)_2NPO_2(C_6H_5)_2$ at $-20°$	[9]
13	[(n-$C_6H_{13}C\equiv C)_3Cu$]Li_2	only prepared in ethereal solution: I, II (X = CN) gives n-$C_6H_{13}C\equiv CN(CH_3)_2$ with $(CH_3)_2NPO_2(C_6H_5)_2$ at $-20°$	[9]

References:

[1] R. Nast (Z. Naturforsch. **8b** [1953] 381/3).
[2] R. Nast, W. Pfab (Chem. Ber. **89** [1956] 415/21).
[3] R. Nast, P.-G. Kirst, G. Beck, J. Gremm (Chem. Ber. **96** [1963] 3302/5).
[4] R. Nast (Angew. Chem. **77** [1965] 352).
[5] H. O. House, W. F. Fischer Jr. (J. Org. Chem. **34** [1969] 3615/8).
[6] G. M. Whitesides, W. F. Fischer Jr., J. S. Filippo Jr., R. W. Brashe, H. O. House (J. Am. Chem. Soc. **91** [1969] 4871/82).
[7] G. Palmisano, R. Pellegata (J. Chem. Soc. Chem. Commun. **1975** 892/3).
[8] C. Chuit, J. P. Foulon, J. F. Normant (Tetrahedron **37** [1980] 1385/9).
[9] G. Boche, M. Bernheim, M. Nießner (Angew. Chem. Suppl. **1983** 34/8; Angew. Chem. **95** [1983] 48).

1.1.2.9 Salts of the Anion [($C_6H_5C\equiv C)_3Cu^0$]$^{3-}$

Compounds containing the anion [($C_6H_5C\equiv C)_3Cu^0$]$^{3-}$ are very strong reductants and extremely sensitive towards air and humidity. Ignition, fog formation, and carbonization have been observed upon sudden contact with air. Ignition occurs also upon contact with NO.

[($C_6H_5C\equiv C)_3Cu^0$]K_3 is prepared as a red solution on the treatment of [($C_6H_5C\equiv C)_3Cu^I$]K_2 (see Section 1.1.2.8) with metallic K or Li in liquid NH_3. Li is preferable because after reduction the insoluble $LiNH_2$ can be removed [1].

[($C_6H_5C\equiv C)_3Cu^0$]$_2Ba_3$ is prepared from a solution of [($C_6H_5C\equiv C)_3Cu^0$]K_3 (see above) in liquid NH_3 with Ba^{2+} [1] or from [($C_6H_5C\equiv C)_3Cu^I$]K_2, metallic Li, and Ba^{2+} in liquid NH_3 [2]. It is orange-brown and forms chocolate-colored crystals after drying in high vacuum. It is sparingly soluble in liquid NH_3, but soluble in absolute tetrahydrofuran. It is not explosive [1, 2].

Possibly the actual structure of this compound is [($C_6H_5C\equiv C)_3Cu-Cu(C\equiv CC_6H_5)_3$]$Ba_3$ because of its diamagnetic character [1, 2]. The IR spectrum in a Nujol mull under nitrogen shows the bands 690, 750 (both CH), 900 to 1200 (aromatic CH), 1440, 1480, 1575, 1590 (all C-C), 2045 (C$\equiv$C), 3030 (aromatic CH) cm^{-1} [1]. "Handling" for more than two or three minutes (air contact?) produces decomposition products, which show IR bands at 1390 cm^{-1} and at 3450 cm^{-1} (OH). The reaction with traces of oxygen produces $C_6H_5C\equiv CCu$ and [($C_6H_5C\equiv C)_3Cu^I$]Ba [1, 2].

References:

[1] R. Nast, P.-G. Kirst, G. Beck, J. Gremm (Chem. Ber. **96** [1963] 3302/5).
[2] R. Nast (Angew. Chem. **77** [1965] 352).

1.1.3   Carbonyls

1.1.3.1   Compounds with One CO Ligand

The literature up to 1949 is described in "Kupfer" B1, 1958, on pp. 240/1, 362, 410, and B2, 1961, on p. 678.

The following Table 27 contains compounds with one CO group. Those with an uncertain composition, like $(CO)CuX \cdot yNH_3$ where $X = ClO_4^-$, NO_3^-, and IO_3^- and $y = 2.8$ to 3.6 [5], or postulated intermediates, like $[(CO)CuD_x]^+$ (D can be several amines) [22] and $[(CO)Cu(C_5H_{10}NH)_2]Cl$ $(C_5H_{10}NH = piperidine)$ [11], are not included. Compounds of formula $(CO)CuX$ $(X = HSO_4^-, ClO_4^-, O_2CH^-,$ and $O_2CCH_3^-$, Table 27, Nos. 10, 13, 15, and 16) are formulated in [29]. CO stabilizes Cu^I phenoxides by formation of carbonyl complexes [96].

The penta-coordinate complexes $[(CO)Cu^ID]^{n+}$ (Nos. 55 to 63) with tetradentate ligands D are accessible from the $[Cu^ID]^{n+}$ complexes and CO in $(CH_3)_2NCHO$. The equilibrium constants $K = [(CO)Cu^ID^{n+}]/[Cu^ID^{n+}] \cdot [CO]$ given in the Table 27 can be calculated from the influence of CO on the redox potential $[Cu^ID]^{n+}/[Cu^{II}D]^{(n+1)+}$ [57].

In CH_3CN solution, Cu^I complexes with tri- and tetradentate Schiff bases $D = \{RN=C(CH_3)\}_2C_5H_3N$ derived from 2,6-diacetylpyridine and amines RNH_2 absorb CO reversibly with a color change from red or brown to yellow or yellow-brown, $R = n-C_3H_7$, $n-C_4H_9$, $CH_2=CHCH_2$, $CH_2=CHCH_2CH_2$, $C_2H_5SCH_2CH_2$. The IR spectra of the solutions indicate that the CO is terminally bonded to the Cu atom. At 0 °C the molar ratio CO:Cu never exceeded 0.6:1, being largest for $R = n-C_3H_7$ and smallest for $R = C_2H_5SCH_2CH_2$. The variation of the stability constants $K = [(CO)CuD^+]/[CuD^+] \cdot [CO]$ for the different complexes can be related to the degree of competition between CO and the side-arm donor groups in R for the fourth coordination site on the Cu^I according to Scheme I. For complexes with $R = alkyl$ there is, of course, no competition. The difference in K between $R = n-C_3H_7$ and $n-C_4H_9$ probably lies in the greater steric effect of the longer side chain. The K value with $R = CH_2=CHCH_2$ indicates only a minor degree of competition from the pendant alkenyl group. With $R = CH_2=CHCH_2CH_2$, the competition is much greater, as also concluded from both ^1H NMR and UV spectra in solution and consistent with the solid state structure of dimeric $[Cu_2D_2][B(C_6H_5)_4]_2$ which reveals the attachment of one alkenyl group per Cu^I. Stability constants K of $[(CO)Cu\{RN=C(CH_3)\}_2C_5H_3N]^+$ [90] are as follows:

R	K in M^{-1} at 0 °C	ν (CO) in cm^{-1} at 20 °C
$n-C_3H_7$	32.1	2085
$n-C_4H_9$	24.1	2085
$CH_2=CHCH_2$	27.4	2085
$CH_2=CHCH_2CH_2$	1.6	2083
$C_2H_5SCH_2CH_2$	too low for measurement	2095

As in the dinuclear complexes $[(CO)Cu_2^IZ_2D]^{n+}$ (Nos. 65 to 74) only one Cu atom is coordinated by CO; they are also treated in this section. The most likely geometries around the Cu^I atoms are square pyramidal and tetrahedral. The equilibrium constants $K = [(CO)Cu_2Z_2D]/[Cu_2Z_2D] \cdot [CO]$ have been determined spectrophotometrically at 440 nm [70, 87].

For luminescing tripod Cu carbonyls (not in Table 27), see [95].

References on pp. 208/10

Table 27
Compounds Containing One CO Group.
Further information on compounds preceded by an asterisk is given at the end of the table.
For abbreviations and dimensions see p. X.

No.	compound	preparation (yield)	spectra and further remarks	Ref.
*1	(CO)Cu	matrix isolated	IR (Ar): 2010.4	[26]
*2	[(CO)Cu]$^+$	$Cu_2O + CO$ in strong acids	IR: 2140 to 2160; Raman (FSO_3H, − 50°): 2149; ^{13}C NMR: 168.3 to 170.7	[14, 15, 29, 75]
*3	(CO)CuCl	see "Kupfer" B 1, p. 240; or CuCl + CO (1 bar, 25°) in CH_3OH, C_6H_6 (slow), THF, $(CH_2OCH_3)_2$, $O(CH_2CH_2OCH_3)_2$. CuCl on active carbon absorbs 0.88 mol CO reversibly at 20° and 1 atm CO/N_2 (9:1) [94]	white solid; IR: 2128 (1.3% $HClO_4/H_2O$), 2070 (CH_3OH), 2085 (THF), 2120 (Nujol); Raman: 2130; force constant f(CuC) = 1.70 mdyn · Å^{-1}; $P(OC_2H_5)_3$ replaces CO, with $[(CH_3)_2NCH_2]^+Br^-$ in THF at 20° $[(CH_3)_2NCH_2]^+[Cu_2Cl_2Br]^-$ is formed	[19, 29, 41, 68]
*4	(CO)CuCl · $2H_2O$	see "Kupfer" B 1, pp. 240/1, and No. 5	possibly identical with No. 3 [46]; IR (solid): 2126	[29]
5	(CO)CuCl · $2H_2O$ · $MgCl_2$ · $4H_2O$ (?)	CuCl + CO in concentrated $MgCl_2/H_2O$, or No. 4 + $MgCl_2$	the fine crystals have lower CO vapor pressure than No. 4; with dilute $MgCl_2/H_2O$ forms No. 4	[4]
6	(CO)CuBr	see "Kupfer" B 1, p. 362	acts as intermediate in the carbonylation of Cu(OCH$_3$)Br, not isolated; a CuBr solution in CH_3OH absorbs at 25° and 1 bar only 25 mol% CO	[25, 68]
7	(CO)CuI (?)	CuI + CO at 100 atm and 20°; but compare "Kupfer" B 1, p. 410	not isolated; Raman: 2128	[41]

References on pp. 208/10

Table 27 [continued]

No.	compound	preparation (yield)	spectra and further remarks	Ref.
8	(CO)CuCN	CuCN + CO in NH_3 at $-79°$	contains possibly $2NH_3$ dec. at $-33.5°$	[2]
9	[(CO)Cu]AsF$_6$	CuAsF$_6$ + CO in SO_2 soln.	yellow solid IR (Nujol) : 2180 ± 5	[58]
10	[(CO)Cu]HSO$_4$ (?)	$Cu_2O + H_2SO_4/H_2O$	not isolated IR (50 to 80% H_2SO_4) : 2128 to 2136	[29]
11	[(CO)Cu]O$_3$SCH$_3$	$Cu_2O + CH_3SO_3H + CO$ in $(C_2H_5)_2O$	IR (solid) : 2130 CO is replaced by alkenes	[78]
12	[(CO)Cu]O$_3$SCF$_3$	$Cu_2O + CF_3SO_3H + CO$ in $(C_2H_5)_2O$	pale pink solid IR: 2128 (solid) [78], 2143 [53]	[53, 78]
	[(CO)Cu]O$_3$SC$_2$H$_5$	according to an X-ray analysis polymeric, see the corresponding section		
13	[(CO)Cu]ClO$_4$ (?)	$Cu_2O + HClO_4/H_2O$	not isolated adducts are formed with chelating amines IR (60% HClO$_4$) : 2130	[29]
14	[(CO)Cu(H$_2$O)$_2$]ClO$_4$	CuClO$_4 \cdot 6H_2O + Cu + CO$ in CH_3OH	explosive, colorless crystals, m.p. 10° IR (NaCl, polyethylene) : 2130 dec. at 70° with formation of $(CO)_2Cu_2(H_2O)_3(ClO_4)_2$	[30]
15	(CO)CuO$_2$CH (?)	$Cu_2O + HCO_2H/H_2O$	not isolated IR (95% HCO$_2$H) : 2119	[29]
16	(CO)CuO$_2$CCH$_3$	see "Kupfer" B2, p. 678	not isolated IR (95% CH$_3$CO$_2$H) : 2109 acts as intermediate in the carbonylation of $Cu(OCH_3)O_2CCH_3$	[25, 29]

No.	Compound	Preparation	Properties	Ref.
17	$(CO)CuO_2CCF_3$	pumping off CF_3CO_2H from No. 18	colorless crystals IR (Nujol): 2155 prolonged pumping leaves CuO_2CCF_3	[12]
18	$(CO)CuO_2CCF_3 \cdot CF_3CO_2H$	$Cu_2O + CO$ in CF_3CO_2H (35.3%) or $(CF_3CO_2)_2O$ (54.2%)	colorless crystals IR (Nujol): 2155	[12]
19	$(CO)CuOC_4H_9\text{-}t$	$t\text{-}C_4H_9OCu + CO$ in C_6H_6	pale yellow, sublimes at 60°/1 Torr, dec. at 150° IR (Nujol): 2062 ^1H NMR (C_6H_6): 1.41 (s, CH_3) CO is replaced by $P(C_4H_9\text{-}n)_3$, $P(OCH_3)_3$, $t\text{-}C_4H_9NC$	[23]
20	$(CO)CuF_2$	from CuF_2 in Ar/CO matrix	IR (Ar/CO): 704.2 (CuF), 2210.4 (CO), isotopic IR data are given force constant f (CuF) = 4.53 mdyn $\cdot$ Å^{-1} indicates a 100% ionic character of the Cu–F bond	[24, 49]
*21	$[(CO)CuCl_2]^-$	from $CuCl + CO$ in HCl/H_2O	IR (7 or 36% HCl): 2103 Raman: 2100	[29, 47]
*22	$[(CO)Cu(NH_3)_2]^+$	$Cu^I + CO + NH_3$ in weak aqueous acids	—	[8, 34]
*23	$[(CO)Cu(NH_3)_3]^+$	$Cu^I + CO$ in NH_3/H_2O solutions	—	[1, 6, 8, 34]
24	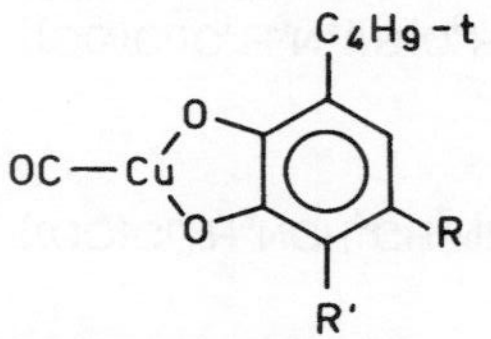 structure (R = t-C_4H_9, R' = H)	$CuCl + CO + Na$ or Tl o-semiquinolate in THF	ESR (THF): g = 2.0032, a(Cu) = 6.1 G	[39]
25	see No. 24 (R = H, R' = t-C_4H_9)	see No. 24	ESR (THF): g = 2.0034, a (Cu) = 5.4 G	[39]
*26	$[(CO)CuH_2NCH_2CH_2NH_2]Cl$	from $(CO)CuCl + (CH_2NH_2)_2$ in CH_3OH at $-30°$	white solid IR (Nujol): 2080 1:1 electrolyte in CH_3OH at $-25°$	[17]

Table 27 [continued]

No.	compound	preparation (yield)	spectra and further remarks	Ref.
*27	$[(CO)CuH_2NCH_2CH_2NH_2][B(C_6H_5)_4]$	see No. 26, in addition of $NaB(C_6H_5)_4$ (25%)	white solid IR (Nujol): 2117	[40, 64]
28	$[(CO)Cu(H_2NCH_2CH_2NH_2)_2]I$	$CuI + (CH_2NH_2)_2 + CO$ in CH_3OH (~52%)	light blue solid IR (Nujol): 2060 $P(OC_2H_5)_3$ replaces CO	[40, 64]
*29	$[(CO)CuC_5H_9N_3][B(C_6H_5)_4]$ ($C_5H_9N_3$ = histamine)	$(CO)CuCl + C_5H_9N_3 + NaB(C_6H_5)_4$ $+ CO$ in CH_3OH at 0°	white solid, thermally very stable IR (Nujol): 2091	[45, 63]
*30	$[(CO)Cu(H_2NCH_2CH_2)_2NH][B(C_6H_5)_4]$	$CuI + (H_2NCH_2CH_2)_2NH +$ $NaB(C_6H_5)_4 + CO$ in CH_3OH (87%)	IR (Nujol): 2080 CO is replaced by $P(C_6H_5)_3$, $P(OC_2H_5)_3$, cyclo-$C_6H_{11}NC$	[42]
31	$(CO)Cu(CF_3CO)_2CH$	$Cu_2O + (CF_3CO)_2CH_2 + CO$ in THF	yellow to green solution IR: ~2100	[83]
32	$[(CO)Cu(H_2NCH_2CH_2CH_2)_2NH][B(C_6H_5)_4]$	—	Cu $2p_{3/2}$ binding energy 933.0 eV (ESCA measurements)	[80]
33	$[(CO)Cu(H_2NCH_2CH_2CH_2)_2NCH_3][B(C_6H_5)_4]$	—	electrochemical behavior as No. 30	[86]
34	$(CO)CuO_3SC_6H_4CH_3\text{-}4$	$Cu_2O + 4\text{-}CH_3C_6H_4SO_3H + CO$ in $(C_2H_5)_2O$	IR: 2113	[78]
35	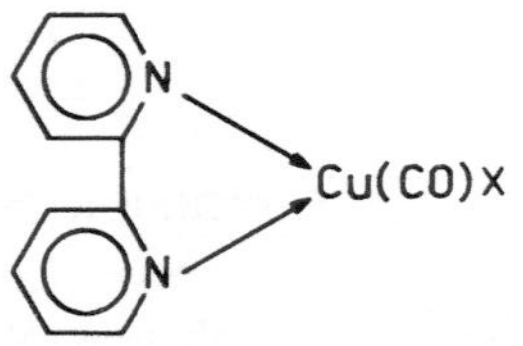 (X = Cl)	not isolated	K = 4.1 (±0.3) atm^{-1} for the reaction $Cu(C_{10}H_8N_2)Cl + CO \rightleftharpoons$ $(CO)Cu(C_{10}H_8N_2)Cl$ in C_2H_5OH at 25°, determined spectrophotometrically at 440 nm	[70]
36	see No. 35 (X = ClO_4)	$CuClO_4 + C_{10}H_8N_2 + CO$ in CH_3OH	^1H NMR (CD_3COCD_3): 7.99 (t, 2H), 8.37 (t, 2H), 8.67 (d, 2H), 9.09 (d, 2H) IR (Nujol): 2110	[84]

References on pp. 208/10

No.	Compound	Preparation	Properties	Ref.
37	see No. 35 (X=O₃SCF₃)	from No. 12 + C₁₀H₈N₂ in (C₂H₅)₂O (quantitative)	purple solid	[78]
38	[(CO)Cu(CH₃)₂NCH₂CH₂N(CH₃)₂]ClO₄	CuClO₄ + (CH₃)₂NCH₂CH₂N(CH₃)₂ + CO in CH₃OH	¹H NMR (CD₃COCD₃): 2.73 (s, CH₃), 2.80 (s, CH₂) IR (Nujol): 2110	[84]
39	[(CO)Cu(CH₃)₂NCH₂CH₂N(CH₃)₂]O₃SCH₃	from No. 11 + (CH₃)₂NCH₂CH₂N(CH₃)₂ in THF	white solid	[78]
40	[(CO)Cu(C₂H₅)₂NCH₂CH₂N(C₂H₅)₂]ClO₄	CuClO₄ + (C₂H₅)₂NCH₂CH₂N(C₂H₅)₂ + CO in CH₃OH	¹H NMR (CD₃COCD₃): 1.15 (t, 12H), 2.95 (m, 12H) IR (Nujol): 2105	[84]
41	[(CO)Cu(CH₃)₂NCH₂CH₂N(CH₃)₂(CH₃OH)]B(C₆H₅)₄	C₆H₅CO₂Cu + CO + NaB(C₆H₅)₄ + (CH₃)₂NCH₂CH₂N(CH₃)₂ in CH₃OH at −30° (35%)	white solid IR (CH₃OH): 2080 dec. at 25° with loss of CH₃OH and CO in CH₃OH at 25° [Cu(CH₃)₂NCH₂CH₂N(CH₃)₂]B(C₆H₅)₄ is formed excess (CH₃)₂NCH₂CH₂N(CH₃)₂ yields [Cu{(CH₃)₂NCH₂CH₂(CH₃)₂}₂]B(C₆H₅)₄ addition of methanolic CH₃CO₂Na and concentrating the solution in a CO stream yields [(CO)₂Cu₂{(CH₃)₂NCH₂CH₂N(CH₃)₂}₂(O₂CCH₃)]B(C₆H₅)₄ after cooling to −80°	[74]
*42	$(CO)CuHB\left(\begin{array}{c}R\\ \diagdown \\ N-N\\ R\end{array}\right)_3$ (R=H)	CuCl + CO + K[HB(N₂C₃HR₂)₃] in polar solvents (63%)	IR (light petroleum): ν(CO) = 2083, ν(BH) = 2465 ¹H NMR: 6.17 (d, 2H), 7.70 (t, 3H) decarbonylation at 165°	[20, 21]

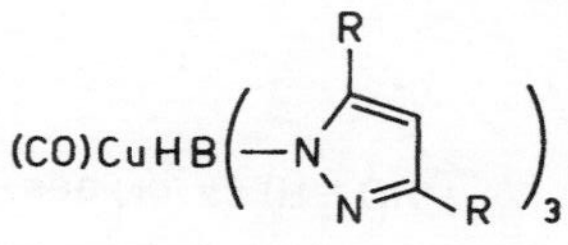

Table 27 [continued]

No.	compound	preparation (yield)	spectra and further remarks	Ref.
43	see No. 42 (R=CH₃)	see No. 42 (32 to 78%)	white crystals, dec. 185 to 190° ^{1}H NMR (CD_3COCD_3): 6.17 (m), 7.70 (m) IR: 2060 to 2066 (s, νCO), 2500 (m, νBH) with D = $P(OCH_3)_3$ or $[(C_6H_5)_2PCH_2]_3CCH_3$ formation of $DCu[HB(N_2C_3H(CH_3)_2)_3]$	[20, 21, 32, 59]
*44	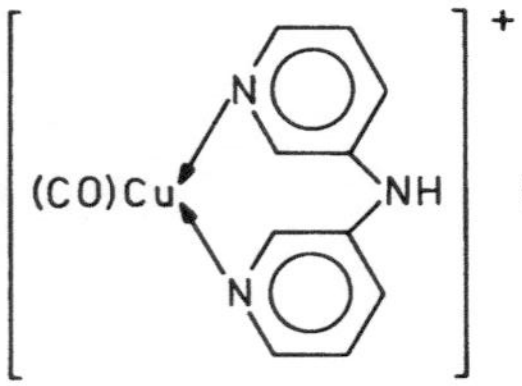	$CuClO_4 + C_{10}H_9N_3 + CO$ in CH_3OH	crystals are stable under N_2 ^{1}H NMR (CD_3COCD_3): 7.23 (q, 4H), 7.98 (t, 2H), 8.50 (d, 2H), 9.65 (s, 1H)	[81]
45	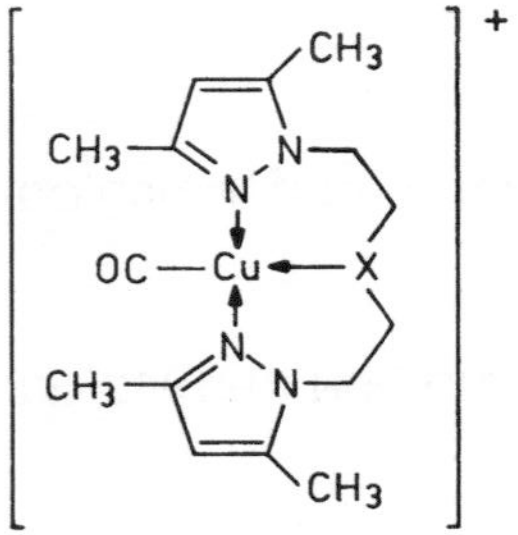	from $[CuC_{14}H_{22}N_5]BF_4 + CO$ in CH_3NO_2, not isolated	IR (CH_3NO_2): 2082 50% formed/decomposed in CH_3NO_2 at an equilibrium CO pressure $p_{1/2}$ = 120 Torr	[79]
46	see No. 45 (X=O)	from $[CuC_{14}H_{21}N_4O]BF_4 + CO$ in CH_3NO_2, not isolated	IR (CH_3NO_2): 2106 $p_{1/2}$ (see No. 45) >1600 Torr X-ray structure given in [79]	[79]
47	see No. 45 (X=S)	from $[CuC_{14}H_{21}N_4S]BF_4 + CO$ in CH_3NO_2, not isolated	IR (CH_3NO_2): 2123 $p_{1/2}$ (see No. 45) = 730 Torr	[79]

References on pp. 208/10

No.	Compound / Structure	Preparation	Properties	Ref.
*48	(CO)CuB(N₂C₃H₃)₄ (N₂C₃H₃ = pyrazol-1-yl)	CuCl + CO + KB(N₂C₃H₃)₄ in (CH₃)₂CO (64%)	m.p. 150° (dec.) ¹H NMR (CDCl₃): 6.16(s), 7.66(s), 7.72(s) of equal intensity IR: 2083 with D = P(C₂H₅)₃, P(OCH₃)₃, or P(OC₆H₅)₃ formation of DCu[B(N₂C₃H₃)₄]	[28, 59]
49	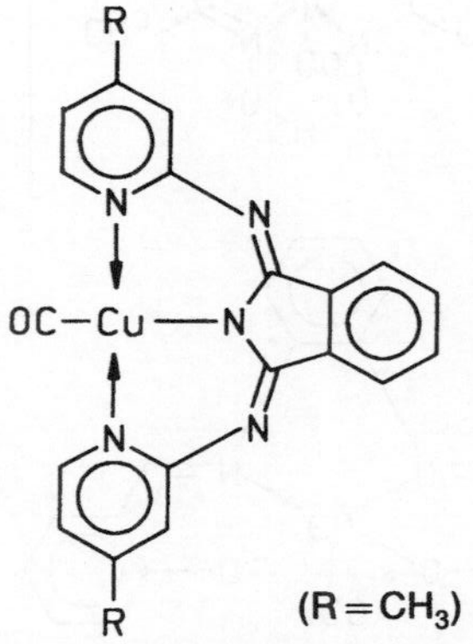 (R = CH₃)	CuCl + C₂₀H₁₆N₅ + CO + N(C₂H₅)₃ in C₂H₅OH (88%); CH₃CO₂CuC₂₀H₁₆N₅ + Na/Hg in presence of CO (65%)	light yellow powder IR: 2068 slightly soluble in organic solvents with O₂ in benzene or toluene formation of Cu^II complexes with CO₃ and OH bridges	[48]
50	see No. 42 (R = t-C₄H₉)	—	IR: 2067 slightly soluble in organic solvents	[48]
51	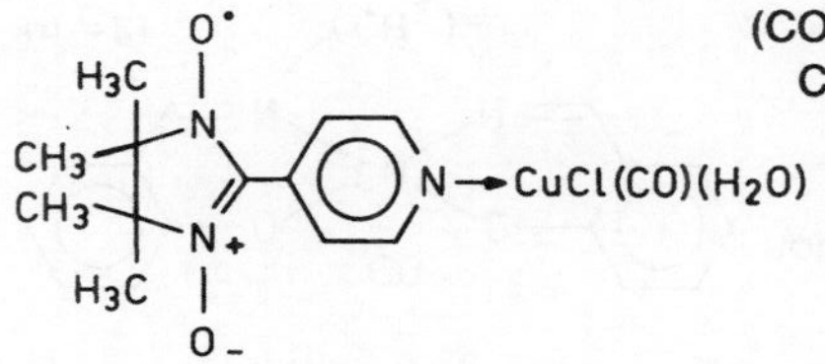	(CO)CuCl·2H₂O + C₁₂H₁₆N₃O₂ (1:1) in C₆H₆	green solid magnetic moment (solid, 295 K): 1.98 μB ESR (solid): g = 2.0067 (CHCl₃): g = 2.0065, aN = 7.5 G	[77]
52	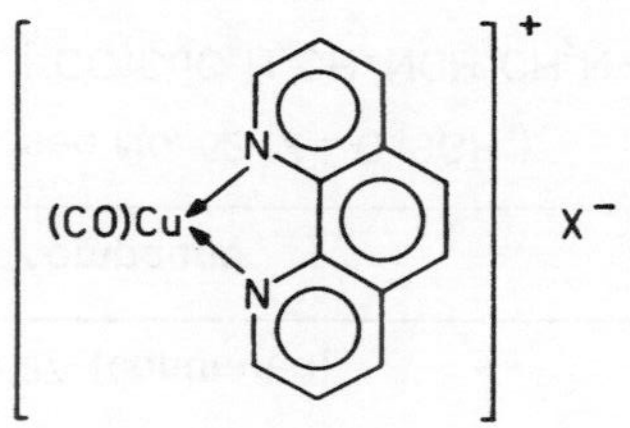 (X = ClO₄)	CuClO₄ + C₁₂H₈N₂ + CO in CH₃OH	¹H NMR (CD₃COCD₃): 8.30 (q, 4H), 8.93 (d, 2H), 9.52 (d, 2H) IR (Nujol): 2115	[84]

Table 27 [continued]

No.	compound	preparation (yield)	spectra and further remarks	Ref.
53	see No. 52 (X = O_3SCH_3)	from No. 12 + $C_{10}H_8N_2$ in $(C_2H_5)_2O$	—	[78]
*54	$[(CO)CuC_6H_5CH=NCH_2CH_2N=CHC_6H_5]I$	$CuI + (C_6H_5CH=NCH_2)_2$ + CO in THF (57%)	yellow crystals IR (Nujol): 1635 (C=N), 2063 (CO)	[85]
55	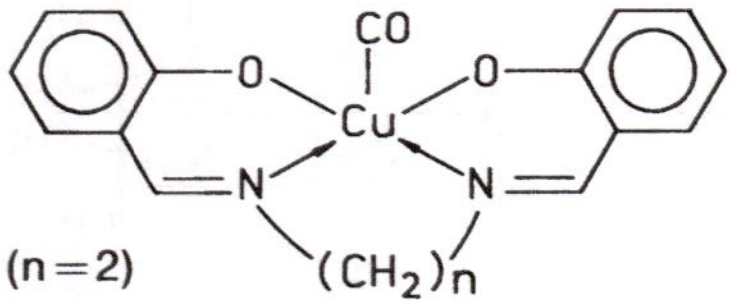 (n=2)	not isolated	$K = 2.6(3) \times 10^3$ M^{-1} in $(CH_3)_2NCHO$ at 22° (see p. 188)	[57]
56	see No. 55 (n=3)	not isolated	$K = 2.1(3) \times 10^3$ M^{-1} in $(CH_3)_2NCHO$ at 22° (see p. 188)	[57]
57	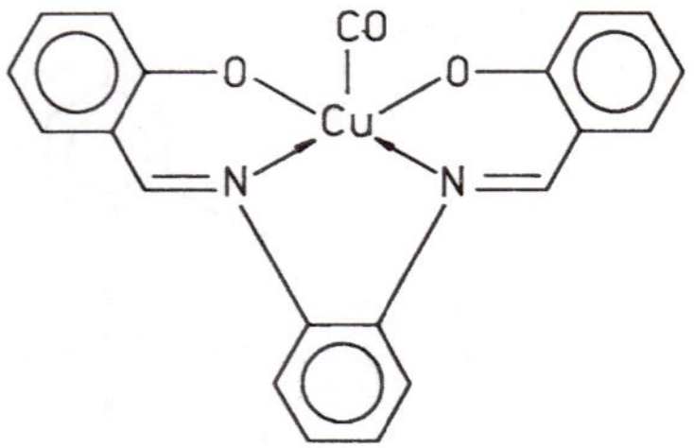	not isolated	$K = 4.7(30) \times 10$ M^{-1} in $(CH_3)_2NCHO$ at 22° (see p. 188)	[57]
58	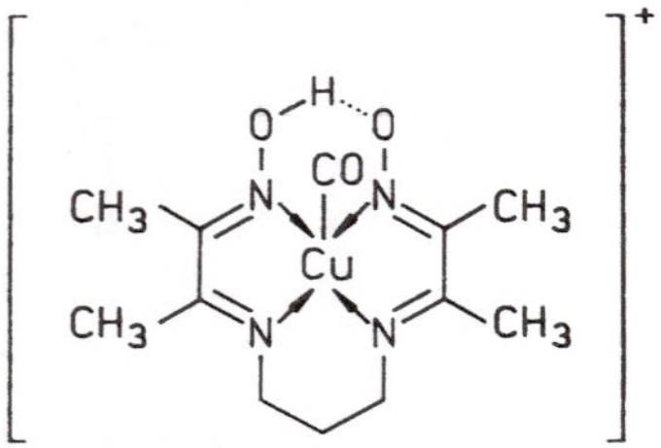	not isolated	$K = 5.8(30) \times 10$ M^{-1} in $(CH_3)_2NCHO$ at 22° (see p. 188)	[57]

References on pp. 208/10

No.	Compound / preparation	Properties	Ref.
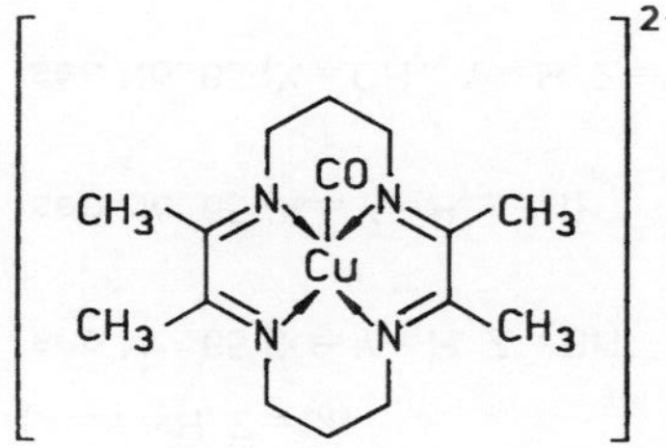 59	[2+] structure; not isolated	$K = 4.2\,(30) \times 10\ M^{-1}$ in $(CH_3)_2NCHO$ at 22° (see p. 188)	[57]
*60	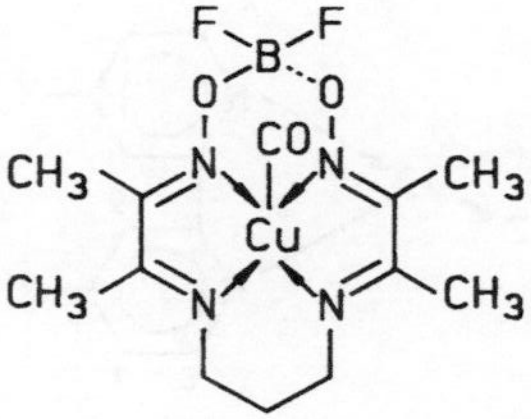 structure; $CuC_{11}H_{18}BF_2N_4O_2 + CO$ in acetone, CH_3CN, CH_2Cl_2, or $C_6H_5CH_3$	orange solid; IR (solid): 2068 (CH_2Cl_2): 2080; X-ray photoelectron spectrum: 285.1 (C1s), ~400 (N1s), 933.1 (Cu $2p_{3/2}$) eV; $K = 1.2 \times 10^5\ M^{-1}$ in $(CH_3)_2NCHO$ at 22° (see p. 188)	[35, 36, 40, 57, 66]
61	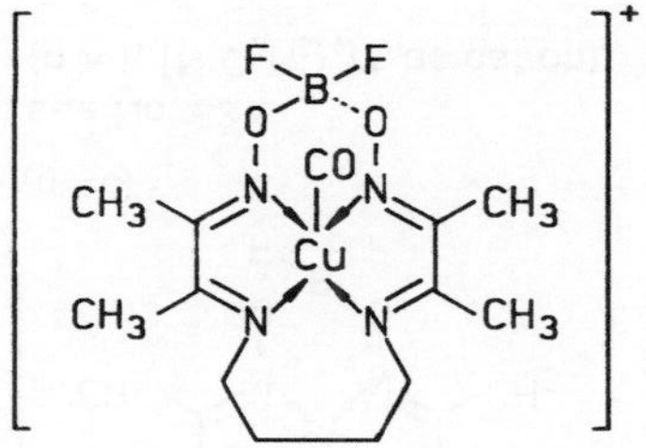 [+] structure; not isolated	$K = 1.2\,(2) \times 10^3\ M^{-1}$ in $(CH_3)_2NCHO$ at 22° (see p. 188)	[57]
62	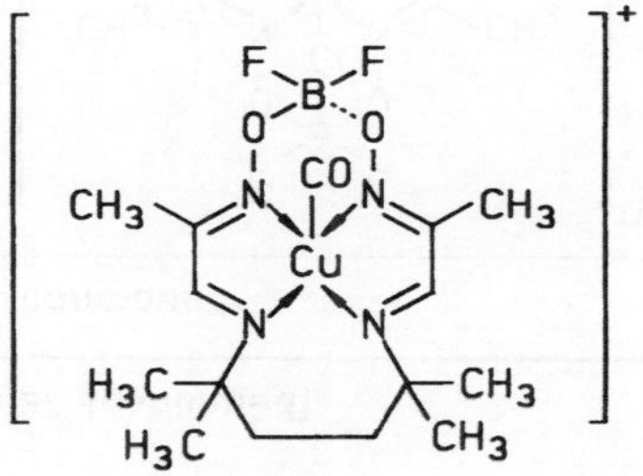 [+] structure; not isolated	$K = 5.3\,(5) \times 10^3\ M^{-1}$ in $(CH_3)_2NCHO$ at 22° (see p. 188)	[57]

Table 27 [continued]

No.	compound	preparation (yield)	spectra and further remarks	Ref.
63	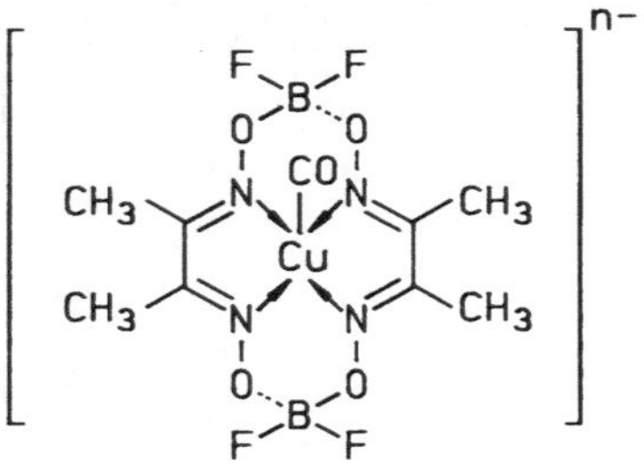 *(structure: Cu complex with N–O–B(F)F bridges, CH3 groups; charge n−)* (n=0)	not isolated	$K = 8.8(9) \times 10^5\ M^{-1}$ in $(CH_3)_2NCHO$ at 22° (see p. 188)	[57]
*64	see No. 63 (n=1, [N(C2H5)4]+ as cation)	$[N(C_2H_5)_4][Cu(C_8H_{12}B_2F_4N_4O_4)] + CO$ in $(CH_3)_2CO$	—	[69]
65	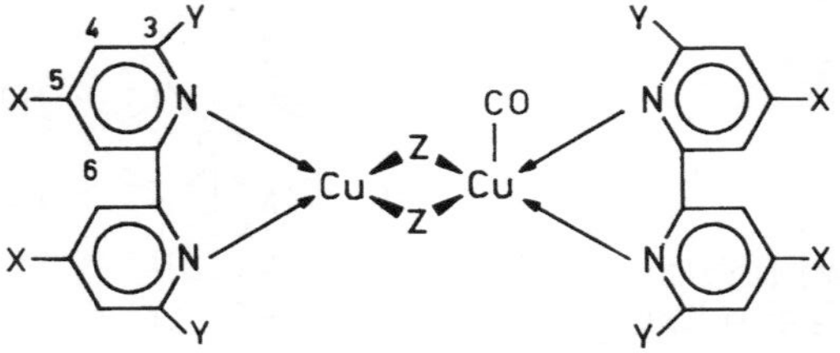 *(structure: bis-pyridyl Cu–CO dimer; X, Y substituents)* (X=Y=H, Z=Cl)	$Cu_2(C_{20}H_{16}Cl_2N_4) + CO$ in $(CH_3)_2CO$	pale yellow crystals, dec. at 25° ^{1}H NMR (acetone): 7.76 (br, H–5), 8.22 (H–4), 8.98 (broad, H–6) IR (solid): 2066 (acetone): 2075 $K = 1.2$ to $1.3\ (\pm 0.1)\ atm^{-1}$ in $(CH_3)_2CO$, $1.7\ (\pm 0.1)\ atm^{-1}$ in $C_2H_5COCH_3$, and $3.8\ (\pm 0.1)\ atm^{-1}$ in $(CH_3)_2CHCH_2COCH_3$ at 25° (see p. 188)	[70, 87]
66	see No. 65 (X=Y=H, Z=Br)	not isolated	$K = 3.5\ (\pm 0.3)\ atm^{-1}$ in $(CH_3)_2CO$ at 25° (see p. 188)	[70]
67	see No. 65 (X=Y=H, Z=I)	not isolated	$K = 6.4\ (\pm 0.3)\ atm^{-1}$ in $(CH_3)_2CO$ at 25° (see p. 188)	[70]
68	see No. 65 (X=CH3, Y=H, Z=Cl)	$Cu_2(C_{24}H_{24}Cl_2N_4) + CO$ in $(CH_3)_2CO$	pale yellow crystals, stable at low temperature	[87]

References on pp. 208/10

No.	Structure / Name	Formula	Properties	Ref.
69	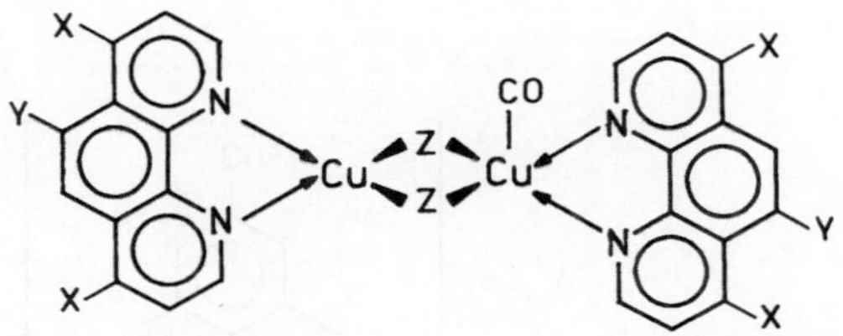 (X=Y=H, Z=Cl)	$Cu_2(C_{24}H_{16}Cl_2N_4) + CO$ in $(CH_3)_2CO$	pale yellow crystals, stable at low temperature IR (KBr): 2063 (acetone): 2076 $K = 1.2 (\pm0.1)$ atm^{-1} in $(CH_3)_2CO$ at 25° (see p. 188)	[70, 87]
70	see No. 69 (X=H, Y=Z=Cl)	$Cu_2(C_{24}H_{14}Cl_4N_4) + CO$ in $(CH_3)_2CO$	pale yellow crystals, stable at low temperature IR (KBr): 2088 (acetone): 2085 $K = 0.9 (\pm0.1)$ atm^{-1} in $(CH_3)_2CO$ at 25° (see p. 188)	[87]
71	see No. 69 (X=H, Y=Br, Z=Cl)	$Cu_2(C_{24}H_{14}Br_2Cl_2N_4) + CO$ in $(CH_3)_2CO$	pale yellow crystals, stable at low temperature IR (KBr): 2082	[87]
72	see No. 69 (X=H, Y=CH$_3$, Z=Cl)	$Cu_2(C_{26}H_{20}Cl_2N_4) + CO$ in $(CH_3)_2CO$	pale yellow crystals, stable at low temperature IR (KBr): 2068 $K = 1.1 (\pm0.1)$ atm^{-1} in $(CH_3)_2CO$ (see p. 188)	[87]
73	see No. 69 (X=CH$_3$, Y=H, Z=Cl)	$Cu_2(C_{28}H_{24}Cl_2N_4) + CO$ in $(CH_3)_2CO$	pale yellow crystals, stable at low temperature $K = 1.6 (\pm0.1)$ atm^{-1} in $(CH_3)_2CO$ (see p. 188)	[87]
74	see No. 69 (X=C$_6$H$_5$, Y=H, Z=Cl)	$Cu_2(C_{48}H_{32}Cl_3N_4) + CO$ in $(CH_3)_2CO$	pale yellow crystals, stable at low temperature	[87]

References on pp. 208/10

Table 27 [continued]

No.	compound	preparation (yield)	spectra and further remarks	Ref.
75	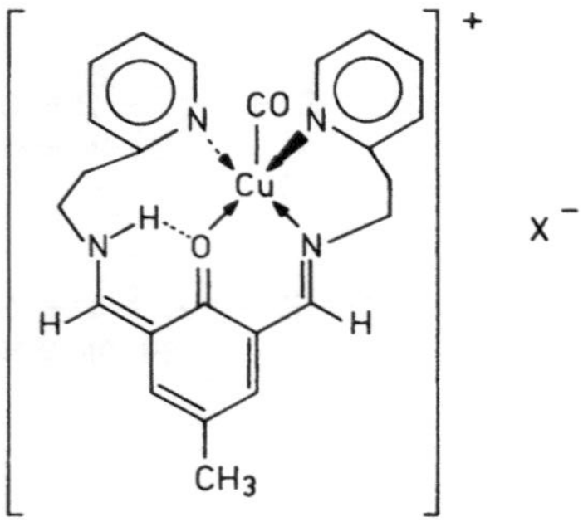	$[CuC_{23}H_{24}N_4O]X + CO$ in CH_3CN at $-10°$	IR: 2075 UV: 23000 cm^{-1} ($\varepsilon = 9300$) in $(CH_3)_2SO$, 23000 cm^{-1} ($\varepsilon = 7900$) in CH_3CN	[65]
76	($X = CuCl_2$) see No. 75 ($X = BF_4$)	see No. 75	yellow crystals IR: 2075 UV: 23000 cm^{-1} ($\varepsilon = 9300$) in $(CH_3)_2SO$, 23000 cm^{-1} ($\varepsilon = 7900$) in CH_3CN	[65]
*77	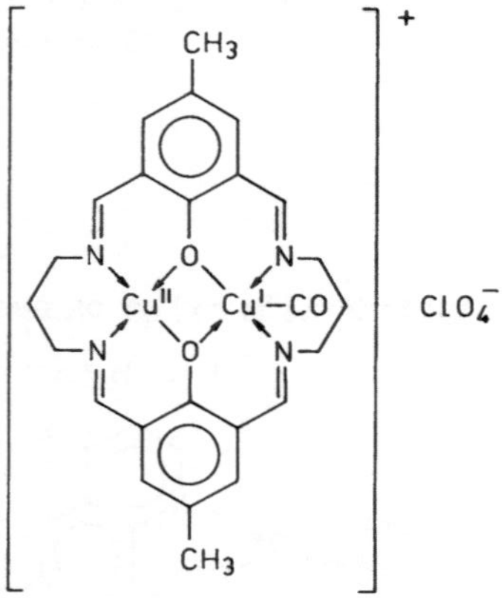	electrolysis of $[Cu_2^{II}C_{24}H_{26}N_4O_2][ClO_4]_2$ in $(CH_3)_2NCHO$, CH_3OH, $(CH_3)_2CO$, CH_3CN or CH_2Cl_2, followed by CO saturation	light brown solid IR: 2065 (solid), 2074 (in $(CH_3)_2NCHO$) X-ray photoelectron spectrum (eV): 399.5 (N 1 s), 932.7, 935.5 (Cu 2 p$_{3/2}$) magnetic moment (25°): $1.94 \pm 0.04\ \mu_B$ $K = 3.1(3) \times 10^4$ M^{-1} in $(CH_3)_2NCHO$ at 22° (see p. 188)	[51, 56, 57, 66]

78	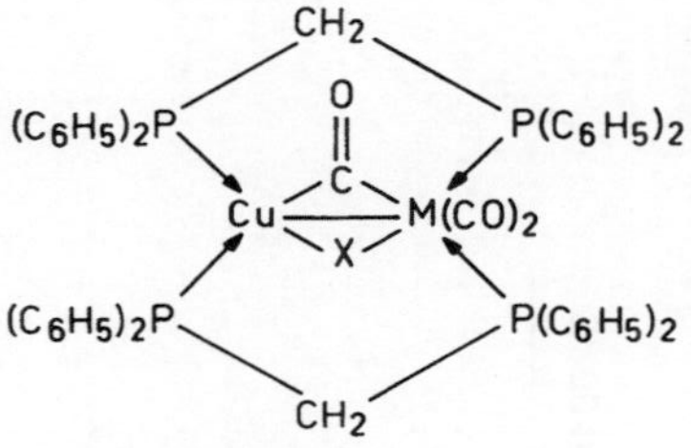 (M = Cr, X = Cl)	$(CO)_3M[(C_6H_5)_2PCH_2P(C_6H_5)_2]_2 + CuX + CH_2Cl_2$ at 20° (50 to 65%)	red solid IR of Nos. 78 to 83: low-frequency bands (≦1810) indicate bridging CO $^1H\{^{31}P\}$NMR of Nos. 78 to 83: at 20° AB patterns for CH_2 indicate rigid systems with pseudo-equatorial and pseudo-axial H	[82]
79	see No. 78 (M = Cr, X = I)	see No. 78	red see No. 78 for spectra	[82]
80	see No. 78 (M = Mo, X = Cl)	see No. 78	orange see No. 78 for spectra	[82]
81	see No. 78 (M = Mo, X = I)	see No. 78	orange see No. 78 for spectra	[82]
*82	see No. 78 (M = W, X = Cl)	see No. 78	orange plates IR (CH_2Cl_2): 1784 (μ-CO), 1838 + 1952 (CO) see also No. 78	[82]
83	see No. 78 (M = W, X = I)	see No. 78	orange see No. 78 for spectra	[82]

References on pp. 208/10

* Further information:

(CO)Cu (Table **27**, No. 1) is obtained by cocondensation of Cu, CO, and Ar ($\leq 1:2:1000$) at 10 to 15 K and identified by its IR isotopic spectra. Upon warming to 35 K it produces IR absorptions of $(CO)_2Cu$ and $(CO)_3Cu$ (see Section 1.1.3.2) [26, 91]. From the $\nu(CO)$ of the ^{63}Cu, ^{65}Cu, $^{12}C^{16}O$, $^{13}C^{16}O$, and $^{12}C^{18}O$ isotopic species the Cotton–Kraihanzel force constant $f(CO) = 16.33$ mdyn $\cdot$ Å^{-1} has been derived [26].

In contrast, cocondensation of Cu atoms from a Knudsen cell with 1/600 of CO/Ar onto a CsI plate cooled to 10 K yields Cu cluster monocarbonyls $(CO)Cu_n$ with $n = 1$ to 4 [31].

Earlier results from cocondensation of Cu atoms from a Knudsen cell with a large excess of pure CO onto a CsI plate cooled to 20 K are given in [20]. But no evidence for the composition of the reported "copper carbonyl" is given. The reported IR data are very similar to those of $(CO)_3Cu$, as given in [26].

An analysis of the ESR data for (CO)Cu indicates that the odd electron is almost entirely localized in a pure 4s orbital [33]. g and hfc tensors of $(CO)^{63}Cu$: $g_{\parallel} = 1.998(1)$, $A_{\parallel}(^{63}Cu) = 4174(2)$ MHz, $g_{\perp} = 1.995(1)$, $A_{\perp}(^{63}Cu) = 4126(2)$ MHz [92], see also [91].

From ab initio restricted Hartree–Fock MO calculations of linear (CO)Cu a binding energy between Cu and CO of -1.64 eV is derived [50, 76].

The interaction of the Cu atom with CO was investigated by MO calculations. Coupling occurs mainly through the HOMO 5σ level of CO with the 4 s and $3d_{z^2}$ orbital of the metal atom. The HOMO level is lowered in energy due to donor–type interaction with the metal [55, 73].

From detailed SCF–X_α–SW calculations a satisfactory interpretation for the thermal lability of the complex in terms of predominant σ–bonding Cu–C interaction with some pπ–pπ* and minimal dπ–dπ* charge transfer from the metal to the CO ligand is possible [72].

The nature of the Cu–CO bonding in (CO)Cu has also been analyzed by the constrained–space–orbital–variation technique [88, 89, 93].

[(CO)Cu]$^+$ (Table **27**, No. 2) is formed by absorption of CO in mixtures of Cu_2O and strong acids like H_2SO_4, $BF_3 \cdot H_2O$, FSO_3H, CF_3SO_3H, and HF. $[(CO)_3Cu]^+$ and $[(CO)_4Cu]^+$ (see Section 1.1.3.2) exist in equilibrium with $[(CO)Cu]^+$ [29]. In dilute H_2SO_4, the monocarbonyl cation is comparatively stable [14, 15].

Carbonyl stretching frequencies and ^{13}C NMR chemical shifts in acid solvents [29]:

solvent	IR $\nu(CO)$ in cm^{-1}	13 C NMR δ in ppm at 30 °C
H_2SO_4	2140	168.5
FSO_3H/H_2SO_4 (1/0.3)	2150	—
CF_3SO_3H	2150	168.3
FSO_3H	2152	169.2
HF	2156	—
$BF_3 \cdot H_2O$	2160	169.0
H_2SO_4/FSO_3H (1/0.3)	—	169.0
HCl/H_2O (36%)	—	170.7

For the significance of $[(CO)Cu]^+$ as an intermediate in the Cu^{II} catalyzed oxidation of CO by O_2, see [9, 10, 13, 18].

(CO)CuCl (Table **27**, No. **3**). IR absorptions at 2112 and 2069 cm⁻¹ were obtained from aqueous or pyridine solutions. But the composition of these compounds is not clear [16]. At 25.0 °C the equilibrium pressure p_E of CO over solid (CO)CuCl/CuCl is 436.6±1 Torr. Thermodynamic data for the reaction CO (g) + CuCl (s)⇌(CO)CuCl (s) are obtained from the temperature dependence ln p_o = ln (436.6/760.0) − 5481.2 (1/T − 1/298.15) of p_o, the standard equilibrium pressure of CO in Torr: ΔG^o_{298} = − 1.374±0.006, ΔH^o_{298} = − 45.58±0.11 kJ · mol⁻¹, ΔS^o_{298} = − 148.3±0.4 J · mol⁻¹ · K⁻¹ [46].

Furthermore, (CO)CuCl is an intermediate in the carbonylation of [Cu(OCH₃)Cl] to give dimethyl carbonate [25]. For the chemical behavior of (CO)CuCl, see Table **27**, Nos. 26, 27, 29, 51. With [(CH₃)₂N=CH₂CuCl]Br (see Section 1.1.5) in THF above 15 °C (CO)Cu₂(CH₂=N(CH₃)₂)Cl₂Br is assumed to be formed [19]. With Na phenoxides, labile dinuclear complexes are formed [96].

(CO)CuCl · 2 H₂O (Table **27**, No. **4**) has been postulated as an intermediate in the carbonylation of aryldiazonium compounds to give the corresponding carboxylic acids [7].

[(CO)CuCl₂]⁻ (Table **27**, No. **21**). Further investigations have been made by gas solubility measurements [3]. The stability constant β = 2 × 10⁶ M⁻² · atm⁻¹ is estimated from potentiometric measurements in aqueous solutions under the assumption that no other carbonyl complex is formed [44].

[(CO)Cu(NH₃)₂]⁺, [(CO)Cu(NH₃)₃]⁺ (Table **27**, Nos. **22, 23**). At a CO pressure of 0.02 to 0.9 atm below 11 °C, No. 23 is more stable than No. 22. The converse is true above 11 °C. For the formation of No. 22 ΔH = 8.7 kcal · mol⁻¹ and ΔS = − 16.4 cal · mol⁻¹ · K⁻¹, for No. 23 ΔH = 15.3 kcal · mol⁻¹ and ΔS = − 38.4 cal · mol⁻¹ · K⁻¹ [34].

Nonempirical LCAO-MO-SCF computations were carried out on the ground and core hole states of Nos. 22 and 23. For No. 22, the computations show that a linear Cu–C–O

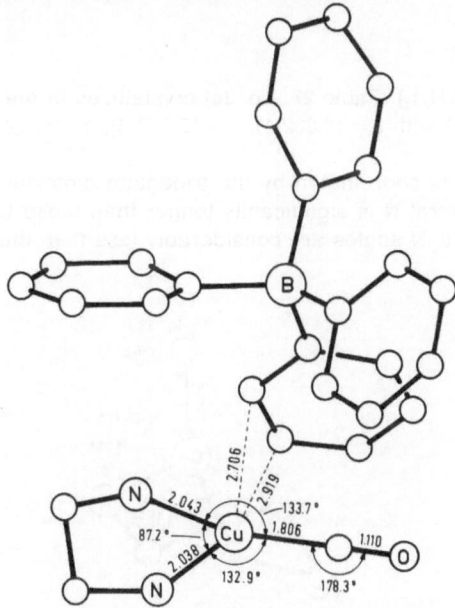

Fig. 7.   Molecular structure of [(CO)CuC₂H₈N₂][B(C₆H₅)₄] (No. 27) with selected bond lengths (in Å) and angles.

arrangement is preferred, the computed coordination energy for CO being -8.1 kcal $\cdot$ mol^{-1}. The bond overlap populations showed that for Nos. 22 and 23 the CO bond length should be somewhat shorter than for the free ligand [71].

[(CO)CuC$_2$H$_8$N$_2$]Cl (Table **27**, No. **26**). When the methanolic solution is evaporated to dryness at $-20\,°C$ [C$_2$H$_8$N$_2$Cu(CO)$_2$CuC$_2$H$_8$N$_2$]Cl$_2$ is obtained. It has been shown that both complexes exist in CH$_3$OH in a temperature dependent equilibrium [17].

[(CO)CuC$_2$H$_8$N$_2$][B(C$_6$H$_5$)$_4$] (Table **27**, No. **27**) crystallizes in the monoclinic space group P2$_1$/c –C$_{2h}^5$ (No. 14) with a $= 10.301\,(1)$, b $= 12.234\,(1)$, c $= 18.390\,(2)$ Å, and β $= 91.11\,(1)°$; Z $= 4$, d$_c = 1.350$ g $\cdot$ cm^{-3}. The structure is shown in **Fig. 7**, p. 203. A rather distant but significant interaction with two carbon atoms of one of the phenyl rings completes the pseudo–trigonal pyramidal coordination around the Cu atom. It is displaced by 0.26 Å from the plane defined by the two N atoms and the C atom of CO toward the coordinated C$_6$H$_5$ [40, 64].

The CO can be replaced by P(OC$_2$H$_5$)$_3$. With H$_2$NCH$_2$CH$_2$NH$_2$ compound No. 27 yields [(CO)$_2$Cu$_2$(C$_2$H$_8$N$_2$)$_3$][B(C$_6$H$_5$)$_4$]$_2$ [40].

[(CO)CuC$_5$H$_9$N$_3$][B(C$_6$H$_5$)$_4$] (Table **27**, No. **29**). A mixture with [(CO)$_2$Cu$_2$(C$_5$H$_9$N$_3$)$_3$]-[B(C$_6$H$_5$)$_4$]$_2$ is obtained after CO absorption of a methanolic suspension of CuI with histamine (molar ratio < 1.0) and NaB(C$_6$H$_5$)$_4$ [45]. The polymeric structure II has been proposed. CO was displaced by P(OC$_2$H$_5$)$_3$ [63].

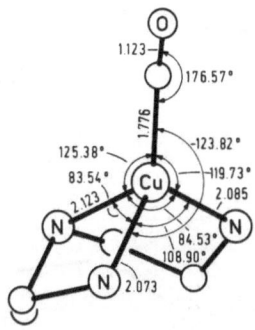

II

[(CO)CuC$_4$H$_{13}$N$_3$][B(C$_6$H$_5$)$_4$] (Table **27**, No. **30**) crystallizes in the monoclinic space group P2$_1$/n (P2$_1$/c)-C$_{2h}^5$ (No. 14) with a $= 14.832(1)$, b $= 18.737(3)$, c $= 9.599(1)$ Å, β $= 90.07(2)°$; Z $= 4$, d$_c = 1.280$ g $\cdot$ cm^{-3}.

As **Fig. 8** shows, Cu is coordinated by the tridentate diethylene triamine and CO. The bond from Cu to the central N is significantly longer than those to the terminal nitrogens. Two intramolecular N–Cu–N angles are considerably less than the ideal tetrahedral angle.

Fig. 8.   Molecular structure of the cation in [(CO)Cu(C$_4$H$_{13}$N$_3$)][B(C$_6$H$_5$)$_4$] (No. 30) with selected bond lengths (in Å) and angles.

References on pp. 208/10

This is due to a combination of the constraints within the ligand and the requirements for normal Cu-N bond distances [42].

The electrochemical behavior has been studied in $(CH_3)_2SO$ solution at Pt electrodes by voltammetry and controlled potential coulometry. The Cu^I center undergoes one-electron charge transfers in both anodic and cathodic processes. Anodic oxidation leads to decarbonylated Cu^{II} species [86].

The compound is quite stable in air. It is slightly soluble in THF or acetone without decomposition. The solid does not lose CO in vacuum or by heating to 100 °C, but it is easily hydrolyzed when $[B(C_6H_5)_4]^-$ is replaced by Cl^- [42].

(CO)CuHB(C₃H₃N₂)₃ (Table **27**, No. **42**) crystallizes in the space group R3c-C_{3v}^6 (No. 161) with a = 13.8616(21) Å, α = 91.37(1)°, Z = 8, d_c = 1.520, and d_m = 1.45(5) g · cm⁻³.

The structure is shown in **Fig. 9**. There are two crystallographically distinct types of molecule within the unit cell. Two molecules of Type 1 (Fig. 5a) have precise C_3 symmetry with O, C, Cu, B, and H atoms lying along the [111] directions. Six molecules of Type 2 (Fig. 5b) are in a distorted C_{3v} symmetry [27].

The air-stable complex is very soluble in light petroleum ether. It reacts with many ligands, e.g., $P(C_6H_5)_3$, $CH_3P(C_6H_5)_2$, $P(OCH_3)_3$, $P(OC_6H_5)_3$, $(C_6H_5)_2MCH_2CH_2M(C_6H_5)_2$ (M = P, As), $As(C_6H_5)_3$, $Sb(C_6H_5)_3$, $t-C_4H_9NC$, with evolution of CO to give the derivatives [20, 21, 59].

The reaction with pure norbornadiene yields $C_7H_8CuHB(C_3H_3N_2)_3$ [61, 62]. It also reacts with $(C_5H_5)_2MoH_2$ to give the Cu-Mo cluster $(C_5H_5)_2Mo(H_2)CuHB(C_3H_3N_2)_3$ [37].

As $(CO)CuHB(C_3H_3N_2)_3$ is effective in sensitizing the light-induced quadricyclene production from dilute norbornadiene solutions (0.1 M in C_2H_5OH, quantum yield ~0.2) a solar energy storage system has been proposed [52, 54, 60].

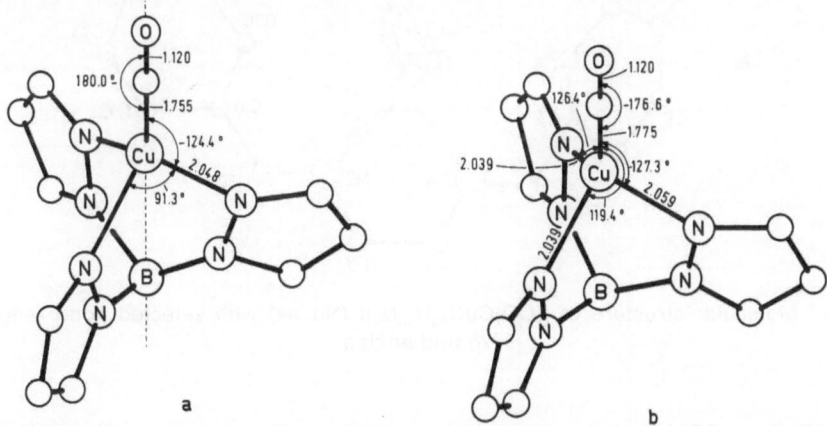

Fig. 9.   Molecular structure of $(CO)CuHB(C_3H_3N_2)_3$ (No. 42) with selected bond lengths (in Å) and angles.

[(CO)CuC₁₀H₉N₃]ClO₄ (Table **27**, No. **44**) crystallizes in the triclinic space group P1̄-C_i^1 (No. 2) with a = 8.922(1), b = 10.755(1), c = 7.379(1) Å, α = 91.51(1)°, β = 99.24(1)°, γ = 81.25(1)°, Z = 2. The molecular structure is shown in **Fig. 10**, p. 206 [81].

References on pp. 208/10

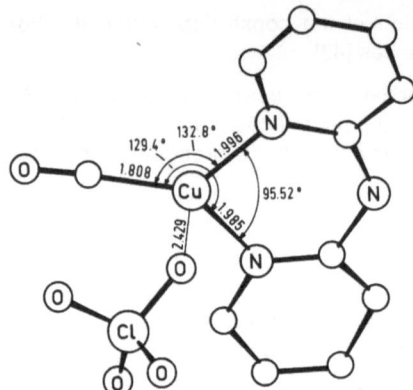

Fig. 10.   Molecular structure of [(CO)CuC$_{10}$H$_9$N$_3$]ClO$_4$ (No. 44) with selected bond lengths
(in Å) and angles.

(CO)CuB(C$_3$H$_3$N$_2$)$_4$ (Table **27**, No. **48**). It is not possible for all pyrazolyl groups in the
ligand to coordinate to one Cu atom. In conjunction with the monomeric osmometric molecular weight and low temperature NMR studies, the authors suggest that the complex is fluxional in solution [28].

[(CO)CuC$_{16}$H$_{16}$N$_2$]I (Table **27**, No. **54**) crystallizes in the monoclinic space group P2$_1$/n
(P2$_1$/c)-C$_{2h}^5$ (No. 14) with a=7.113(2), b=17.204(4), c=14.794(4) Å, and β=97.87(3)°; Z=4,
d$_c$=1.68 g·cm^{-3}. The structure is given in **Fig. 11** [85].

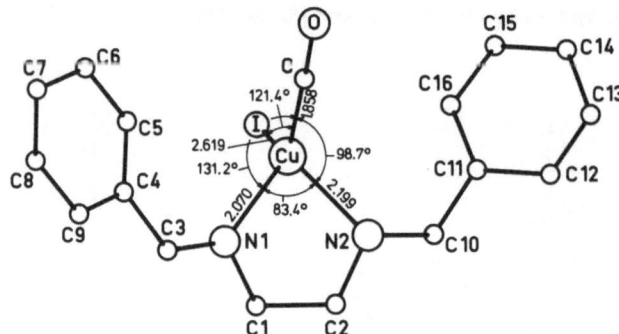

Fig. 11.   Molecular structure of [(CO)CuC$_{16}$H$_{16}$N$_2$]I (No. 54) with selected bond lengths (in
Å) and angles.

(CO)CuC$_{11}$H$_{18}$BF$_2$N$_4$O$_2$ (Table **27**, No. **60**) crystallizes in the space group Pbca-D$_{2h}^{15}$ (No. 61)
with a=13.926(1), b=14.209(1), c=16.297(1) Å, Z=8, d$_c$=1.56, and d$_m$=1.55(3) g·cm^{-3}.

The molecular structure is shown in **Fig. 12**. The Cu atom is situated in an asymmetrical
square pyramid and is displaced 0.96 Å from the basal plane of the four N atoms. Two
significantly different sets of Cu–N bond lengths, 2.164 and 2.104 Å, characterize the equatori-
al asymmetry. The carbonyl ligand coordinates at the apex of the square pyramid with
a Cu–CO distance of 1.780(3) Å and a Cu–C–O angle of 177.5(3)° [36].

References on pp. 208/10

From molecular orbital calculations it is found that upon CO coordination, the $x^2 - y^2$ metal orbital has lower energy than the ligand π^* orbitals. CO is attached to the metal largely by interactions between ligand σ orbitals and metal 4s and 4p orbitals, augmented by stabilization via unoccupied $CO\pi^*$ orbitals of two metal d orbitals. The contribution of the metal d orbitals to the Cu–CO σ linkage is zero [67].

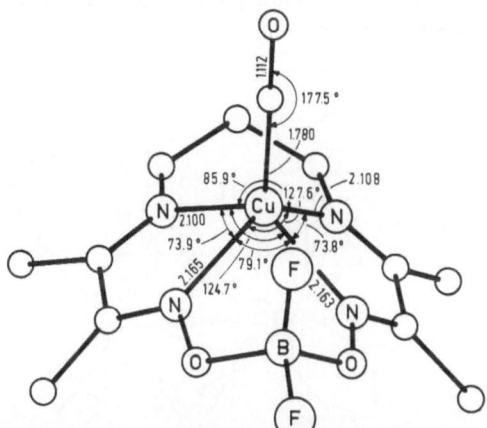

Fig. 12.   Molecular structure of $(CO)CuC_{11}H_{18}BF_2N_4O_2$ (No. 60) with selected bond lengths (in Å) and angles.

$[N(C_2H_5)_4][(CO)CuC_8H_{12}B_2F_4N_4O_4]$ (Table **27**, No. **64**) crystallizes in the monoclinic space group $P2_1/c\text{-}C_{2h}^5$ (No. 14) with a = 13.681(1), b = 25.634(3), c = 15.687(2) Å, β = 115.35(3)°, Z = 8, and d = 1.463 g · cm⁻³. The structure of the anion is shown in **Fig. 13** [69].

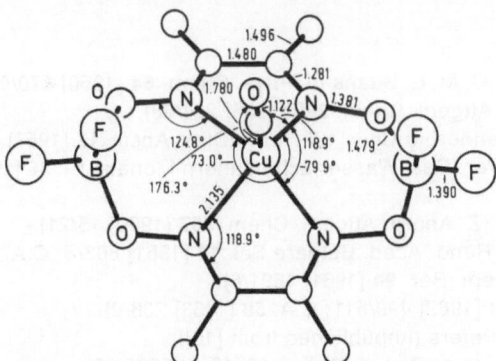

Fig. 13.   Structure of the $[(CO)CuC_8H_{12}B_2F_4N_4O_4]^-$ anion of No. 64 with selected bond lengths (in Å) and angles.

$[(CO)Cu^ICu^{II}C_{24}H_{26}N_4O_2]ClO_4$ (Table **27**, No. **77**). The unusual coordination number of 5 is expected for Cu^I. The four line ESR spectrum in CH_2Cl_2 at 25 °C indicates the localization of the odd electron on a single Cu, at least on the ESR time scale. The electronic absorption spectrum shows a weak band at ~600 mm, which is assigned to a ligand field → Cu^{II} transition [51].

References on pp. 208/10

Cu(μ-CO)(μ-Cl)(μ-(C$_6$H$_5$)$_2$PCH$_2$P(C$_6$H$_5$)$_2$)$_2$W(CO)$_2$ (Table **27**, No. **82**) shows an AA'XX' type ^{31}P{^{1}H} NMR pattern for the P nuclei bonded to W. Eight of the ten lines possible have been observed. For the P nuclei bonded to Cu the resonance is broad at 20 °C but sharpens when the temperature is lowered and is of the AA'XX' type below −60 °C [82].

The compound crystallizes in the monoclinic space group P2$_1$/c-C$^5_{2h}$ (No. 14) with a = 10.760(2), b = 25.132(4), c = 20.675(3) Å and β = 112.62(1)°, Z = 4, d$_c$ = 1.57 g · cm^{-3} [82]. The molecular structure is shown in **Fig. 14**.

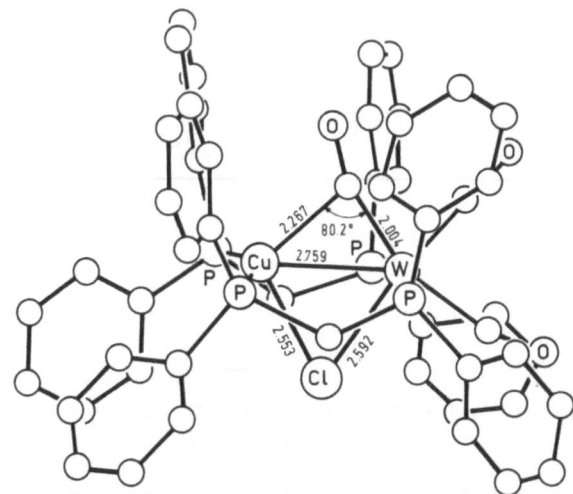

Fig. 14.   Molecular structure of Cu(μ-CO)(μ-Cl)(μ-(C$_6$H$_5$)$_2$PCH$_2$P(C$_6$H$_5$)$_2$)$_2$W(CO)$_2$ (No. 82) with selected bond lengths (in Å) and angles.

References:

[1]   D. W. van Krevelen, C. M. E. Baans (J. Phys. Chem. **54** [1950] 370/90).
[2]   H. Weiss (Z. Anorg. Allgem. Chem. **280** [1955] 284/93).
[3]   W. D. Bond (Diss. Vanderbilt Univ. 1957 from Diss. Abstr. **17** [1957] 1671).
[4]   M. Peter (Schweiz. Ver. Gas-Wasserfachmännern Monats-Bl. **36** [1956] 58/63; C.A. **1958** 8814).
[5]   R. Nast, C. Schultze (Z. Anorg. Allgem. Chem. **307** [1960] 15/21).
[6]   C. Balarew (Compt. Rend. Acad. Bulgare Sci. **14** [1961] 803/6; C.A. **56** [1962] 14987).
[7]   G. N. Schrauzer (Chem. Ber. **94** [1961] 1891/8).
[8]   G. Toselli (Calore **33** [1962] 499/511; C.A. **59** [1963] 238/9).
[9]   R. T. McAndrew, E. Peters (unpublished from [10]).
[10] J. J. Byerley, J. Y. H. Lee (Can. J. Chem. **45** [1967] 3025/30).

[11] W. Brackman (Discussions Faraday Soc. No. 46 [1968] 122/6).
[12] A. F. Scott, L. L. Wilkening, B. Rubin (Inorg. Chem. **8** [1969] 2533/4).
[13] J. J. Byerley, E. Peters (Can. J. Chem. **47** [1969] 313/21).
[14] Y. Souma, H. Sano (Nippon Kagaku Zasshi **91** [1970] 625/30; C.A. **73** [1970] No. 94192).
[15] H. Sano (Nippon Kagaku Zasshi **91** [1970] 630/6; C.A. **73** [1970] No. 94193).
[16] J. O. Alben, L. Yen, N. J. Farrier (J. Am. Chem. Soc. **92** [1970] 4475/6).
[17] G. Rucci, C. Zanzottera, M. P. Lachi, M. Camia (Chem. Commun. **1971** 652).
[18] J. S. Ogden (Chem. Commun. **1971** 978/9).

[19] R. Mason, G. Rucci (Chem. Commun. **1971** 1132).

[20] M. I. Bruce, A. P. P. Ostazewski (J. Chem. Soc. Chem. Commun. **1972** 1124/5).

[21] M. I. Bruce, A. P. P. Ostazewski (J. Chem. Soc. Dalton Trans. **1973** 2433/6).

[22] J. F. Knifton (J. Org. Chem. **38** [1973] 3296/301).

[23] T. Tsuda, H. Habu, S. Horiguchi, T. Saegusa (J. Am. Chem. Soc. **96** [1974] 5930/1).

[24] D. A. van Leirsburg, C. W. De Kock (J. Phys. Chem. **78** [1974] 134/42).

[25] P. Koch, G. Cipriani, E. Perrotti (Gazz. Chim. Ital. **104** [1974] 599/605).

[26] H. Huber, E. P. Kuendig, M. Moskovits, G. A. Ozin (J. Am. Chem. Soc. **97** [1975] 2097/106).

[27] M. R. Churchill, B. G. DeBoer, F. J. Rotella, O. M. Abu Salah, M. I. Bruce (Inorg. Chem. **14** [1975] 2051/6).

[28] O. M. Abu Salah, M. I. Bruce (J. Organometal. Chem. **87** [1975] C15/C18).

[29] Y. Souma, J. Iyoda, H. Sano (Inorg. Chem. **15** [1976] 968/70).

[30] T. Ogura (Inorg. Chem. **15** [1976] 2301/3).

[31] M. Moskovits, J. E. Hulse (Surface Sci. **61** [1976] 302/5; C. A. **86** [1977] No. 79168).

[32] C. Mealli, C. S. Arcus, J. L. Wilkinson, T. J. Marks, J. A. Ibers (J. Am. Chem. Soc. **98** [1976] 711/8).

[33] E. P. Kündig, D. McIntosh, B. McGarvey, G. A. Ozin (unpublished from D. McIntosh, G. A. Ozin, J. Am. Chem. Soc. **98** [1976] 3167/75, ref. 6).

[34] N. V. Ksandrov, N. V. Fedorova (Izv. Vysshikh Uchebn. Zavedenii Khim. Khim. Tekhnol. **19** [1976] 1147 from C.A. **85** [1976] No. 167455).

[35] R. R. Gagné (J. Am. Chem. Soc. **98** [1976] 6709/10).

[36] R. R. Gagné, J. L. Allison, R. S. Gall, C. A. Koval (J. Am. Chem. Soc. **99** [1977] 7170/8).

[37] O. M. Abu Salah, M. I. Bruce (Australian J. Chem. **30** [1977] 2293/5).

[38] V. A. Golodov, Yu. L. Sheludyakov, R. I. Di, V. K. Fokanov (Kinetika Kataliz **18** [1977] 234/7; Kinet. Catal. [USSR] **18** [1977] 193/6).

[39] G. A. Razuvaev, V. K. Cherkasov, G. A. Abakumov (J. Organometal. Chem. **160** [1978] 361/71).

[40] M. Pasquali, C. Floriani, A. G. Manfredotti (J. Chem. Soc. Chem. Commun. **1978** 921/2).

[41] M. Bigorgne (J. Organometal. Chem. **160** [1978] 345/52).

[42] M. Pasquali, F. Marchetti, C. Floriani (Inorg. Chem. **17** [1978] 1684/8).

[43] R. R. Gagné, J. L. Allison, G. C. Lisensky (Inorg. Chem. **17** [1978] 3563/71).

[44] M. A. Busch, T. C. Franklin (Inorg. Chem. **18** [1979] 521/4).

[45] M. Pasquali, C. Floriani, A. Gaetani-Manfredotti, C. Guastini (J. Chem. Soc. Chem. Commun. **1979** 197/9).

[46] W. Backén, R. Vestin (Acta Chem. Scand. A **33** [1979] 85/91).

[47] J. J. Rafalko (unpublished from [44]).

[48] R. R. Gagné, R. S. Gall, G. C. Lisensky, R. E. Marsh, L. M. Speltz (Inorg. Chem. **18** [1979] 771/81).

[49] R. H. Hange, S. E. Gransden, J. L. Margrave (J. Chem. Soc. Dalton Trans. **1979** 745/8).

[50] H. Itoh, A. B. Kunz (Z. Naturforsch. **34a** [1979] 114/6).

[51] R. R. Gagné, C. A. Koval, T. J. Smith (J. Am. Chem. Soc. **99** [1977] 8367/8).

[52] C. Kutal, D. P. Schwendiman, P. Grutsch (Sol. Energy **19** [1977] 651/5).

[53] M. B. Dines (unpublished from P. F. Rodesiler, E. L. Amma, J. Inorg. Nucl. Chem. **39** [1977] 1227/9).

[54] R. Sterling, C. Kutal (unpublished from C. Kutal, Advan. Chem. Ser. No. 168 [1978] 158/73, ref. 33).

[55] H. Itoh (Fundam. Res. Homogeneous Catal. **3** [1979] 73/82; C.A. **92** [1980] No. 65105).

[56] R. R. Gagné, C. A. Koval, T. J. Smith, M. C. Cimolino (J. Am. Chem. Soc. **101** [1979] 4571/80).

[57] R. R. Gagné, J. L. Allison, M. D. Ingle (Inorg. Chem. **18** [1979] 2767/74).

[58] C. D. Desjardins, D. B. Edwards, J. Passmore (Can. J. Chem. **57** [1979] 2714/5).

[59] O. M. Abu Salah, M. I. Bruce, J. D. Walsh (Australian J. Chem. **32** [1979] 1209/18).

[60] R. F. Sterling (Abstr. Papers 179th Natl. Meeting Am. Chem. Soc., Houston, Texas, 1980, Abstr. No. INOR 47).

[61] R. F. Sterling (Diss. Univ. Georgia 1979; Diss. Abstr. Intern. B **40** [1980] 3160/1).

[62] R. F. Sterling, C. Kutal (Inorg. Chem. **19** [1980] 1502/5).

[63] M. Pasquali, G. Marini, C. Floriani, A. Gaetani-Manfredotti, C. Guastini (Inorg. Chem. **19** [1980] 2525/31).

[64] M. Pasquali, C. Floriani, A. Gaetani-Manfredotti (Inorg. Chem. **19** [1980] 1191/7).

[65] J. J. Grzybowski, F. L. Urbach (Inorg. Chem. **19** [1980] 2604/8).

[66] R. R. Gagné, J. L. Allison, C. A. Koval, W. S. Mialki, T. J. Smith, R. A. Walton (J. Am. Chem. Soc. **102** [1980] 1905/9).

[67] J. K. Burdett, W. P. Douglas (Inorg. Chem. **19** [1980] 2779/84).

[68] M. Pasquali, C. Floriani, A. Gaetani-Manfredotti (Inorg. Chem. **20** [1981] 3382/8).

[69] M. W. McCool, R. E. Marsh, D. M. Ingle, R. R. Gagné (Acta Cryst. B **37** [1981] 935/7).

[70] S. Kitagawa, M. Munakata (Inorg. Chem. **20** [1981] 2261/7).

[71] D. T. Clark, A. Sgamellotti, F. Tarantelli (Inorg. Chem. **20** [1981] 2602/7).

[72] D. F. McIntosh, G. A. Ozin, R. P. Messmer (Inorg. Chem. **20** [1981] 3640/50).

[73] H. Itoh, G. Ertl (Z. Naturforsch. **37a** [1982] 346/52).

[74] M. Pasquali, C. Floriani, G. Venturi, A. Gaetani-Manfredotti, A. Chiesi-Villa (J. Am. Chem. Soc. **104** [1982] 4092/9).

[75] A. Neppel, J. P. Hickey, I. S. Butler (J. Raman Spectrosc. **8** [1979] 57/9).

[76] H. Itoh, A. B. Kunz (Fundam. Res. Homogenous Catal. **3** [1979] 73/82).

[77] A. Dorfer, K. E. Schwarzhans (Z. Naturforsch. **38b** [1983] 265/7).

[78] G. Doyle, K. A. Eriksen, D. van Engen (Inorg. Chem. **22** [1983] 2892/5).

[79] T. M. Sorrell, M. R. Malachowski (Inorg. Chem. **22** [1983] 1883/7).

[80] D. T. Clark, W. J. Brennan, R. S. Allaker, M. Pasquali, A. Sgamellotti, F. Tarantelli (Inorg. Chim. Acta **87** [1984] 67/9).

[81] J. S. Thompson, J. F. Whitney (Inorg. Chem. **23** [1984] 2813/9).

[82] A. Blagg, A. T. Hutton, B. L. Slaw, M. Thornton-Pett (Inorg. Chim. Acta **100** [1985] L33/L34).

[83] G. Doyle, K. A. Eriksen, D. van Engen (Organometallics **4** [1985] 830/5).

[84] J. S. Thompson, R. M. Swiatek (Inorg. Chem. **24** [1985] 110/3).

[85] A. Toth, C. Floriani, M. Pasquali, A. Chiesi-Villa, A. Gaetani-Manfredotti, C. Guastini (Inorg. Chem. **24** [1985] 648/53).

[86] P. Zanello, P. Leoni (Can. J. Chem. **63** [1985] 922/7).

[87] S. Kitagawa, M. Munakata, N. Miyaji (Bull. Chem. Soc. Japan **56** [1983] 2258/62).

[88] P. S. Bagus, K. Hermann, C. W. Bauschlicher Jr. (J. Chem. Phys. **81** [1984] 1966/74).

[89] C. W. Bauschlicher Jr. (J. Chem. Phys. **84** [1986] 260/7).

[90] S. M. Nelson, A. Lavery, M. G. B. Drew (J. Chem. Soc. Dalton Trans. **1986** 911/20).

[91] G. A. Ozin (Appl. Spectrosc. **30** [1976] 573/85).

[92] P. H. Kasai, P. M. Jones (J. Am. Chem. Soc. **107** [1985] 813/8).

[93] C. W. Bauschlicher Jr., P. S. Bagus, C. J. Nelin, B. O. Roos (J. Chem. Phys. **85** [1986] 354/64).

[94] H. Hirai, K. Wada, M. Komiyama (Bull. Chem. Soc. Japan **59** [1986] 2217/23).

[95] T. N. Sorrell, A. S. Borovik, C. C. Shen (Inorg. Chem. **25** [1986] 589/90).

[96] P. Fiaschi, C. Floriani, M. Pasquali, A. Chiesi-Villa, C. Guastini (Inorg. Chem. **25** [1986] 462/9).

1.1.3.2 Compounds with More Than One CO Ligand

Only a few unstable compounds are known for this group.

(CO)$_2$Cu is prepared by cocondensation of Cu, CO, and Ar($\leq 1:12.5:1000$) at 10 to 15 K. The existence of (CO)$_2$Cu in addition to that of (CO)Cu and (CO)$_3$Cu is shown by IR spectroscopy together with CO mixed isotope experiments. The ν(CO) of ($^{12}C^{16}O)_n(^{13}C^{16}O)_{2-n}$Cu (n$=0$ to 2) are given for two different matrix sites, ν(CO) $=1891.5$ and 1976.1 cm^{-1} for ($^{12}C^{16}O)_2$Cu. The Cotton-Kraihanzel force constants are f(CO) $=15.41$ and 15.29, f(CO,CO) $=0.97$ and 1.08 mdyn/Å for the two sites [17, 38].

MO calculations by the self-consistent-field X$_\alpha$ scattered-wave method have been carried out for (CO)$_2$Cu. As for (CO)Cu, major bonding interaction arise from σ donation of charge density from the 3σ orbital of the CO ligands to the metal [31]. For extended Hückel MO calculations see [28].

[(CO)$_2$Cu(C$_5$H$_{10}$NH)$_2$]Cl$_2$ is believed to be an intermediate in the carbonylation of piperidine with CO in the presence of CuCl$_2 \cdot$ 2H$_2$O [15].

[(CO)$_2$CuC$_{10}$H$_8$N$_2$]ClO$_4$ (C$_{10}$H$_8$N$_2 =2,2'$-bipyridine) is formed from [(Cu(CH$_3$CN)$_4$]ClO$_4$ and C$_{10}$H$_8$N$_2$ in (CD$_3$)$_2$CO under CO atmosphere at 25 °C. ^{1}H NMR (acetone-d$_6$, 25 °C): 7.89 (H-5), 8.35 (H-4), 8.67 (H-3), 9.13 (H-6) ppm. ^{13}C NMR (acetone-d$_6$, 31 °C): 123.31 (C-5), 127.89 (C-3), 141.47 (C-4), 151.66 (C-6), 152.65 (C-2) ppm [33].

(CO)$_3$Cu is prepared by cocondensation of Cu with CO (1:1000 to 100000) at 10 to 15 K [17]; see also [3]. It is also formed when Cu atoms and CO are allowed to react on adamantane or cyclohexane at 77 K in a rotating cryostat. The compound is remarkably stable, surviving up to 253 K in adamantane [34, 37, 40].

The IR spectrum shows two CO stretching modes of approximately equal intensity at 1976.8 and 1990.0 cm^{-1} together with two modes at 323 and 376 cm^{-1}. From IR frequencies of ($^{12}C^{16}O)_n(^{13}C^{16}O)_{3-n}$Cu (n$=0$ to 3) in Ar matrices the force constants f(CO) $=16.51$ and 16.43, f(CO,CO) $=0.59$ and 0.66 mdyn/Å are calculated for two different matrix sites. The electronic spectrum shows absorptions at 375 and 562 nm assigned to $^2A_2'' \rightarrow {}^2E''$ and $^2A_2'' \rightarrow {}^2A_1'$, weak shoulders at 344 and 495 nm, as well as a weak feature at 262 nm. A qualitative molecular orbital energy level scheme for (CO)$_3$Cu has been given for D$_{3h}$ symmetry [17, 38]. The mean metal-carbon bond dissociation energy has been estimated from the Cotton-Kraihanzel force constants to be 36 kcal $\cdot$ (mol CO)$^{-1}$ [25].

At 4 to 100 K the EPR spectrum of (^{12}CO)$_3$^{63}Cu in cyclo-C$_6$D$_{12}$ or adamantane consisted of four equally spaced (80 G) lines centered at g$=2.0010$. The hyperfine structure was that of ^{63}Cu (I$=3/2$). In addition there was a central asymmetric triplet at g$=2.0028$ to 2.0029 which carried no ^{63}Cu hyperfine structure. On warming a ^{13}C containing sample to 250 K, a dramatic change occurred which revealed immediately the number of carbon atoms in the molecule, the spectrum consisted of four 1:3:3:1 quartets. On recooling to 77 K the main features were regenerated, proving that the two spectra are associated with the same molecule. EPR parameters [37, 40], see also [34]:

matrix	temperature (K)	axis	g	a(^{63}Cu) (MHz)	a(^{13}C) (MHz)
cyclo-C$_6$D$_{12}$	100	x, y	2.0028(2)	0(10)	29(2)
		z (threefold)	2.0010(2)	225(5)	8(2)
adamantane	250	isotropic	2.0014(2)	93.9(5)	22.0(5)

References on pp. 213/4

The low isotropic ^{63}Cu hyperfine interaction seems to indicate a planar, trigonal π radical whose ground state is $^2A_2'$ in D_{3h} symmetry. The unpaired electron occupies an orbital which is largely Cu $4p_z$ (z=threefold axis) and which is bonding in each Cu-C distance [19, 34, 37, 40]. Other authors [17, 39] regard the unpaired electron as residing mostly at the C atoms.

SCF$-X_\alpha$-SW calculations indicate that, as for (CO)Cu and (CO)$_2$Cu, the major interaction arises from the 3 σ orbital of the CO ligand with the metal [31]. For extended Hückel MO calculations see [28].

[(CO)$_3$Cu]$^+$. For formation see [(CO)Cu]$^+$ in Section 1.1.3.2. The equilibrium between [(CO)$_3$Cu]$^+$ and [(CO)Cu]$^+$ depends on the concentration of acid, the temperature, and the CO pressure. The [(CO)$_3$Cu]$^+$/[(CO)Cu]$^+$ ratio gradually increases with an increase in the H$_2$SO$_4$ concentration above 80%. [(CO)$_3$Cu]$^+$ ist not formed in H$_2$SO$_4$ concentrations below 80% [7, 91]. At -10 °C and a CO pressure of 7 atm the mole ratio of CO/Cu$^+$ reaches 3 in 100% H$_2$SO$_4$ [1, 7]. By addition of a small amount of another solvent, such as H$_2$O, CH$_3$OH, C$_2$H$_5$OH, acetic anhydride, or ether, the ratio rapidly becomes 1 [2].

The carbonyl stretching frequency appears at 2175 cm^{-1} in H$_2$SO$_4$. A ^{13}C NMR chemical shift of 169.2 ppm is obtained from H$_2$SO$_4$/FSO$_3$H (1.0/0.3) solutions [18].

(C$_6$H$_5$)$_3$PCu(CO)$_3$Mo(C$_3$H$_3$N$_2$)$_3$GaCH$_3$, [(C$_3$H$_3$N$_2$)$_3$GaCH$_3$]$^-$=tridentate methyltris(pyrazol-1-yl)gallate, has been prepared from [CuP(C$_6$H$_5$)$_3$Cl]$_4$ and Na$^+$[(CO)$_3$Mo(C$_3$H$_3$N$_2$)$_3$GaCH$_3$]$^-$ (1:4) in THF in form of golden-yellow air-stable crystals ($\sim$60% yield). ^{1}H NMR (acetone-d$_6$): 0,58 (s, CH$_3$), 6.31 (t, H-4 in C$_3$H$_3$N$_2$), 7.55 (m, C$_6$H$_5$), 7.81 (d, H-5 in C$_3$H$_3$N$_2$), 7.98 (d, H-3 in C$_3$H$_3$N$_2$) ppm. IR: ν(CO)=1798, 1898 in CH$_2$Cl$_2$; 1780, 1805, 1890 in Nujol. The X-ray crystallographic analysis shows semibridging CO groups with mean bond angles Mo-C-O of 170.1(5)°, 170.7(1)°, and 172.1(6)°, Cu-C-O of 117.0(1)°, 117.5(3)°, and 118.5(1)°, and mean bond lengths Mo-C of 1.973(7), 1.964(6), and 1.966(6)Å, Cu-C of 2.247(13), 2.298(23), and 2.415(5)Å for two crystallographically independent molecules. The exact nature of the Cu-CO interaction is not clear [36].

[(CO)$_4$Cu]$^+$ has been formed by CO absorption of Cu$_2$O in BF$_3$·H$_2$O [10, 18], FSO$_3$H [18], FSO$_3$H·SbF$_5$ [23], CF$_3$SO$_3$H [18], and HF [18] solutions.

IR and ^{13}C NMR spectra in strong acids:

solvent	IR ν(CO) in cm^{-1}	^{13}C NMR δ in ppm at -30 to -60 °C	Ref.
BF$_3$·H$_2$O	2186	169.5	[18]
	2190	—	[10]
FSO$_3$H	2183	171.1	[18]
FSO$_3$H/SbF$_5$	2183	—	[23]
CF$_3$SO$_3$H	2180	168.4	[18]
HF	2184	—	[18]

Raman (FSO$_3$H, -50 °C):2181 cm^{-1} (t$_2$) [30].

[(CO)$_6$Cu]$^{2+}$. The Hartree-Fock-Roothan method in the CNDO/2 approximation has been used to carry out calculations for the octahedral complexes of the ions Fe^{2+}, Co^{2+}, Ni^{2+}, Cu^{2+}, and Zn^{2+} with the ligands F$^-$, H$_2$CO, H$_2$O, NH$_3$, CH$_3$CN, and CO [35].

Catalytical Use of [(CO)$_n$Cu]$^+$ (n = 3, 4). The [(CO)$_n$Cu]$^+$ are active intermediates for carboxylation of alcohols [5, 9, 11, 16, 22, 26, 27], olefins [4, 6, 7, 11, 13, 14, 22, 24, 26, 29], C$_6$ to C$_{12}$ dienes and diols [21], and saturated hydrocarbons without [23] or with alcohols or olefins [8] as sources of alkyl cations in strong acids like H$_2$SO$_4$, BF$_3$/H$_2$O, BF$_3$/H$_3$PO$_4$, BF$_3$/H$_2$SO$_4$, HF/H$_2$O, HF/H$_2$O/BF$_3$, H$_2$SO$_4$/HSO$_3$F, and HSO$_3$F/SbF$_5$. The rate constant for the carbonylation of CH$_2$O by CO was increased 30 times by [(CO)$_n$Cu]$^+$ [32]. The reuse of the catalytic solutions has been tested [20]. They are stabilized by metallic Cu [12].

References:

[1] Y. Souma, H. Sano (Nippon Kagaku Zasshi **91** [1970] 625/30; C.A. **73** [1970] No. 94192).

[2] H. Sano (Nippon Kagaku Zasshi **91** [1970] 630/6; C.A. **73** [1970] No. 94193).

[3] J. S. Ogden (Chem. Commun. **1971** 978/9).

[4] Y. Souma, H. Sano (Japan. Kokai 74-61092 [1974] from C.A. **83** [1975] No. 137525).

[5] Y. Souma, H. Sano (Nikkakyo Geppo **26** [1973] 220/4 from C.A. **80** [1974] No. 14510).

[6] Y. Souma, H. Sano (Kagaku Kogyo **24** [1973] 1342/7 from C.A. **79** [1973] No. 125765).

[7] Y. Souma, H. Sano, J. Iyoda (J. Org. Chem. **38** [1973] 2016/20).

[8] Y. Souma, H. Sano (J. Org. Chem. **38** [1973] 3633/5).

[9] Y. Souma, H. Sano (Bull. Chem. Soc. Japan **46** [1973] 3237/40).

[10] Y. Matsushima, T. Koyano, T. Kitamura, S. Wada (Chem. Letters **1973** 433/4).

[11] T. Koyano, Y. Matsushima, T. Kitamura (Ger. Offen. 2339947 [1974] from C.A. **81** [1974] No. 54838).

[12] Y. Matsushima, T. Koyano, K. Shiozawa, S. Tomita (Japan. Kokai 75-134992 [1975] from C. A. **84** [1976] No. 121162).

[13] H. Sano, Y. Souma (Japan. 74-03511 [1974] from C.A. **81** [1974] No. 151548).

[14] N. Yoneda, T. Fukuhara, Y. Takahashi, A. Suzuki (Chem. Letters **1974** 607/10).

[15] B. K. Nefedov, A. A. Slinkin, A. V. Kucherov, Ya. T. Eidus (Izv. Akad. Nauk SSSR Ser. Khim. **1974** 2044/8; Bull. Acad. Sci. USSR Div. Chem. Sci. **1974** 1962/5).

[16] K. V. Puzitskii, S. D. Pirozhkov, T. N. Myshenkova, K. G. Ryabova, Ya. T. Eidus (Izv. Akad. Nauk SSSR Ser. Khim. **1975** 443/5; Bull. Acad. Sci. USSR Div. Chem. Sci. **1975** 371/3).

[17] H. Huber, E. P. Kündig, M. Moskovits, G. A. Ozin (J. Am. Chem. Soc. **97** [1975] 2097/106).

[18] Y. Souma, J. Iyoda, H. Sano (Inorg. Chem. 15 [1976] 968/70).

[19] E. P. Kündig, D. McIntosh, B. McGarvey, G. A. Ozin (unpublished from D. McIntosh, G. A. Ozin, J. Am. Chem. Soc. **98** [1976] 3167/75, ref. 6).

[20] Y. Souma, H. Sano (Osaka Kogyo Gijutsu Shikensho Kiho **27** [1976] 52/6; C.A. **85** [1976] No. 142559).

[21] Y. Souma, J. Iyoda, H. Sano (Bull. Chem. Soc. Japan 49 [1976] 3291/5).

[22] Y. Souma, H. Sano (Bull. Chem. Soc. Japan **49** [1976] 3296/9).

[23] Y. Souma, H. Sano (Bull. Chem. Soc. Japan **49** [1976] 3335/6).

[24] N. Yoneda, T. Fukuhara, Y. Takahashi, A. Suzuki (Bull. Chem. Soc. Japan **51** [1978] 2347/53).

[25] R. G. Behrens (J. Less-Common Metals **58** [1978] 47/54).

[26] Y. Souma, H. Sano (Osaka Kogyo Gijutsu Shikensho Kiho **29** [1978] 106/9; C.A. **89** [1978] No. 108019).

[27] J. M. Bregeault, C. Jarjour, S. Yolou (J. Mol. Catal. **4** [1978] 225/9).

[28] H. Itoh, A. B. Kunz (Fundam. Res. Homogeneous Catal. **3** [1979] 73/82).

[29] S. Yolou, J. M. Bregeault (Bull. Soc. Chim. France **1979** 485/95).

[30] A. Neppel, J. P. Hickey, I. S. Butler (J. Raman. Spectrosc. **8** [1979] 57/9).

[31] D. F. McIntosh, G. A. Ozin, R. P. Messmer (Inorg. Chem. **20** [1981] 3640/50).

[32] Y. Souma, H. Sano (Nippon Kagaku Kaishi **1982** 263/7).

[33] S. Kitagawa, M. Munakata, N. Miyaji (Inorg. Chem. **21** [1982] 3842/3).

[34] J. A. Howard, B. Mile, J. R. Morton, K. F. Preston, R. Sutcliffe (Chem. Phys. Letters **117** [1985] 115/7).

[35] V.M. Pinchuk, V.A. Korobskii, L.B. Shevardina (Zh. Neorg. Khim. **29** [1984] 1111/3; Russ. J. Inorg. Chem. **29** [1984]

[36] G.A. Banta, B.M. Louie, E. Onyiriuka, S.J. Rettig, A. Storr (Can J. Chem. **64** [1986] 373/86).

[37] J.A. Howard, B. Mile, J.R. Morton, K.F. Preston (J. Phys. Chem. **90** [1986] 2027/9).

[38] G. A. Ozin (Appl. Spectrosc. **30** [1976] 573/85).

[39] P. H. Kasai, P. M. Jones (J. Am. Chem. Soc. **107** [1985] 813/8).

[40] J. A. Howard, B. Mile, J. R. Morton, K. F. Preston, R. Sutcliffe (J. Phys. Chem. **90** [1986] 1033/6).

1.1.4 Isocyanide Compounds

General Literature:

E. Singleton, H. E. Oosthuizen, Metal Isocyanide Complexes, Advan. Organometal. Chem. **22** [1983] 209/310.

L. P. Yur'eva, Isocyanide Complexes of Transition Metals, Metody Elementoorg. Khim. Tipy Metalloorg. Soedin. Perekhodnykh Metal. **1** [1975] 162/216; C.A. **84** [1976] No. 11597.

L. Malatesta, F. Bonati, Isocyanide Complexes of Metals, London – New York – Sydney – Toronto 1969, pp. 37/42, 54.

L. Malatesta, Isocyanide Complexes of Metals, Progr. Inorg. Chem. **1** [1959] 283/379, 291/7.

The literature before 1949 is also reviewed in "Kupfer" B2, 1961, p. 855.

Most of the copper isocyanide compounds contain one, two, three, or four isocyanides and ligands like halogens or pseudohalogens. Others contain N, P, or As donor ligands. Some have halides, BF_4, $B(C_6H_5)_4$, or PF_6 as counterions. The number of the isocyanide ligands depends on the nature of the additional ligands or counterions, X, and for a particular X two or more complexes with different numbers of isocyanides can often be synthesized. While complexes of the $[(RNC)_4Cu]X$ type are readily formed from Cu^I halides and excess isocyanide, they have not been isolated for X = CN. And although $(t-C_4H_9NC)_3CuBr$ has been isolated, $(t-C_4H_9NC)_3CuCN$ has not [2].

The compounds RNCCuX most likely are polymeric; see Chapter 1.1.4.1.1.

Copper isocyanides $(RNC)_nCu$ have not yet been isolated. However, their existence is indicated by IR and ESR data. Metallic copper dissolves in $cyclo-C_6H_{11}NC$. The solution has an IR band at 2180 cm^{-1}, which can be assigned to a coordinated isocyanide, in addition to the band at 2140 cm^{-1} for the free ligand [4, 5]. Its ESR spectrum exhibits a g value of 2.0041, which was tentatively assigned to a zero valent copper isocyanide complex [6, 7, 9]. A soluble complex of unknown composition which forms from Cu_2O and $cyclo-C_6H_{11}NC$ also has an IR band at 2181 cm^{-1} [5].

Complexes of metallic Cu or Cu_2O and $t-C_4H_9NC$, not fully characterized, show Δv (NC) = +40 cm^{-1} from free $t-C_4H_9NC$. They react in C_6H_6 with $1,3,5-C_6H_3(NO_2)_3$ to give unstable crystalline products with the molar proportions $Cu:t-C_4H_9NC:C_6H_3(NO)_3:C_6H_6 = 1:5:1:0.33$ and $Cu_2O:t-C_4H_9NC:C_6H_3(NO_2)_3:C_6H_6 = 1:8:2:0.66$ [15]. Solutions of 2,3,4,6-tetra-O-acetyl-β-D-glucopyranos-1-ylisocyanide in CH_2Cl_2 or CH_3CN show Δv (NC) = +46 to 47 cm^{-1} after addition of Cu, CuO, Cu_2O, or CuCl [19].

Catalytic systems like RNC/Cu, RNC/Cu$_2$O, and RNC/CuCl are frequently used very efficiently for the syntheses of a great variety of organic compounds because the copper isocyanide complexes so prepared are soluble in organic solvents. Reactions occur with activated C–H bonds or alkyl halides followed by additions to C=C and C=O double bonds and concomitant cyclizations as well as insertions of the isocyanide ligand into the hetero-atom H–bond of amines, alcohols, thiols, phosphanes, and silanes. In all these reactions Cu isocyanide complexes are formed as byproducts [8]. The system cyclo-C$_6$H$_{11}$/Cu exhibits mono–electronic reducing character and catalyzes the dimerization of alkyl halides with excellent yields [13], cf. [14]. The formation of radical anions by electron transfer from the systems t-C$_4$H$_9$NC/Cu or t-C$_4$H$_9$NC/Cu$_2$O to π substrates such as C$_6$H$_5$NO$_2$, m- and p-C$_6$H$_4$(NO$_2$)$_2$, benzoquinone, fluorenone, and (NC)$_2$C=C(CN)$_2$ is shown by ESR spectroscopy [18].

Nonstoichiometric (t-C$_4$H$_9$NC)$_n$Cu$_2$CO$_3$ is prepared by hydrolysis of (t-C$_4$H$_9$NC)$_3$CuO-CO$_2$C$_4$H$_9$-t (see 1.1.4.3) with H$_2$O (mole ratio 2:1) in THF at 0 °C (94% yield) or from (t-C$_4$H$_9$NC)$_3$CuOCO$_2$H (see 1.1.4.3) with proton abstractors such as CuOC$_4$H$_9$-t in THF at 0 °C. It is colorless and extremely sensitive to H$_2$O. IR (Nujol): 1511 (CO), 2164+2189 (NC). The reaction with C$_2$H$_5$I in (CH$_3$)$_2$NCHO at room temperature yields 57% (C$_2$H$_5$O)$_2$CO, 12% CO$_2$, and 5% C$_2$H$_5$OH. Treatment with t-C$_4$H$_9$NC, CO$_2$, and H$_2$O (mole ratio 1:3:1.5:1.2) gives (t-C$_4$H$_9$NC)$_3$CuOCO$_2$H [16].

The nonstoichiometric product (t-C$_4$H$_9$NC)$_n$CuNCO (n ≈ 2.5 by iodometry, 2.7 from NMR), possibly a mixture of different isocyanide complexes, is prepared either from t-C$_4$H$_9$NC, Cu$_2$O, and RNHCO$_2$C$_2$H$_5$ in 93% (R = H) and 90% (R = (CH$_3$)$_3$Si) yield [10], or from t-C$_4$H$_9$NC, [(CH$_3$)$_3$Si]$_2$NCu, and CO$_2$ with 91% yield of the additionally formed [(CH$_3$)$_3$Si]$_2$O in toluene at 0 °C [11]. The product forms colorless crystals, m.p. 103 °C [10]. IR: 610 (δ,γNCO), 1350 (v$_s$NCO), 2175 (v$_{as}$NCO), and 2225 (NC) cm^{-1}, suggesting that the NCO group may be coordinated to Cu through N [11], cf. [10]. ^{1}H NMR (CDCl$_3$): 1.20 ppm (CH$_3$) [10]. It is soluble in most organic solvents, not very sensitive to air in solid state, but readily oxidized in solution [10]. In benzene its degree of association of 1.3 was measured by cryoscopy [11]. When treated with aqueous H$_2$SO$_4$, it evolves CO$_2$ [11]. With C$_2$H$_5$OH/HCl it yields NH$_2$CO$_2$C$_2$H$_5$ [10]; with alkyl bromides in the presence of alcohol it produces N-alkylcarbamates [10, 11]. However, with n-C$_4$H$_9$I [11] or with acyl chlorides [10] the corresponding isocyanides are formed. Addition of n-C$_4$H$_9$NH$_2$ to the reaction mixture with n-C$_4$H$_9$I yields (n-C$_4$H$_9$NH)$_2$CO [11].

Nonstoichiometric carbamato complexes (t-C$_4$H$_9$NC)$_n$CuO$_2$CNRR' (n = 1.4 for R = R' = C$_2$H$_5$; n = 1.6 for R = H, R' = C$_6$H$_5$) were isolated by evaporation of the C$_6$H$_6$ solution of CuOC$_4$H$_9$-t, t-C$_4$H$_9$NC (1:3), HNRR', and CO$_2$ under reduced pressure and identified by their Cu content, CO$_2$ evolution on acidolysis, and IR spectra (not given). Complexes of this type are the key intermediates in the copper/isocyanide–promoted formation of urethanes from CO$_2$, amines, and alkyl halides [12], cf. [17].

A complex of C$_2$H$_5$NC with the copper protein hemocyanin, obtained from the hemolymph of the whelk Murex trunculus, seems to be formed by the reversible displacement of O$_2$ in oxyhemocyanin, as was indicated by UV spectroscopy [3].

References:

[1] F. Klages, K. Mönkemeyer, R. Heinle (Chem. Ber. **85** [1952] 109/22).

[2] S. Otsuka, K. Mori, K. Yamagami (J. Org. Chem. **31** [1966] 4170/4).

[3] E. J. Wood, W. H. Bannister (Nature **215** [1967] 1091).

[4] T. Saegusa, Y. Ito, S. Kobayashi, K. Hirota, H. Yoshioka (Bull. Chem. Soc. Japan **42** [1969] 3310/3).

[5] T. Saegusa, Y. Ito, S. Kobayashi, K. Hirota, N. Takeda (Can. J. Chem. **47** [1969] 1217/22).

[6] T. Saegusa, K. Yonezawa, Y. Ito (Syn. Commun. **2** [1972] 431/9).

[7] T. Saegusa, K. Yonezawa, I. Murase, T. Konoike, S. Tomita, Y. Ito (J. Org. Chem. **38** [1973] 2319/28).

[8] T. Saegusa, Y. Ito (Synthesis **1975** 291/300).

[9] Y. Ito, T. Saegusa (Organotransition Met. Chem. Proc. 1st Japan Am. Semin., Honolulu 1974 [1975], pp. 49/55; C.A. **85** [1976] No. 160272).

[10] Y. Ito, Y. Inubushi, S. Matsumura, T. Saegusa (Bull. Chem. Soc. Japan **49** [1976] 573/4).

[11] T. Tsuda, H. Washita, T. Saegusa (J. Chem. Soc. Chem. Commun. **1977** 468/9).

[12] T. Tsuda, H. Washita, K. Watanabe, M. Miwa, T. Saegusa (J. Chem. Soc. Chem. Commun. **1978** 815/6).

[13] A. Ballatore, M. P. Crozet, J.-M. Surzur (Tetrahedron Letters **1979** 3073/6).

[14] M. P. Crozet, J.-M. Surzur, R. Jauffred, C. Ghiglione (Tetrahedron Letters **1979** 3077/8).

[15] A.-T. Hansson, M. Nilsson (Acta Chem. Scand. B **34** [1980] 119/23).

[16] T. Tsuda, Y. Chujo, T. Saegusa (J. Am. Chem. Soc. **102** [1980] 431/3).

[17] T. Tsuda, K. Watanabe, K. Miyata, H. Yamamoto, T. Saegusa (Inorg. Chem. **20** [1981] 2728/30).

[18] Y. Ito, T. Konoike, T. Saegusa (Tetrahedron Letters **1974** 1287/90).

[19] D. Marmet, P. Boullanger, G. Descotes (Tetrahedron Letters **21** [1980] 1459/62).

1.1.4.1 Compounds Containing One Isocyanide Ligand

Cu^I semiquinolato complexes of the type $RNCCuO_2C_6H_3R'$ ($R = t-C_4H_9$; nature and position of R' not given) have been found in equilibrium with $(RNC)_2CuO_2C_6H_3R'$ and $CuO_2C_6H_3R'$ on dilution of $(RNC)_2CuO_2C_6H_3R'$ solutions.

Reference:

G. A. Abakumov, A. V. Lobanov, K. K. Cherkasov (Abstr. 2nd All-Union Conf. Organometal. Chem., Gorkii 1982, p. 168) from E. A. Kashtanov, V. K. Cherkasov, L. V. Gorbunova, G. A. Abakumov (Izv. Akad. Nauk SSSR Ser. Khim. **1983** 2121/5; Bull. Acad. Sci. USSR Div. Chem. Sci. **1983** 1917/20).

1.1.4.1.1 Compounds of the Type RNCCuX

In this chapter the compounds RNCCuX are described for R = alkyl or aryl and X = Cl, Br, I, CN, alkoxy, or acyloxy. Most of the compounds in Table 28 are prepared from CuX and the corresponding isocyanide.

Method I: In an exothermic reaction without solvent.

Method II: In CH_3CN (boiling or at room temperature) or at room temperature in C_6H_6, $CHCl_3$, or aqueous C_2H_5OH.

Another method is the alkylation of HCN with olefins like isobutene in the presence of CuBr. Although $t-C_4H_9NCCuCN$ was obtained in this manner, no $t-C_4H_9NCCuBr$ could be detected in the reaction products. Because the isocyanide is easily liberated from RNCCuX with KCN, this method can be used for preparing t-alkyl isocyanides from olefins [6]. An X-ray crystallographic study of CH_3NCCuI (Table 28, No. 2) revealed a polymeric structure, and it is very likely that other compounds are also polymeric in the solid state. However, for compounds No. 17 and 18 a monomeric structure in solution was inferred from the osmometric molecular mass in dimethylformamide. Since only one $\nu(NC)$ was observed in the solid state IR spectra for Nos. 17 to 19, they are assumed to be two-coordinate [9].

References on p. 219

Table 28

Compounds of the Type RNCCuX.

Further information on compounds preceded by an asterisk is given at the end of the table. For abbreviations and dimensions see p. X.

No.	RNCCuX R	X	method of preparation (solvent, yield) remarks	Ref.
1	CH_3	Br	II (CH_3CN) colorless crystals	[4]
2	CH_3	I	according to X-ray studies polymeric, therefore reported in the corresponding chapter	[2, 4, 5]
*3	CH_3	CN	light gray solid	[4]
4	C_2H_5	I	II (CH_3CN) colorless needles	[4]
5	C_2H_5	CN	I colorless prisms from ethanol loss of isocyanide at room temperature	[1]
6	$n-C_3H_7$	I	II (CH_3CN, low) dirty white needles	[4]
7	$n-C_3H_7$	CN	I colorless rhombic plates from ethanol loss of isocyanide at room temperature	[1]
8	$(CH_3)_2CHCH_2$	CN	I colorless rhombic prisms from ethanol loss of isocyanide at room temperature	[1]
9	$t-C_4H_9$	Cl	preparation not given IR: 2189	[11, 13]
*10	$t-C_4H_9$	CN	colorless prisms from CH_3CN, m.p. 196 to 198° IR (Nujol): 2140 (CN, cyanide), 2182 (NC, isocyanide)	[6]
*11	$t-C_4H_9$	OC_4H_9-t	II (C_6H_6) yellow crystals, sublimes at 90°/1 Torr ^1H NMR (C_6D_6): 0.99 $(t-C_4H_9N)$, 1.90 $(t-C_4H_9O)$ IR (Nujol): 945 $(t-C_4H_9O)$, 1195 + 1220 $(t-C_4H_9N)$, 2102 (NC)	[11, 13]
*12	$t-C_4H_9$	O_2CCH_3	colorless crystals from benzene ^1H NMR (CD_3CN): 1.49 $(O_2CCH_3 + t-C_4H_9)$ IR (KBr): 1210 + 1235 $(t-C_4H_9)$, 1410 + 1560 + 1585 (O_2C), 2169 (NC)	[10]
13	$t-C_4H_9$	$O_2CC_6H_5$	preparation similar to No. 12 ^1H NMR (CD_3CN): 1.45 (CH_3), 7.4 (C_6H_5) IR (KBr): 1570 + 1600 $(C_6H_5 + O_2C)$, 2168 (NC), 3060 (C_6H_5)	[10]

References on p. 219

Table 28 [continued]

No.	RNCCuX		method of preparation (solvent, yield)	Ref.
	R	X	remarks	
14	t-C$_4$H$_9$	OCO$_2$C$_4$H$_9$-t	from No. 11+CO$_2$ (C$_6$H$_6$) exists according to IR studies in C$_6$H$_6$ at room temperature in equilibrium with No. 11 and CO$_2$; attempted isolation even at 0° gives only No. 11	[13]
15	C$_6$H$_5$	Cl	II (ether, almost quantitative); or from (C$_6$H$_5$NC)$_2$CuCl (boiling CH$_3$OH, 96%) colorless needles, m.p. 184 to 185° insoluble in ordinary solvents	[3]
16	cyclo-C$_6$H$_{11}$	Cl	I colorless crystals from CHCl$_3$/ether, m.p. 95 to 96° IR: 2192 (NC)	[7, 8]
17	4-CH$_3$OC$_6$H$_4$	Cl	II (CHCl$_3$, 85%) colorless crystals from CHCl$_3$, m.p. 188 to 190° stable in air, even in solution IR (Nujol): 2155 (NC)	[9]
18	4-CH$_3$OC$_6$H$_4$	Br	II (CHCl$_3$) colorless crystals from CHCl$_3$, m.p. 190 to 193° stable in air, even in solution IR (Nujol): 2173 (NC)	[9]
19	4-CH$_3$OC$_6$H$_4$	I	II (H$_2$O/C$_2$H$_5$OH, 40%) colorless crystals from CHCl$_3$, m.p. 148 to 150° stable in air, even in solution IR (Nujol): 2158 (NC)	[9]
20	C$_6$H$_5$CO$_2$CH$_2$NC	Cl	II (CH$_3$CN, 25°, 41%) m.p. 147 to 148° (dec.) ^{1}H NMR (acetone-d$_6$): 6.02 (CH$_2$)	[14]

*Further information:

CH$_3$NCCuCN (Table **28**, No. **3**) is probably obtained as a noncrystalline solid if CH$_3$I and CuCN in CH$_3$CN are heated at 100 °C for less than 12 h. It is insoluble in water and all common solvents. According to the elemental analysis it seemed to be contaminated with CuCN. When warmed with aqueous KCN, it releases CH$_3$NC [4].

t-C$_4$H$_9$NCCuCN (Table **28**, No. **10**) is prepared from (t-C$_4$H$_9$NC)$_3$CuBr (Table **32**, No. **2**) in boiling water in 84% yield; in benzene the yield is very low. It is also formed by repeated recrystallization of (t-C$_4$H$_9$NC)$_2$CuCN (Table **30**, No. **2**) from CHCl$_3$/ether and by reaction of CuX (X=Cl, Br), HCN, and isobutene either without a solvent or in water, tetrahydrofuran, and benzene. It is insoluble in hydrocarbons, alcohols, and ether, and only slightly soluble in hot acetonitrile. The ν(NC) of the isocyanide ligand is 39 cm^{-1} higher than that of the free ligand; the ν(CN) of the cyano group is 32 cm^{-1} lower than for CuCN. On treatment with aqueous KCN the isocyanide is liberated [6].

t-C$_4$H$_9$NCCuOC$_4$H$_9$-t (Table **28**, No. **11**) is also prepared from t-C$_4$H$_9$NC and Cu(OC$_4$H$_9$-t)$_2$ (33%) or (CO)CuOC$_4$H$_9$-t (Table **27**, No. **19**) [11]. It is soluble in most organic solvents. In cyclohexane it has a degree of association of about 4 or 5 [13]. That the $t\text{-}C_4H_9NC$ ligand functions as a good π acceptor is inferred from the 36 cm^{-1} lowering of $\nu(NC)$ upon complexation [11, 13].

Reaction with cyclopentadiene gives $\pi\text{-}C_5H_5CuCNC_4H_9\text{-}t$ in good yield. For the reaction with fluorene followed by treatment with CH_3I in tetrahydrofuran at 90 °C the proposed mechanism is shown in Scheme I [11]. The compound is further assumed to be an intermediate in the formation of $C_6H_5CO_2CH_2C_6H_5$ from C_6H_5CHO in the presence of $t\text{-}C_4H_9NC$ and $CuOC_4H_9\text{-}t$ [12].

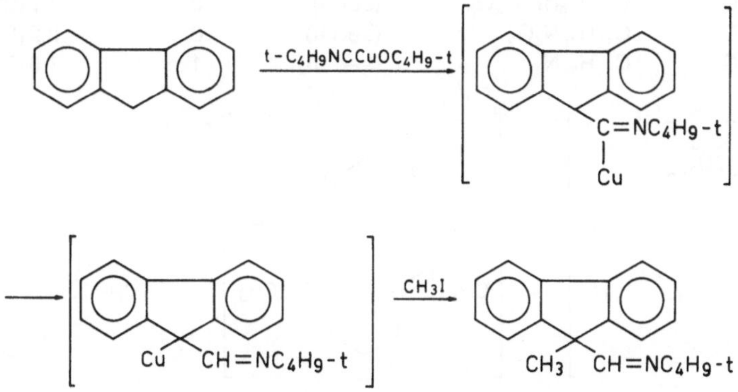

I

$t\text{-}C_4H_9NCCuO_2CCH_3$ (Table **28**, No. **12**) is obtained by heating Cu_2O, $t\text{-}C_4H_9NC$, and CH_3CO_2H (1:2:2 mole ratio) in benzene in 53% yield. It is air–sensitive, especially in benzene solution. In the presence of $t\text{-}C_4H_9NC$ it is readily soluble in benzene even at room temperature. The compound reacts with alkyl halides such as $(+)\text{-}(R)\text{-}C_6H_5CHBrCH_3$, to give the corresponding acetates. The reaction proceeds quantitatively with 75% inversion of configuration. The isocyanide ligand greatly influences the stereochemistry because with CH_3CO_2Cu, subjected to the same conditions, racemization takes place [10].

References:

[1] H. Guillemard (Ann. Chim. Phys. [8] **14** [1908] 311/432, 429/30).

[2] E. G. J. Hartley (J. Chem. Soc. **1928** 780/1).

[3] F. Klages, K. Mönkemeyer, R. Heinle (Chem. Ber. **85** [1952] 109/22).

[4] H. Irving, M. Jonason (J. Chem. Soc. **1960** 2095/7).

[5] P. J. Fisher, N. E. Taylor, M. M. Harding (J. Chem. Soc. **1960** 2303/9).

[6] S. Otsuka, K. Mori, K. Yamagami (J. Org. Chem. **31** [1966] 4170/4).

[7] T. Saegusa, Y. Ito, S. Kobayashi, K. Hirota, H. Yoshioka (Bull. Chem. Soc. Japan **42** [1969] 3310/3).

[8] T. Saegusa, Y. Ito, S. Kobayashi, K. Hirota, N. Takeda (Can. J. Chem. **47** [1969] 1217/22).

[9] J. Bailey, M. J. Mays (J. Organometal. Chem. **47** [1973] 217/24).

[10] T. Saegusa, J. Murare, Y. Ito (J. Org. Chem. **38** [1973] 1753/5).

[11] T. Tsuda, H. Habu, S. Horiguchi, T. Saegusa (J. Am. Chem. Soc. **96** [1974] 5930/1).

[12] T. Tsuda, H. Habu, T. Saegusa (J. Chem. Soc. Chem. Commun. **1974** 620).

[13] T. Tsuda, S. Sanada, K. Meda, T. Saegusa (Inorg. Chem. **15** [1976] 2329/32).

[14] W. P. Fehlhammer, K. Bartel, B. Weinberger, U. Plaia (Chem. Ber. **118** [1985] 2220/34).

1.1.4.1.2 Compounds of the Type $RNCCu^nD_mX$ Including Ionic Compounds

Complexes of the Type $[RNCCu^I{}^8D]^{n+}$. Binding constants $K = [RNCCu^I{}^8D^{n+}]/[RNC] \cdot [Cu^I{}^8D^{n+}]$ in $(CH_3)_2NCHO$ at 22 ± 1 °C are calculated from the influence of RNC on the polarographic half-wave potentials of the redox couples $[Ci^I{}^8D]^{n+}/[Cu^{II}{}^8D]^{(n+1)+}$ [6]:

RNC	^8D		n	$K \, [M^{-1}]$
$p-NCC_6H_4NC$	$C_{11}H_{18}BF_2N_4O_2$	(see I)	0	$1.9(4) \times 10^7$
$p-O_2NC_6H_4NC$	$C_{11}H_{18}BF_2N_4O_2$	(see I)	0	$1.7(2) \times 10^7$
$p-O_2NC_6H_4NC$	$C_{11}H_{19}N_4O_2$	(see II)	0	$8.2(8) \times 10^3$
$p-O_2NC_6H_4NC$	$C_{14}H_{24}N_4$	(see III)	1	$7.6(7) \times 10^4$

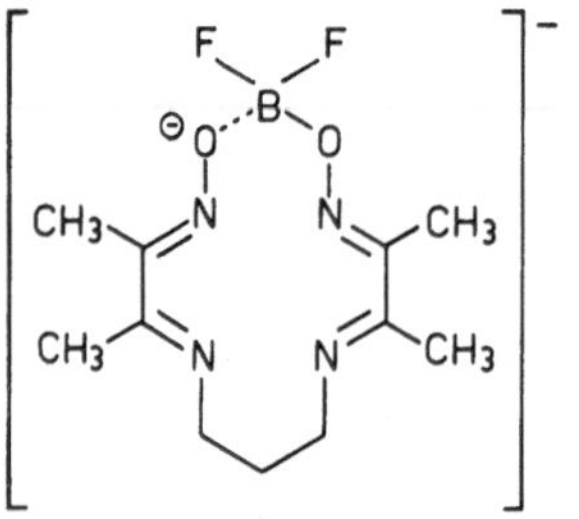

I 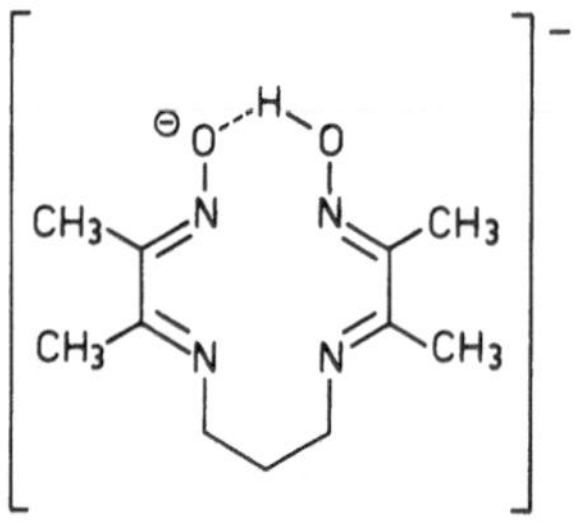   II 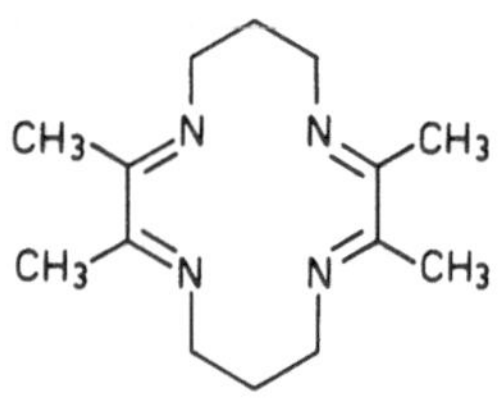   III

Most of the **compounds in Table 29** are prepared by the following methods:

Method I: From $(C_6H_5)_3PCuCl$ and RNC (1:1 mole ratio) in $CHCl_3$. After precipitation with light petroleum, the compound is recrystallized from $CHCl_3$.

Method II: From RNCCuX and D (1:1 mole ratio, 1:3 for No. 5) in $CHCl_3$ or/and dimethylformamide.

All compounds except for No. 8 are crystalline materials. The complexes No. 1, 2, and 9 to 16 are air-stable in the solid state and in solution, soluble to some extent in $CHCl_3$, and apparently stable with respect to disproportionation [1].

The solid state structures for Nos. 9 to 12 are assumed to be dimeric halogen-bridged cis structures in accordance with the two $\nu(NC)$ bands in the IR spectra. For Nos. 2 and 13 to 16 four-coordinate Cu atoms are presumed. Osmometric molecular weights for Nos. 9, 11, and 15 in $CHCl_3$ correspond to the monomeric formulations, whereas No. 2 in $CHCl_3$ and No. 16 in $(CH_3)_2CO$ are ionized. Conductivity data in CH_3NO_2 indicate Nos. 9 and 12 to be only slightly ionized, Nos. 13 to 15 to be partly ionized, and Nos. 2 and 16 to be extensively ionized [1].

In Table 29 all IR data except those for Nos. 10, 11, and 12 (no details given) are measured in Nujol, $\nu(NC) = 2094$ to 2184 cm^{-1}.

References on p. 223

Table 29

Compounds of the Type RNCCu^{n}D$_m$X Including Ionic Compounds.

Further information on compounds preceded by an asterisk is given at the end of the table.
For abbreviations and dimensions see p. X.

No.	compound	method of preparation (yield) remarks IR: ν(NC)	Ref.
1	C$_2$H$_5$NCCuP(C$_6$H$_5$)$_3$Cl	I (90%) colorless, m.p. 191 to 195° IR: 2168, 2184 reacts with P(C$_6$H$_5$)$_3$ + AgPF$_6$ to give No. 2	[1]
2	[C$_2$H$_5$NCCu(P(C$_6$H$_5$)$_3$)$_3$]PF$_6$	see No. 1 (89%) m.p. 161 to 165° IR: 2193	[1]
*3	t-C$_4$H$_9$NCCu(C$_3$H$_3$N$_2$)$_2$B(C$_6$H$_5$)$_2$ (C$_3$H$_3$N$_2$ = 2-η-pyrazol-1-yl)	colorless, m.p. >280° IR: 2180 ^{1}H NMR (CDCl$_3$): 1.46 (t-C$_4$H$_9$), 6.97 + 7.50 + 7.70 (2H each, pyrazol-yl), 7.27 (C$_6$H$_5$)	[5]
*4	t-C$_4$H$_9$NCCu(C$_3$H$_3$N$_2$)$_3$BC$_3$H$_3$N$_2$ (C$_3$H$_3$N$_2$ = pyrazol-1-yl-, three of them 2-η-coordinated to Cu)	colorless, m.p. 137 to 138° IR: 2140 ^{1}H NMR (CDCl$_3$): 1.57 (t-C$_4$H$_9$), 6.34 + 7.34 + 7.80 (4H each, pyrazolyl)	[5]
*5	t-C$_4$H$_9$NCCu(C$_3$H$_3$N$_2$)$_3$BH (C$_3$H$_3$N$_2$ = 2-η-pyrazol-1-yl)	colorless, m.p. 156 to 158° IR: 2155 (NC), 2480 (BH) ^{1}H NMR: 1.60 (s, t-C$_4$H$_9$), 5.94 + 7.31 + 7.38 (t, d, d, C$_3$H$_3$N$_2$)	[2]
*6	[cyclo-C$_6$H$_{11}$NCCu(NH$_2$CH$_2$CH$_2$)$_2$NH]B(C$_6$H$_5$)$_4$	colorless, air stable IR: 2150	[3]
*7	4-CH$_3$C$_6$H$_4$NCCu(C$_6$H$_5$CH=NCH$_2$)$_2$I	yellow crystals IR: 1630 (C=N), 2135 (NC of isocyanide)	[7]
*8	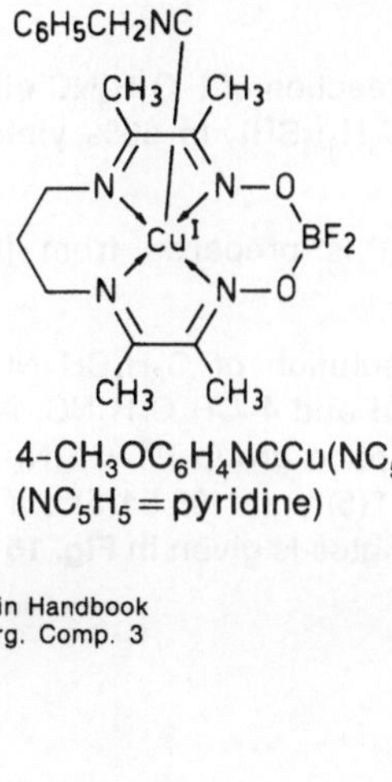	not isolated	[4]
9	4-CH$_3$OC$_6$H$_4$NCCu(NC$_5$H$_5$)Cl (NC$_5$H$_5$ = pyridine)	II (95%) colorless, m.p. 131 to 137° IR: 2098, 2133	[1]

Table 29 [continued]

No.	compound	method of preparation (yield) remarks IR: ν(NC)	Ref.
10	4-CH$_3$OC$_6$H$_4$NCCuP(C$_6$H$_5$)$_3$Cl	I, II (82%) colorless, m.p. 192 to 195° IR: 2094, 2130	[1]
11	4-CH$_3$OC$_6$H$_4$NCCuP(C$_6$H$_5$)$_3$Br	II colorless, m.p. 201 to 205° IR: 2096, 2133	[1]
12	4-CH$_3$OC$_6$H$_4$NCCuAs(C$_6$H$_5$)$_3$Br	II colorless, m.p. 175 to 177° IR: 2099, 2138	[1]
13	4-CH$_3$OC$_6$H$_4$NCCu(N$_2$C$_{12}$H$_8$)Cl (N$_2$C$_{12}$H$_8$ = 1,10-phenanthroline)	II (90%) yellow, m.p. 202 to 205° IR: 2127 reacts with P(C$_6$H$_5$)$_3$ + AgPF$_6$ to give No. 16	[1]
14	4-CH$_3$OC$_6$H$_4$NCCu(N$_2$C$_{12}$H$_8$)Br (N$_2$C$_{12}$H$_8$ = 1,10-phenanthroline)	II yellow, m.p. 206 to 211° IR: 2131	[1]
15	4-CH$_3$OC$_6$H$_4$NCCu(CH$_2$P(CH$_3$)$_2$)$_2$Br	II yellow, m.p. 230 to 235° IR: 2138	[1]
16	[4-CH$_3$OC$_6$H$_4$NCCu(N$_2$C$_{12}$H$_8$)P(C$_6$H$_5$)$_3$]PF$_6$ (N$_2$C$_{12}$H$_8$ = 1,10-phenanthroline)	see No. 13 (73%) yellow, m.p. 255 to 260° (dec.) IR: 2161	[1]

*Further information:

t-C$_4$H$_9$NCCu(C$_3$H$_3$N$_2$)$_2$B(C$_6$H$_5$)$_2$ (Table 29, No. 3) is prepared from t-C$_4$H$_9$NC, CuCl, and Na[(C$_3$H$_3$N$_2$)$_2$B(C$_6$H$_5$)$_2$] (2:1:1 mole ratio) in ether with 36% yield [5].

t-C$_4$H$_9$NCCu(C$_3$H$_3$N$_2$)$_3$BC$_3$H$_3$N$_2$ (Table 29, No. 4) is prepared from t-C$_4$H$_9$NC, CuCl, and K[B(C$_3$H$_3$N$_2$)$_4$] in an aqueous 2M KI solution with 47% yield [5].

t-C$_4$H$_9$NCCu(C$_3$H$_3$N$_2$)$_3$BH (Table 29, No. 5) is prepared by the reaction of t-C$_4$H$_9$NC either with OCCu(N$_2$C$_3$H$_3$)BH in benzene in 85% yield or with [Cu(N$_2$C$_3$H$_3$)$_3$BH]$_2$ in 66% yield. It is recrystallized from petroleum ether [2].

[cyclo-C$_6$H$_{11}$NCCu(NH$_2$CH$_2$CH$_2$)$_2$NH]B(C$_6$H$_5$)$_4$ (Table 29, No. 6) is prepared from [OC-Cu(NH$_2$CH$_2$CH$_2$)$_2$NH]B(C$_6$H$_5$)$_4$ and cyclo-C$_6$H$_{11}$NC in THF [3].

4-CH$_3$C$_6$H$_4$NCCu(C$_6$H$_5$CH=NCH$_2$)$_2$I (Table 29, No. 7). A THF solution of C$_6$H$_5$CH=NCH$_2$-CH$_2$N=CHC$_6$H$_5$ was reacted with nearly equimolar amounts of CuI and 4-CH$_3$C$_6$H$_4$NC. After 10 h, the title compound precipitated in a 37% yield. Triclinic, space group P$\bar{1}$-C$_i^1$ (No. 2); a = 10.139(7), b = 13.427(9), c = 9.830(4) Å, α = 91.56(5)°, β = 111.31(5)°, γ = 108.54(5)°; Z = 4, d$_c$ = 1.55 g · cm^{-3}. The structure with selected bond lengths and angles is given in **Fig. 15** [7].

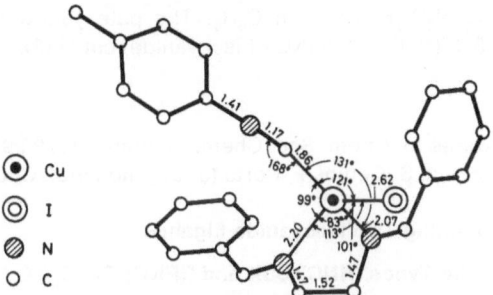

Fig. 15. Molecular structure of 4-CH$_3$C$_6$H$_4$NCCu(C$_6$H$_5$CH=NCH$_2$)$_2$I (No. 9) with selected bond lengths (in Å) and angles.

C$_6$H$_5$CH$_2$NCCu^{I}C$_{11}$H$_{18}$N$_4$O$_2$BF$_2$ (Table 29, No. 8) seems to be formed from [Cu^{II}C$_{11}$H$_{18}$N$_4$O$_2$-BF$_2$]ClO$_4$ · H$_2$O and an excess of C$_6$H$_5$CH$_2$NC by electrochemical reduction in CH$_3$O(CH$_2$)$_2$OH at 25 °C. The formation constant K = [RNCCu(D–X)]/[RNC] · [Cu(D–X)] is estimated to be 3 × 10^6 M^{-1} and to be much greater than the K of the corresponding Cu^{II} species [4].

References:

[1] J. Bailey, M. J. Mays (J. Organometal. Chem. **47** [1973] 217/24).
[2] M. I. Bruce, A. P. P. Ostazewski (J. Chem. Soc. Dalton Trans. **1973** 2433/6; see also J. Chem. Soc. Chem. Commun. **1972** 1124/5).
[3] M. Pasquali, F. Marchetti, C. Floriani (Inorg. Chem. **17** [1978] 1684/8).
[4] A. W. Addison, M. Carpenter, L. K.-M. Lau, M. Wicholas (Inorg. Chem. **17** [1978] 1545/52).
[5] O. M. Abu Salah, M. I. Bruce, J. D. Walsh (Australian J. Chem. **32** [1979] 1209/18).
[6] R. R. Gagné, J. L. Allison, D. M. Ingle (Inorg. Chem. **18** [1979] 2767/74).
[7] A. Toth, C. Floriani, M. Pasquali, A. Chiesi-Villa, A. Gaetani-Manfredotti, C. Guastini (Inorg. Chem. **24** [1985] 648/53).

1.1.4.1.3 Compounds of the Types RNCCuR′ and RNCCuC≡CR′(⁴D)

Two compounds have been described:

cyclo-C$_6$H$_{11}$NCCuC(=NC$_6$H$_{11}$-cyclo)C$_6$H$_3$(CH$_3$-5)CH$_2$N(CH$_3$)$_2$-2 (see Formula I). The red-brown solid is formed by a 1:1 reaction of CuC(=NC$_6$H$_{11}$-cyclo)C$_6$H$_3$(CH$_3$-5)CH$_2$N(CH$_3$)$_2$-2 with cyclo-C$_6$H$_{11}$NC in C$_6$H$_6$ at 60 °C in 75% yield. IR (Nujol): 2146 cm^{-1} (NC) [1].

I

cyclo-C$_6$H$_{11}$NCCuC≡CCO$_2$C$_2$H$_5$(N$_2$C$_{12}$H$_4$(CH$_3$)$_4$) (N$_2$C$_{12}$H$_4$(CH$_3$)$_4$ = 3,4,7,8-tetramethyl-1,10-phenanthroline) has been prepared from [C$_2$H$_5$O$_2$CC≡CCuN$_2$C$_{12}$H$_4$(CH$_3$)$_4$]$_4$ · H$_2$O (see Section 1.1.2.4) and cyclo-C$_6$H$_{11}$NC in C$_6$H$_6$ as well as from [cyclo-C$_6$H$_{11}$NCCuC≡

$CCO_2C_2H_5(P(C_6H_5)_3)]_2$ and $N_2C_{12}H_4(CH_3)_4$ in C_6H_6. The pale yellow solid melts at 130 °C. IR (Nujol): 1670 (C=O), 2055 (C≡C), 2145 (NC of isocyanide) cm^{-1} [2].

References:

[1] G. van Koten, J. G. Noltes (J. Chem. Soc. Chem. Commun. **1972** 59).
[2] G. la Monica, G. Ardizzoia, S. Cenini, F. Porta (J. Organometal. Chem. **273** [1984] 263/73).

1.1.4.2 Compounds Containing Two Isocyanide Ligands

1.1.4.2.1 Compounds of the Types $(RNC)_2CuX$ and $[(RNC)_2Cu^nD_m]X$

In Table 30 the compounds $(RNC)_2CuX$, $[(RNC)_2Cu^2D_2]X$, and $[(RNC)_2Cu^4D]X$ are described. Most of them are prepared by the following methods:

Method I: From CuX or CuX_2 and an excess of RNC (usually 1:3 mole ratio). The solvents are given in parentheses in the table.

Method II: A suspension of $(RNC)_2CuX$, such as Nos. 7 or 8, and $AgPF_6$ in acetone is treated with the equivalent amount of D. The compound is isolated by adding ether or petroleum ether [4].

Method III: A suspension of CuI in $CH_3OH/(CH_2NR_2)_2$ is reacted first with CO, then with RNC and $Na[B(C_6H_5)_4]$.

Method IV: From $[(RNC)_4Cu]X$ and an excess of ^4D by refluxing in n-C_3H_7OH or C_2H_5OH for 2 to 5 d. After cooling to 0 °C, $[(RNC)_2Cu^4D]X$ precipitates in good yields. Even with a large excess of ^4D, only two isocyanides are replaced.

Method V: From $[Cu(NCCH_3)_4]X$, ^4D, and RNC (1:1:2) in C_2H_5OH. Because good yields have been obtained after only 30 min at room temperature, this method is superior to Method IV.

The $(RNC)_2CuX$ compounds No. 7 and 8 are probably dimeric and halogen-bridged in the solid state as there are two $\nu(NC)$ bands in the IR spectrum. In $CHCl_3$, their osmometric molecular masses are very close to those required for the monomer. However, in CH_3NO_2 conductivity measurements show some ionization [4].

The $[(RNC)_2Cu^nD_m]PF_6$ compounds No. 9 to 13 and 17 to 19 are stable against air [4, 9]. Nos. 9 to 13 and 20 are not affected by moisture and light [9]. Nos. 18 and 19 exist in $CHCl_3$ as ion pairs, but in acetone they are fully ionized according to osmometric measurements. Also conductivity data on Nos. 17 to 19 in CH_3NO_2 indicate extensive ionization [4].

Table 30
Compounds of the Types $(RNC)_2CuX$, $[(RNC)_2Cu^2D_2]X$, or $[(RNC)_2Cu^4D]X$.
Further information on compounds preceded by an asterisk is given at the end of the table.
For abbreviations and dimensions see p. X.

No.	compound	method of preparation (yield) remarks IR: $\nu(NC)$	Ref.
	compounds of the type $(RNC)_2CuX$:		
1	$(C_2H_5NC)_2CuCN$	I (ether) unstable, gradual loss of C_2H_5NC	[1]
*2	$(t\text{-}C_4H_9NC)_2CuCN$	I m.p. 173 to 175°	[3]
*3	$(t\text{-}C_4H_9NC)_2CuOCO_2C_4H_9\text{-}t$	exists only in benzene solution	[5]
*4	$(C_6H_5NC)_2CuCl$	I ($CHCl_3$/ether, 92%) colorless needles, m.p. 128°	[2]
5	$(C_6H_5NC)_2CuCN$	addition of ether to $(C_6H_5NC)_3CuCN$ in $CHCl_3$ (74%) colorless flakes from $CHCl_3$/ether, m.p. 158 to 162°	[2]
*6	$(4\text{-}CH_3C_6H_4NC)_2CuOC_6H_3(C_4H_9\text{-}t)_2\text{-}2,6$	$CuCl+RNC+NaX$ in THF (~40%) IR (Nujol): 2145, 2170	[8, 10]
7	$(4\text{-}CH_3OC_6H_4NC)_2CuCl$	I (ether, 72%) colorless, m.p. 139 to 142° IR (Nujol): 2141, 2163	[4]
8	$(4\text{-}CH_3OC_6H_4NC)_2CuBr$	I m.p. 129 to 131° IR (Nujol): 2143, 2161	[4]
	compounds of the type $[(RNC)_2Cu^nD_m]X$:		
9	$[(CH_3NC)_2CuN_2C_{10}H_8]PF_6$ $(N_2C_{10}H_8=2,2'\text{-bipyridyl})$	IV (86%), V (91%) pale yellow needles ^1H NMR (acetone-d_6): 3.40 (s, CH_3), 7.80 (td) +8.21 (td) +8.54 (d) +8.96 (d; each 2H of $N_2C_{10}H_8$) IR (Nujol): 2164 (sh), 2193, 2213	[9]
10	$[(t\text{-}C_4H_9NC)_2Cu(t\text{-}C_4H_9N\!=\!CHCH\!=\!NC_4H_9\text{-}t)]PF_6$	IV (80%) orange-yellow ^1H NMR (acetone-d_6): 1.42 (s, CH_3 of RNC), 1.49 (s, CH_3 of ^4D), 8.30 ("NH") IR (Nujol): 2176, 2197 (CH_2Cl_2): 2168, 2190	[9]

References on p. 228

Table 30 [continued]

No.	compound	method of preparation (yield) remarks IR: ν(NC)	Ref.
11	[(t-C_4H_9NC)$_2$Cu$N_2C_{10}H_8$]PF$_6$ ($N_2C_{10}H_8$ = 2,2′-bipyridyl)	IV (82%) yellow ^{1}H NMR (acetone-d$_6$): 1.40 (s, CH$_3$), 7.70 (td) + 8.15 (td) + 8.50 (d) + 8.88 (d; each 2H of ^{4}D) IR (Nujol): 2138, 2168, 2188 UV (CH$_2$Cl$_2$): 383 (sh)	[9]
12	[(t-C_4H_9NC)$_2$Cu$N_2C_{12}H_8$]PF$_6$ ($N_2C_{12}H_8$ = 1,10-phenanthroline)	IV (85%) pale yellow ^{1}H NMR (acetone-d$_6$): 1.48 (s, CH$_3$), 8.12 (m) + 8.17 (s) + 8.78 (dd) + 9.32 (dd; each 2H of ^{4}D) IR (Nujol): 2170, 2189	[9]
13	[(t-C_4H_9NC)$_2$Cu(CH$_2$P(C_6H_5)$_2$)$_2$]PF$_6$	IV (75% in presence of KF$_6$) colorless IR (Nujol): 2188	[9]
14	[(cyclo-$C_6H_{11}NC$)$_2$Cu(CH$_2$NH$_2$)$_2$][B(C_6H_5)$_4$]	III (52%) colorless solid, sensitive to O$_2$ and H$_2$O IR (Nujol): 2143, 2177 suggests pseudo- tetrahedral coordination around Cu	[6]
*15	[(cyclo-$C_6H_{11}NC$)$_2$Cu(CH$_2$N(CH$_3$)$_2$)$_2$][B(C_6H_5)$_4$]	III (53%) IR (Nujol): 2160, 2180	[6]
*16	[(cyclo-$C_6H_{11}NC$)$_2$Cu$N_2C_{10}H_8$]ClO$_4$ ($N_2C_{10}H_8$ = 2,2′bipyridyl)	not isolated	[7]
17	[(4-CH$_3$OC_6H_4NC)$_2$Cu(P(C_6H_5)$_3$)$_2$]PF$_6$	II (65%) colorless, m.p. 93 to 95° IR: 2156, 2168	[4]
18	[(4-CH$_3$OC_6H_4NC)$_2$Cu(CH$_2$P(C_6H_5)$_2$)$_2$]PF$_6$	II colorless, m.p. 176 to 179° IR: 2152, 2162	[4]
19	[(4-CH$_3$OC_6H_4NC)$_2$Cu$N_2C_{10}H_8$]PF$_6$ ($N_2C_{10}H_8$ = 2,2′-bipyridyl)	II (93%) yellow, m.p. 131 to 138° IR: 2142, 2166	[4]
20	[(2,6-(CH$_3$)$_2C_6H_3NC$)$_2$Cu$N_2C_{12}H_6$(CH$_3$)$_2$]PF$_6$ ($N_2C_{12}H_6$(CH$_3$)$_2$ = 2,9-dimethyl-1,10- phenanthroline)	IV (78%) off-white ^{1}H NMR (acetone-d$_6$): 2.39 (s, CH$_3$ of RNC), 3.32 (s, CH$_3$ of ^{4}D) IR (Nujol): 2135, 2147 (CH$_2$Cl$_2$): 2143	[9]

References on p. 228

*Further information:

(t-C₄H₉NC)₂CuCN (Table **30**, No. **2**) is also formed in low yield from CuCl, HCN, and isobutene (1:20:5 mole ratio) at 100 °C, if the ethanol soluble products of the reaction are repeatedly recrystallized from CH₃CN/ether. Upon treatment with hot CHCl₃/ether or water, the compound is readily transformed into the more stable complex t-C₄H₉NCCuCN (see 1.1.4.1.1) [3].

(t-C₄H₉NC)₂CuOCO₂C₄H₉-t (Table **30**, No. **3**) is formed by treating t-C₄H₉NCCuOC₄H₉-t (see 1.1.4.1.1) with 1 mol t-C₄H₉NC and CO₂ or by treating CuOC₄H₉-t with 2 mol t-C₄H₉NC and CO₂ in benzene. The presence of the compound is suggested from IR studies. Although No. 3 is stable in benzene at room temperature, removal of the solvent causes partial decarboxylation and disproportionation to produce an equimolar mixture of t-C₄H₉NCCuOC₄H₉-t (see 1.1.4.1.1) and (t-C₄H₉NC)₃CuOCO₂C₄H₉-t (see 1.1.4.3), from which it can be regenerated by carboxylation in benzene [5].

(C₆H₅NC)₂CuCl (Table **30**, No. **4**) is also formed by treating [(C₆H₅NC)₄Cu]Cl (see 1.1.4.4.1) with boiling water or by the disproportionation of (C₆H₅NC)₃CuCl (see 1.1.4.3) in hot water. In boiling methanol the compound is transformed to C₆H₅NCCuCl (see 1.1.4.1.1) [2].

(4-CH₃C₆H₄NC)₂CuOC₆H₃(C₄H₉-t)₂-2,6 (Table **30**, No. **6**) is monoclinic, space group P2₁/n (P2₁/c)-C⁵₂ₕ (No. 14); a = 23.179(4), b = 9.456(1), c = 12.509(2) Å, β = 94.02(2)°; Z = 4, d_c = 1.22 g · cm⁻³. The structure is shown in **Fig. 16**. The complex is monomeric with Cu^I in a trigonal planar coordination [8, 10].

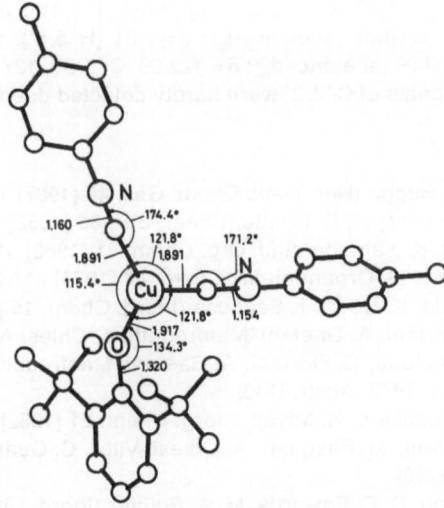

Fig. 16.   Molecular structure of (4-CH₃C₆H₄NC)₂CuOC₆H₃(C₄H₉-t)₂-2,6 (No. 6) with selected bond lengths (in Å) and angles.

[(cyclo-C₆H₁₁NC)₂Cu(CH₂N(CH₃)₂)₂][B(C₆H₅)₄] (Table **30**, No. **15**). Monoclinic, space group P2₁/n(P2₁/c)-C⁵₂ₕ (No. 14); a = 15.626(2), b = 14.753(3), c = 18.024(4) Å, β = 97.47(2)°; Z = 4, d_c = 1.157 g · cm⁻³. The crystals show discrete cations and anions; the structure of the cation is shown in **Fig. 17**, p. 228. A pseudotetrahedral Cu coordination is provided by the two C-bonded isocyanides and the chelating diamine. The C₆H₁₁ ring of one isocyanide shows

References on p. 228

statistical distribution between two conformations. The five–membered CuN_2C_2 ring with the diamine has practically a gauche conformation [6].

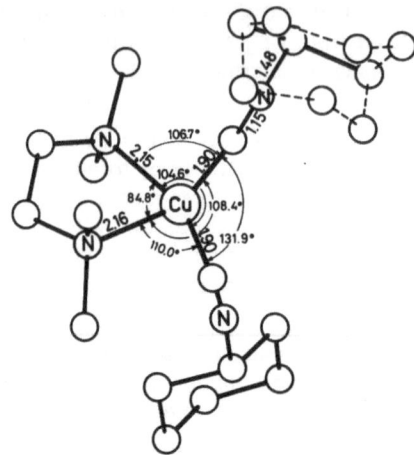

Fig. 17.   Cation structure of $[(cyclo-C_6H_{11}NC)_2Cu(CH_2N(CH_3)_2)_2][B(C_6H_5)_4]$ (No. 7) with selected bond length (in Å) and angles.

[(cyclo-$C_6H_{11}NC$)$_2$CuN$_2$C$_{10}$H$_8$]ClO$_4$ (Table 16, No. 16). Formation in $(CD)_3CO$ solution from cyclo-$C_6H_{11}NC$, $[Cu(CH_3CN)_4]ClO_4$, and 2,2′–bipyridyl (2:1:1) is indicated by ^1H and ^{13}C NMR of the $N_2C_{10}H_8$ group. ^1H NMR (acetone-d_6): $\delta = 7.81$ (H–5,5′), 8.27 (H–4,4′), 8.63 (H–3,3′), 8.98 (H–6,6′) ppm. ^{13}C NMR (acetone-d_6): $\delta = 122.99$ (C–5,5′), 127.04 (C–3,3′), 139.72 (C–4,4′), 150.64 (C–6,6′) ppm, signals of C–2,2′ were hardly detected due to low signal/noise ratio [7].

References:

[1] K. A. Hofmann, G. Bugge (Ber. Deut. Chem. Ges. **40** [1907] 1772/8, 1777).
[2] F. Klages, K. Mönkemeyer, R. Heinle (Chem. Ber. **85** [1952] 109/22).
[3] S. Otsuka, K. Mori, K. Yamagami (J. Org. Chem. **31** [1966] 4170/4).
[4] J. Bailey, M. J. Mays (J. Organometal. Chem. **47** [1973] 217/24).
[5] T. Tsuda, S. Sanada, K. Ueda, T. Saegusa (Inorg. Chem. **15** [1976] 2329/32).
[6] M. Pasquali, C. Floriani, A. Gaetani-Manfredotti, A. Chiesi-Villa (Inorg. Chem. **18** [1979] 3535/42), cf. M. Pasquali, C. Floriani, A. Gaetani-Manfredotti (9th Intern. Conf. Organo-
     metal. Chem., Dijon 1979, Abstr. D42).
[7] S. Kitagawa, M. Munakata, N. Miyaji (Inorg. Chem. **21** [1982] 3842/3).
[8] P. Fiaschi, C. Floriani, M. Pasquali, A. Chiesi-Villa, C. Guastini (J. Chem. Soc. Chem.
     Commun. **1984** 888/90).
[9] A. Bell, R. A. Walton, D. E. Edwards, M. A. Poulter (Inorg. Chim. Acta **104** [1985] 171/8).
[10] P. Fiaschi, C. Floriani, M. Pasquali, A. Chiesi-Villa, C. Guastini (Inorg. Chem. **25** [1986]
     462/9).

1.1.4.2.2 Compounds of the Type (RNC)$_2$Cu(X–^{2}D) or (RNC)$_2$CuR′

The compounds in Table 31 are mainly prepared by the following methods:

Method I: By reaction of Cu_2O, RNC, and the neutral ligand (1:4:2 mole ratio) in benzene at 50 °C and precipitation with ether in almost quantitative yield. Use of metallic Cu instead of Cu_2O decreases the yield.

Method II: From Cu_2O, RNC, and R'CONHX (1:8:2 mole ratio) in refluxing C_6H_6 according to reaction Scheme I,

   a: with X=H and removal of H_2O by azeotropic distillation,

   b: with $X=Si(CH_3)_3$, which results in a shorter reaction time than with X=H and avoids any hydrolysis.

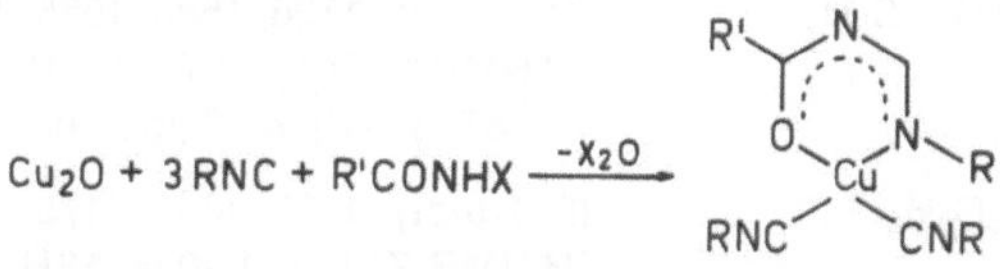

$$Cu_2O + 3\,RNC + R'CONHX \xrightarrow{-X_2O}$$

I

For o-semiquinolato Cu^I complexes of the type $(RNC)_2CuO_2C_6H_3R'$ ($R=t-C_4H_9$; nature and position of R' not given), the dissociation equilibria $(RNC)_2CuO_2C_6H_3R' \rightleftharpoons RNCCuO_2C_6H_3R' + RNC \rightleftharpoons CuO_2C_6H_3R' + 2RNC$ have been observed on dilution of solutions [4].

For most compounds in Table 31 coordination by a chelating $X-^2D$ ligand via O and N atoms is possible as is a direct metal-C bond by substitution of the acidic H. The latter has been demonstrated for Au^I acetylacetonates, see "Organogold Compounds", 1980, Section 1.1.1.4. On the basis of the IR and ^1H NMR data for compounds Nos. 1 and 6 to 10 chelate structures are assumed [1 to 3].

All compounds are sensitive to air, but stable under N_2 [1 to 3]. Compounds No. 1 to 5 are colorless crystalline solids, which decompose at about 150 °C to deposit metallic Cu [2]. They are soluble in most organic solvents [2], as is No. 10 [1]. Nos. 1 to 4 react with excess RNC to give enamines via insertion, $(RNC)_2CuCHR^1R^2 + nRNC \rightarrow [(RNC)_{n+1}CuC(=NR)-CHR^1R^2] \xrightarrow{H_2O} RNHCH=CR^1R^2$ [2]. I_2 causes oxidative coupling of Nos. 1 and 2, $(RNC)_2CuCHR_2^1 + I_2 \rightarrow R_2^1CHCHR_2^1 + 2(RNC)_2CuI$ [2]. Treatment of No. 7 with C_2H_5OH/HCl gives $CH_3CONHCH=NC_4H_9-t$ in 43% yield. Compounds No. 6 to 9 react with alkylating agents R^1X to furnish $RCONR^1CH=NC_4H_9-t$ and isomeric N-alkyl products. For the alkylation it is not necessary to isolate the complexes. They can be generated in situ [3].

Table 31
Compounds of the Type $(RNC)_2Cu(X-^2D)$ or $(RNC)_2CuR'$.
For abbreviations and dimensions see p. X.

No.	R	$X-^2D$ or R'	method of preparation (% yield)	remarks	Ref.
1	$t-C_4H_9$	$CH(COCH_3)_2$	I	IR (Nujol): 1525, 1610, 2142, 2166 ^1H NMR ($CDCl_3$): 1.42 (s, 18H), 1.87 (s, 6H), 5.18 (s, 1H)	[2]
2	$t-C_4H_9$	$CH(CO_2C_2H_5)_2$	I	IR (Nujol): 1580, 1690, 2172 ^1H NMR (C_6D_6): 1.25 (t, 6H), 1.47 (s, 18H), 4.19 (q, 4H)	[2]

References on p. 230

Table 31 [continued]

No.	R	X–^{2}D or R′	method of preparation (% yield)	remarks	Ref.
3	t-C$_4$H$_9$	CH$_3$COCHCO$_2$C$_2$H$_5$ I		IR (Nujol): 1520, 1580, 1655, 1690, 2144, 2167 ^{1}H NMR (CDCl$_3$): 1.21 (t, 3H), 1.43 (s, 18H), 1.87 (s, 3H), 4.06 (q, 2H)	[2]
4	t-C$_4$H$_9$	NCCHCO$_2$C$_2$H$_5$ I		IR (Nujol): 1620, 1640, 2175 ^{1}H NMR (C$_6$D$_6$): 1.20 (s, 18H), 3.18 (s, 3H)	[2]
5	t-C$_4$H$_9$	C$_2$H$_5$C(CO$_2$C$_2$H$_5$)$_2$ I		IR (Nujol): 1585, 1690, 2172 ^{1}H NMR (C$_6$D$_6$): 0.94 (s, 18H), 1.21 (t, 6H), 1.40 (t, 3H), 1.60/1.80 (2H), 4.30 (q, 4H)	[2]
6	t-C$_4$H$_9$	OCHNCHNC$_4$H$_9$-t	(see reaction IIa (70) Scheme I)	IR (KBr): 1200, 1270, 1595, 2150 ^{1}H NMR (C$_6$D$_6$): 1.10 (s, 18H), 1.50 (s, 9H), 8.30 (s, 1H), 9.30 (s, 1H, not exchangeable in D$_2$O)	[3]
7	t-C$_4$H$_9$	CH$_3$CONCHNC$_4$H$_9$-t	(see reaction IIb (90) Scheme I)	IR (KBr): 1295, 1362, 1590, 1630, 2150 ^{1}H NMR (C$_6$D$_6$): 1.45 (s, 9H), 1.78 (s, 18H), 2.48 (s, 3H), 9.26 (s, 1H)	[3]
8	t-C$_4$H$_9$	C$_6$H$_5$CONCHNC$_4$H$_9$-t	(see reaction IIa (80), Scheme I) IIb (83)	IR (KBr): 1200, 1550, 1590, 1640, 2150 ^{1}H NMR (CD$_3$CN): 7.20 to 7.57 (m, 3H), 8.05 to 8.30 (m, 2H), 8.52 (s, 18H), 8.70 (s, 9H), 8.78 (s, 1H)	[3]
9	t-C$_4$H$_9$	C$_4$H$_3$OCONCHNC$_4$H$_9$-t	(see reaction IIb Scheme I, C$_4$H$_3$O = fur-2-yl)	not isolated	[3]
10	C$_6$H$_5$	CH(COCH$_3$)$_2$	—	from CuC$_5$H$_7$O$_2$ · 2.5NH$_3$ + C$_6$H$_5$NC in THF (40%) slightly yellow, sinters at 107°, m.p. 132° (dec.) IR (CsI): 1535 (CO), 1600 (CC), 2120 + 2150 (NC); C$_{2v}$ symmetry proposed $\chi_{mol} = -92.0 \times 10^{-6}$ cm^3 · mol^{-1}	[1]

References:

[1] R. Wast, W. H. Lepel (Chem. Ber. **102** [1969] 3224/8).

[2] Y. Ito, T. Konoike, T. Saegusa (J. Organometal. Chem. **85** [1975] 395/401), see also Y. Ito, T. Saegusa (Organotransition Met. Chem. Proc. 1st Japan. Am. Semin., Honolulu 1974 [1975], pp. 49/55 from C.A. **85** [1976] No. 160272).

[3] Y. Ito, Y. Inobushi, T. Sugaya, T. Saegusa (J. Organometal. Chem. **137** [1977] 1/9).

[4] G. A. Abakumov, A. V. Lobanov, V. K. Cherkasov (Abstr. 2nd All-Union Conf. Organometal. Chem., Gorkii 1982, p. 168) from E. A. Kashtanov, V. K. Cherkasov, L. V. Gorbunova, G. A. Abakumov (Izv. Akad. Nauk SSSR Ser. Khim. **1983** 2121/5; Bull. Acad. Sci. USSR Div. Chem. Sci. **1983** 1917/20).

1.1.4.3 Compounds Containing Three Isocyanide Ligands

Most of the compounds in Table 32 are prepared by the following methods:

Method I: From CuX or CuR′ and RNC

        a: without a solvent [1, 2]; in [2] cooling is recommended,

        b: in $C_2H_5OH/H_2O/NH_3$ [2],

        c: in pyridine/ether at $-10\ °C$ [2].

Method II: A solution of $[(RNC)_4Cu]X$ (see 1.1.4.4.1) in $CHCl_3$ is treated with ether to give $(RNC)_3CuX$ in good yield [2].

$(RNC)_3CuX$ is also formed by alkylation of HCN with olefins like $CH_2=C(CH_3)_2$ in the presence of CuX, cf. compound No. 2. This reaction can be used for preparing t-alkyl isocyanides from olefins and HCN because of the easy elimination in aqueous KCN of the isocyanide from the complex, which can be regenerated in situ [3].

Table 32
Compounds Containing Three Isocyanide Ligands.
Further information on compounds preceded by an asterisk is given at the end of the table.
For abbreviations and dimensions see p. X.

No.	compound	method of preparation (% yield)	remarks IR (Nujol): $\nu(NC)$	Ref.
compounds of the $(RNC)_3CuX$ type				
1	$(C_2H_5NC)_3CuCN$	I a	brown crystals; air-sensitive; soluble in C_2H_5OH, low solubility in other common solvents	[1]
*2	$(t-C_4H_9NC)_3CuBr$	I	colorless needles, m.p. 151 to 153° IR: 2182	[3]
*3	$(t-C_4H_9NC)_3CuOCO_2H$	see No. 4	colorless, soluble in H_2O and organic solvents IR (Nujol): 1618 (CO), 2162 (NC), 2602 (OH)	[8]
*4	$(t-C_4H_9NC)_3CuOCO_2C_4H_9-t$	—	colorless solid IR: 2160 ^{1}H NMR (C_6D_6): 0.95 (s, $t-C_4H_9NC$), 1.75 $(CO_2C_4H_9-t)$	[6, 7]
5	$(n-C_5H_{11}NC)_3CuCN$	I a	properties similar to No. 1	[1]
6	$\{(CH_3)_2CH(CH_2)_2NC\}_3CuCN$	I a	properties similar to No. 1	[1]
7	$(C_6H_5NC)_3CuCl$	II (96)	colorless, m.p. 156° ^{13}C NMR $(CHCl_3)$: 125.9 + 129.1 + 129.5 (C_6H_5), 148.0 (NC), $J(^{14}N, ^{13}C) = 12.9$ disproportionates in hot H_2O into $(RNC)_2CuCl$ and $[(RNC)_4Cu]Cl$	[2, 5]

References on p. 233

Table 32 [continued]

No. compound	method of preparation (% yield)	remarks IR (Nujol): ν(NC)	Ref.
8 $(C_6H_5NC)_3CuCN$	I a (99), I b	colorless, m.p. 126 to 128°; properties similar to No. 1 loss of 1 isocyanide during attempted recrystallization	[1, 2]
9 $(4-CH_3C_6H_4NC)_3CuCl$	II; see also No. 11	colorless needles, m.p. 170 to 172° forms a 1:1 adduct with C_2H_5OH	[2]

compounds of the $(RNC)_3CuR'$ type

*10 $(t-C_4H_9NC)_3CuC_9H_7$ $(C_9H_7 = inden-1-yl)$	—	greenish yellow solid ^1H NMR (C_6D_6): $A_2B_2X_2Y$ pattern for C_9H_7, 0.9 (s, C_4H_9), 6.50 (d, H-1,3), 7.0 (q, H-5,6), 7.35 (t, H-2), 7.78 (q, H-4,7), J(H-1,3, H-2)=3	[4]
*11 $(4-CH_3C_6H_4NC)_3CuC \equiv CC_6H_5$ I c (high)		colorless needles, dec. 100°; air-sensitive	[2]

*Further information:

$(t-C_4H_9NC)_3CuBr$ (Table **32**, No. **2**) is also synthesized from CuBr, HCN, and $(CH_3)_2C=CH_2$ in moderate yield in C_2H_5OH, and in lower yields in other solvents or without a solvent. It is soluble in most organic solvents and is monomeric in CH_3NO_2 according to cryoscopic measurements. Upon treatment with aqueous KCN the isocyanide ligand is easily released. The reaction with $(CH_3)_2CO$ and HBr gives $t-C_4H_9NHCOC(CH_3)_2OH$ [3].

The reaction with $(t-C_4H_9NC)_4Mo(SC_4H_9-t)_2$ or with $Mo(SC_4H_9-t)_4$ in acetone gives $(t-C_4H_9NC)_4Mo(\mu-SC_4H_9-t)_2CuBr$, in which $(t-C_4H_9NC)_4Mo(SC_4H_9-t)_2$ acts as a bidentate ligand through bridging S atoms [9, 10]. It exists in two conformational isomers in the solid state [10].

$(t-C_4H_9NC)_3CuOCO_2H$ (Table 32, No. 3) is also formed in the reaction of $t-C_4H_9NC$ and CO_2 at room temperature, either with CuOH in $(CH_3)_2NCHO$ (mole ratio 3:1:1) or with "$(t-C_4H_9NC)_nCu_2CO_3$" (see 1.1.4, p. 215) and H_2O in THF (mole ratio 3:1.5:1:1.2). Its thermal decarboxylation yields Cu, CO_2, and CO. The last is formed by Cu-induced reaction of CO_2 with $t-C_4H_9NC$. At 120 °C in mesitylene formation of CO is quantitative. Acidolysis gives quantitative amounts of CO_2. It reacts with equimolar amounts of C_2H_5I in $(CH_3)_2NCHO$ at room temperature to give C_2H_5OH (57% yield), $(C_2H_5O)_2CO$ (16%), and CO_2 (80%). The title compound is suitable for carboxylation reactions such as reaction I (82% yield at 130 °C). Proton abstraction by $CuOC_4H_9-t$ in THF at 0 °C gives the "$(t-C_4H_9NC)_nCu_2CO_3$" mentioned above and $t-C_4H_9OH$; with LiC_4H_9-n in THF n-butane is formed [8].

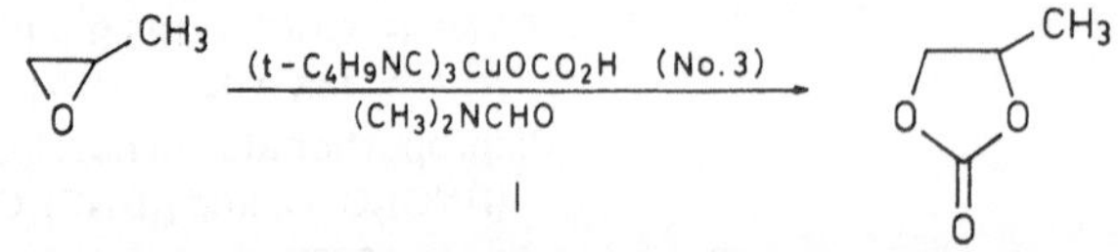

(t-C₄H₉NC)₃CuOCO₂C₄H₉-t (Table 32, No. 4) is synthesized by treating t-C₄H₉NCCuOC₄H₉-t (see 1.1.4.1.1) or CuOC₄H₉-t and t-C₄H₉NC in C₆H₆ with an excess of CO_2 at 10 °C. The complex is soluble in common organic solvents; in C₆H₆ the cryoscopic degree of association is 0.90. It is air-sensitive but stable under N_2 at 10 °C. At 30 °C slight decarboxylation occurs. Acidolysis in 18N H_2SO_4 releases CO_2 completely. In boiling C₆H₆ decomposition into t-C₄H₉NCCuOC₄H₉-t (see 1.1.4.1.1) takes place [6]. By heating in mesitylene or tetraline to 130 °C t-C₄H₉NCO and CO are slowly produced, representing a deoxygenation of CO_2 by t-C₄H₉NC, whereas at 150 °C metallic Cu, CO, and CO_2 are formed [7]. Hydrolysis with H_2O in THF at 0 °C produces a 94% yield of "(t-C₄H₉NC)ₙCu₂CO₃" (see 1.1.4, p. 215) and t-C₄H₉OH at mole ratio 2:1, but quantitative amounts of (t-C₄H₉NC)₃CuOCO₂H (Table 32, No. 3) at mole ratio 1:1 [8]. Reaction with CH₂=CHCH₂Br yields CH₂=CHCH₂OCO₂C₄H₉-t [6]. The ν(NC) indicates that the RNC ligand acts as a σ-donor in the title compound, unlike in t-C₄H₉NCCuOC₄H₉-t (see 1.1.4.1.1) where its π-acceptor character is dominating [6].

(t-C₄H₉NC)₃CuC₉H₇ (Table 32, No. 10) is synthesized in 30% yield from Cu₂O, t-C₄H₉NC, and indene (1:10:10 mole ratio). It is sensitive to air and heat and decomposes in polar solvents. Compared with π-C₅H₅CuCNC₄H₉-t its instability is notable. A σ-bond structure with rapid interconversion between bonding through C–1 and C–3 of indenyl is assumed on the basis of the ¹H NMR spectrum and a D₂O quenching experiment. The compound loses indene upon treatment with dilute acids. It catalyzes the reaction of indene with acetone to give 1-isopropylidene indene [4].

(4-CH₃C₆H₄NC)₃CuC≡CC₆H₅ (Table 32, No. 11) reacts with CHCl₃ in ether to give (4-CH₃C₆H₄NC)₃CuCl (Table 32, No. 9). If washed several times with ether, it dissociates into the initial reactants. If treated with aqueous pyridine, C₆H₅C≡CCu, CO, and 4-CH₃C₆H₄-N=CHNHC₆H₄CH₃-4 · HCl are formed [2].

References:

[1] L. Malatesta (Gazz. Chim. Ital. **77** [1947] 240/7).
[2] F. Klages, K. Mönkemeyer, R. Heinle (Chem. Ber. **85** [1952] 109/22).
[3] S. Otsuka, K. Mori, K. Yamagami (J. Org. Chem. **31** [1966] 4170/4).
[4] T. Saegusa, Y. Ito, S. Tomita (J. Am. Chem. Soc. **93** [1971] 5656/61).
[5] D. L. Cronin, J. R. Wilkinson, L. J. Todd (J. Magn. Resonance **17** [1975] 353/61).
[6] T. Tsuda, S. Sanada, K. Ueda, T. Saegusa (Inorg. Chem. **15** [1976] 2329/32).
[7] T. Tsuda, S. Sanada, T. Saegusa (J. Organometal. Chem. **116** [1976] C10/C12).
[8] T. Tsuda, Y. Chujo, T. Saegusa (J. Am. Chem. Soc. **102** [1980] 431/3).
[9] S. Otsuka, N. Okura, N. C. Payne (J. Chem. Soc. Chem. Commun. **1982** 531/2).
[10] N. C. Payne, N. Okura, S. Otsuka (J. Am. Chem. Soc. **105** [1983] 245/51).

1.1.4.4 Compounds Containing Four Isocyanide Ligands

1.1.4.4.1 Compounds of the Type [(RNC)₄Cu]X

Most compounds described in this section have been formed only for ¹³C or ⁶³Cu NMR investigations, but not isolated.

The compounds listed in Table 33 are obtained by the following two methods:

Method I: An excess [10, 14, 15] or exactly 4 equivalents [6] of RNC are added to [(CH₃CN)₄Cu]X; without a solvent yields are up to 60%, but in a solvent (not reported) quantitative yields are obtained [6]. CH₃OH and (CH₃)₂CO are suitable solvents [15].

References on p. 239

Method II: RNC in C_2H_5OH is added to an ammoniacal solution of CuCl containing NH_4Cl to give the compounds in almost quantitative yield [1].

The compounds No. 2, 3, 17, and 30 are stable with respect to oxidation and decomposition in light and air at room temperature. With an excess of chelating ligands [4]D such as 2,2'-bipyridyl, t-C_4H_9N=CHCH=NC$_4H_9$-t, 1,10-phenanthroline, 2,9-dimethyl-1,10-phenanthroline, and $(C_6H_5)_2PCH_2CH_2P(C_6H_5)_2$ even in refluxing n-C_3H_7OH only two isocyanides are replaced in Nos. 3, 17, and 30, giving [(RNC)$_2$Cu^{4}D]PF$_6$ complexes; see 1.1.4.2.1 [15].

The aryl isocyanide compounds No. 18, 20, 24, and 28 catalyze the insertion of RNC into C_2H_5OH to give RN=CHOC$_2H_5$. A reaction mechanism was proposed from kinetic data. Aliphatic isocyanides do not react [8].

^{13}C NMR Spectra. At room temperature the resonance of the NC group is extremely weak and broad [7, 10]. For some complexes, however, the 1:1:1 triplet from N coupling can be obtained either by adding small amounts of [(CH$_3$CN)$_4$Cu]BF$_4$ or by running the spectra in (CH$_3$)$_2$SO (=DMSO in Table 33) at higher temperatures. No fine structure from Cu-C coupling could be observed at any temperature [10].

The spectra of the alkenyl and aryl isocyanide complexes are measured at lower temperatures to avoid decomposition [10].

On coordination of the RNC ligands, the NC resonances shifts upfield 15 to 20 ppm and J(^{14}N^{13}C) shows an almost threefold increase. The linewidth of the NC resonance is found to be mainly due to ligand exchange and possibly in terms of quadrupole N and Cu relaxation [10]. In Table 33 counting of the C atoms of R begins at the N-bonded C; the isocyanide C is designed by C-0. On coordination C-1 shows a 1 to 3 ppm shift to lower fields for saturated isocyanides, but a 0.3 to 3 ppm upfield shift for unsaturated isocyanides, probably due to conjugation effects. The linewidth of C-1 is found to be controlled by the N relaxation [10].

[63]Cu NMR spectra at 25 °C exhibit a single resonance line with shifts of 450 to 550 ppm downfield from a 0.05 M CH$_3$CN solution of [Cu(NCCH$_3$)$_4$]ClO$_4$; see Table 33. The linewidth $\Delta v_{1/2}$ ranges from 230 to 320 Hz for the complexes No. 19, 22, and 26 with aryl isocyanides. For [(C$_4$H$_9$NC)$_4$Cu]ClO$_4$ (No. 14), however, it amounts to 4000 Hz. Compared with other tetrahedral Cu^I complexes, the ^{63}Cu downfield shift increases in the order of ligands RCN < pyridine < CN$^-$ $\approx$ RNC. This order corresponds well with that in π-acceptor capability, which is obtained from IR spectra of mixed-ligand metal complexes [14].

Secondary ion mass spectra of [(RNC)$_4$Cu]PF$_6$ (R = CH$_3$, t-C_4H_9, cyclo-C_6H_{11}; Table 33, Nos. 3, 17, 23). The absence of [(RNC)$_4$Cu]$^+$ and [(RNC)$_3$Cu]$^+$ — except for the CH$_3$NC complex No. 3 — and the presence of [(RNC)$_2$Cu]$^+$ ions attest to the gas phase stability of two-coordinate Cu^I. Cluster ion formation as well as α-cleavage have been observed. Ligand exchange occurs when mixtures are examined. Cross labeling and isotopic labeling studies have provided insights into the fragmentation mechanisms [16].

Table 33

Compounds of the Type [(RNC)$_4$Cu]X.

Further information on compounds preceded by an asterisk is given at the end of the table.
For abbreviations and dimensions see p. X.

No.	compound	method of preparation (yield) remarks (for NMR see p. 234)	Ref.

*1	[(CH$_3$NC)$_4$Cu]BF$_4$	I
		^{13}C NMR (DMSO, 125°): 28.5 (C–1), 140.7 (NC),
		  J(^{14}N,^{13}C–1) = 7.3, J(^{14}N,^{13}C–0) = 15.5,
		  ^{1}J(^{13}C–1,^{1}H) = 146.8
			[10]

*2	[(CH$_3$NC)$_4$Cu]ClO$_4$	I (85%)
		colorless needles
		slightly soluble in polar solvents such as CHCl$_3$,
		  CH$_2$Cl$_2$, CH$_3$CN, (CH$_3$)$_2$CO, CH$_3$OH, C$_2$H$_5$OH
		^{1}H NMR (CDCl$_3$): 3.35 (s)
		IR (Nujol): 2181 (sh) + 2212 (NC)
		  (CH$_2$Cl$_2$): 2210 (NC)
			[15]

3	[(CH$_3$NC)$_4$Cu]PF$_6$	I (90 to 95%) colorless crystals ^{1}H NMR (acetone-d$_6$): 3.44 (t), ^{3}J(^{14}NH) = 2.4 IR (Nujol): 2176 + 2213 (NC) (CH$_2$Cl$_2$): 2176 + 2211 (NC)	[15]

4	[(CH$_2$=CHNC)$_4$Cu]BF$_4$	I
		^{13}C NMR (CDCl$_3$): 119.0 (C–1), 124.2 (C–2), 145.6 (NC),
		  J(^{14}N,^{13}C–1) = 12.4, J(^{14}N,^{13}C–0) = 16.5 upon
		  addition of [(CH$_3$CN)$_4$Cu]BF$_4$
			[10]

5	[(C$_2$H$_5$NC)$_4$Cu]BF$_4$	I ^{13}C NMR (DMSO, 125°): 14.2 (C–2), 38.1 (C–1), 140.4 (NC), J(^{14}N,^{13}C–1) = 5.9, J(^{14}N,^{13}C–0) = 14.4	[10]
*6	[(CH$_2$=C=CHNC)$_4$Cu]BF$_4$	I m.p. 58 to 61° (dec.) ^{1}H NMR (CD$_3$CN): 5.67 (d, CH$_2$), 6.33 (t, CH), ^{4}J(H,H) = 6.5 IR (Nujol): 1975 (C=C=C), 2170 (NC), 3045 (CH$_2$)	[6]

*7	[(CH$_2$=C=CHNC)$_4$Cu]ClO$_4$	I
		m.p. 79 to 82° (dec.)
		^{13}C NMR (CCl$_4$/CD$_3$CN): 86.4 (C–1), 87.3 (C–3),
		  210.7 (C–2)
		^{1}H NMR (CD$_3$CN): 5.65 (d, CH$_2$), 6.32 (t, CH),
		  ^{4}J(H,H) = 6.5
		IR (Nujol): 1975 (C=C=C), 2167.8 (NC), 3040 (CH$_2$)
			[6, 7]

*8	[(CH≡CCH$_2$NC)$_4$Cu]BF$_4$	I m.p. 83 to 86° (dec.) ^{1}H NMR (CD$_3$CN): 2.88 (t, CH), 4.55 (d, CH$_2$), ^{4}J(H,H) = 2.5 IR (Nujol): 2140 (C≡C), 2202.3 (NC), 3260 (HC≡)	[6]

References on p. 239

Table 33 [continued]

No.	compound	method of preparation (yield) remarks (for NMR see p. 234)	Ref.
*9	$[(CH \equiv CCH_2NC)_4Cu]ClO_4$	I m.p. 91 to 97° (dec.) ^1H NMR (CD_3CN): 2.87 (t, CH), 4.55 (d, CH_2), $^4J(H,H) = 2.5$ IR (Nujol): 2140 $(C \equiv C)$, 2199.7 (NC), 3260 $(HC \equiv)$	[6]
10	$[(CH_2=CHCH_2NC)_4Cu]BF_4$	I ^{13}C NMR (DMSO, 125°): 45.0 (C-1), 117.8 (C-3), 128.7 (C-2), 141.8 (NC), $J(^{14}N,^{13}C-1) = 6.5$, $J(^{14}N,^{13}C-0) = 14.4$	[10]
11	$[\{(Z)-CH_3CH=CHNC\}_4Cu]BF_4$	I ^{13}C NMR (DMSO, 90°): 12.9 (C-3), 112.9 (C-1), 135.7 (C-2), 148.3 (NC), $J(^{14}N,^{13}C-1) = 11.7$, $J(^{14}N,^{13}C-0) = 16.5$	[10]
12	$[(n-C_3H_7NC)_4Cu]BF_4$	I ^{13}C NMR (DMSO, 125°): 10.2 (C-3), 21.6 (C-2), 44.5 (C-1), 140.6 (NC), $J(^{14}N,^{13}C-1) = 5.2$, $J(^{14}N,^{13}C-0) = 14.3$	[10]
13	$[(i-C_3H_7NC)_4Cu]BF_4$	I ^{13}C NMR (DMSO, 125°): 22.8 (C-2), 47.4 (C-1), 139.5 (NC), $J(^{14}N,^{13}C-0) = 12.5$	[10]
14	$[(C_4H_9NC)_4Cu]ClO_4$	I ^{63}Cu NMR (C_4H_9NC): 451 $(\Delta v_{1/2} = 4000)$	[14]
15	$[(t-C_4H_9NC)_4Cu]BF_4$	I ^{13}C NMR (DMSO, 125°): 29.7 (C-2), 57.1 (C-1), 139.2 (NC) IR (CCl_4/CD_3CN): 2177.4	[7, 9, 10]
16	$[(t-C_4H_9NC)_4Cu]ClO_4$	from $[(t-C_4H_9NC)_4Cu(H_2O)_2][ClO_4]_2$ in C_2H_5OH at 45° (79%) colorless crystals, m.p. 129° (dec.), insoluble in water ^1H NMR (acetone-d_6): 1.54 IR (Nujol): 2180.6 (NC)	[4]
17	$[(t-C_4H_9NC)_4Cu]PF_6$	I (90 to 95%) colorless crystals ^1H NMR (acetone-d_6): 1.51 (s) IR (Nujol): 2147 (sh) + 2182 (NC) (CH_2Cl_2): 2144 (sh) + 2180 (NC)	[15]
18	$[(4-ClC_6H_4NC)_4Cu]BF_4$	I ^{13}C NMR (DMSO): 123.9 (C-1), 128.9 (C-2,6), 130.1 (C-3,5), 135.8 (C-4), 147.6 (NC)	[10]

References on p. 239

Table 33 [continued]

No.	compound	method of preparation (yield) remarks (for NMR see p. 234)	Ref.
19	$[(4\text{-}ClC_6H_4NC)_4Cu]ClO_4$	I ^{63}Cu NMR $(CH_2Cl_2/4\,ClC_6H_4NC)$: 553 $(\Delta\nu_{1/2}=$ 230 Hz)	[14]
20	$[(C_6H_5NC)_4Cu]BF_4$	I ^{13}C NMR: 126.7 (C-2,6), 129.8 (C-3,5), 130.8 (C-4), 146.1 (NC) in DMSO at 105°, similar with 125.4 (C-1) in CDCl$_3$ ionization constant $K_c = (2.8 \pm 0.5) \times 10^{-2}$ M (conductometrically)	[8, 10]
*21	$[(C_6H_5NC)_4Cu]Cl \cdot 6H_2O$	II needles, m.p. 102° ^{14}N NMR $(CHCl_3)$: 198 (aqueous NaNO$_3$ as external standard) IR: 2170 ionized according to conductivity in H$_2$O	[1, 3]
22	$[(C_6H_5NC)_4Cu]ClO_4$	I ^{63}Cu NMR (C_6H_5NC): 549 $(\Delta\nu_{1/2}=300$ Hz)	[14]
23	$[(cyclo\text{-}C_6H_{11}NC)_4Cu]ClO_4$	I ^{63}Cu NMR $(C_6H_{11}NC)$: 468 $(\Delta\nu_{1/2}=880$ Hz)	[14]
24	$[(4\text{-}CH_3C_6H_4NC)_4Cu]BF_4$	I ^{13}C NMR (DMSO): 21.0 (CH$_3$), 122.5 (C-1), 126.7 (C-2,6), 130.5 (C-3,5), 141.4 (C-4), 146.5 (NC)	[10]
25	$[(4\text{-}CH_3C_6H_4NC)_4Cu]Cl \cdot H_2O$	II m.p. 127°	[1]
*26	$[(4\text{-}CH_3C_6H_4NC)_4Cu]ClO_4$	I [14]; from No. 25 + HClO$_4$ in C$_2$H$_5$OH [2] colorless needles, m.p. 175° [2] soluble in CHCl$_3$ and C$_6$H$_5$NO$_2$, insoluble in C$_6$H$_6$ [2] ^{63}Cu NMR $(4\text{-}CH_3C_6H_4NC)$: 547 $(\Delta\nu_{1/2}=320$ Hz) [14]	[2, 14]
27	$[(C_6H_5CH_2NC)_4Cu]Cl$	from CuCl + RNC (1:7) in pyridine/24 h refluxing in piperidine yields C$_6$H$_5$CH$_2$N= CHNC$_5$H$_{10}$; tetrahydropterines give no analogous reactions	[13]
28	$[(4\text{-}CH_3OC_6H_4NC)_4Cu]BF_4$	I ^{13}C NMR (DMSO): 55.8 (CH$_3$), 115.2 (C-3,5), 117.5 (C-1), 128.6 (C-2,6), 145.9 (NC), 160.7 (C-4)	[10]
*29	$[(4\text{-}CH_3OC_6H_4NC)_4Cu]PF_6$	m.p. 183 to 184° IR (Nujol): 2144, 2169 fully ionized according to conductivity in CH$_3$NO$_2$	[5]

References on p. 239

Table 33 [continued]

No.	compound	method of preparation (yield) remarks (for NMR see p. 234)	Ref.
30	[(2,6-(CH$_3$)$_2$C$_6$H$_3$NC)$_4$Cu]PF$_6$ I	colorless crystals	[15]

^{1}H NMR (acetone-d$_6$): 2.56 (s, CH$_3$), 7.40 (s, C$_6$H$_3$)
IR (Nujol): 2157 (NC)(CH$_2$Cl$_2$): 2163 (NC)

*Further information:

[(CH$_3$NC)$_4$Cu]BF$_4$ (Table **33**, No. **1**). Orthorhombic, space group Pna2$_1$-C$_{2v}^9$ (No. 33); a = 24.05(1), b = 8.401(5), c = 20.72(1) Å; Z = 12, d$_c$ = 1.497 g · cm^{-3}. The Cu atom is tetrahedrally coordinated by four almost linear CH$_3$NC molecules via C atoms; bond distances Cu↔C are 1.94 to 2.01 Å. A comparison of the bond distances and angles with corresponding values of the acetonitrile complex [(CH$_3$CN)$_4$Cu]BF$_4$ did not reveal significant differences between the two essentially isomorphous structures [12].

[(CH$_3$NC)$_4$Cu]ClO$_4$ (Table **33**, No. **2**). Solid state IR and Raman spectra have been assigned by comparison with those of the analogous CH$_3$CN and CD$_3$CN complexes [15]:

class	description	IR (cm^{-1})	Raman (cm^{-1})	class	description	IR (cm^{-1})	Raman (cm^{-1})
ClO$_4^-$ fundamentals (T$_d$):				(C–N≡C)$_4$Cu fundamentals (T$_d$):			
A$_1$	ClO sym. str.	inactive	928(38)	A$_1$	N≡C str.	inactive	2205(100)
E	OClO deg. def.	inactive	452(7)	A$_1$	CCu str.	inactive	236(2)
T$_2$	ClO deg. str.	1089 vs, br	1091(3)	A$_1$	C–N str.	inactive	906(5)
T$_2$	OClO deg. def.	622 vs	619(9)	E	NCCu def.	inactive	377(2)
				E	CCuC def.	inactive	not observed
CH$_3$ fundamentals (C$_{3v}$):							
A$_1$	CH$_3$ sym. str.	2950 s	2948(52)	T$_2$	N≡C str.	2213 vs 2182 sh	2224(6)
A$_1$	CH$_3$ sym. bend	1402 m	1398(19)				
E	CH$_3$ asym. str.	3010 s	3014(5)	T$_2$	CCu str.	258 m	257(12)
E	CH$_3$ asym. bend	1430 m 1415 m	1431(3) 1408(20)	T$_2$	NCCu def.	367 w 325 w	not observed
E	CH$_3$ rock	hidden by A$_1$ ClO str.		T$_2$	CCuC def.	not observed	
				T$_2$	C–N≡C def.	200 s	not observed
				T$_2$	C–N str.	948 sh 934 m	hidden by A$_1$ ClO str.
				E	C–N≡C def.	inactive	not observed

[(CH$_2$=C=CHNC)$_4$Cu]X and [(CH≡CCH$_2$NC)$_4$Cu]X (X=BF$_4$, ClO$_4$; Table 33, Nos. 6 to 9) are odorless, cream white microcrystalline compounds, soluble in CH$_2$Cl$_2$ and CH$_3$CN but insoluble in most other organic solvents. The propargyl compounds No. 8 and 9 are air-stable at room temperature, whereas the allenyl compounds No. 6 and 7 decompose slowly at room temperature but are stable at −40 °C. The perchlorates No. 7 and 9 detonate violently when exposed to a heavy shock. Nos. 8 and 9 are readily isomerized to Nos. 6 and 7 in the presence of a base like 1,5-diazabicyclo[4.3.0]non-5-en in CH$_3$CN. If treated with aqueous NaCN, Nos. 6 to 9 release the corresponding isocyanides quantitatively [6].

[(C$_6$H$_5$NC)$_4$Cu]Cl · 6H$_2$O (Table 33, No. 21) is transformed by boiling water into (C$_6$H$_5$NC)$_2$CuCl. Treating a chloroform solution of No. 21 with dry ether produces (C$_6$H$_5$NC)$_3$CuCl [1].

[(4-CH$_3$C$_6$H$_4$NC)$_4$Cu]ClO$_4$ (Table 33, No. 26) is also prepared from bis(2-nitropropandiolato-O,O')copper by treatment first with 4-CH$_3$C$_6$H$_4$NC in C$_2$H$_5$OH, then with LiClO$_4$, and recrystallization from water-free C$_2$H$_5$OH (70% yield). IR (solid): 2168 cm^{-1}; IR (CH$_2$Cl$_2$): 2156, 2172 cm^{-1}. ^{1}H NMR (CH$_2$Cl$_2$): δ=2.42 and 2.17 ppm, the latter supposedly due to decomposition products [11].

[(4-CH$_3$OC$_6$H$_4$NC)$_4$Cu]PF$_6$ (Table 33, No. 29) is prepared by treating a suspension of (4-CH$_3$OC$_6$H$_4$NC)$_2$CuBr and AgPF$_6$ (1:1 mole ratio) in acetone with two moles of 4-CH$_3$OC$_6$H$_4$NC. It is air-stable in the solid state and in solution. The IR spectrum reveals two bands in the ν(N≡C) region rather than the one band expected for T$_d$ symmetry. Therefore, the isocyanide ligands are assumed to be bent at the N atom [5].

References:

[1] F. Klages, K. Mönkemeyer, R. Heinle (Chem. Ber. 85 [1952] 109/22).

[2] A. Sacco (Gazz. Chim. Ital. 85 [1955] 989/92).

[3] W. Becker, W. Beck, R. Rieck (Z. Naturforsch. 25b [1970] 1332/7).

[4] R. W. Stephany, W. Drenth (Rec. Trav. Chim. 89 [1970] 305/12).

[5] J. Bailey, M. J. Mays (J. Organometal. Chem. 47 [1973] 217/24).

[6] J. W. Zwikker, R. W. Stephany (Syn. Commun. 3 [1973] 19/23).

[7] R. W. Stephany, M. J. A. de Bie, W. Drenth (Org. Magn. Resonance 6 [1974] 45/7).

[8] D. Knol, C. P. A. van Os, W. Drenth (Rec. Trav. Chim. 93 [1974] 314/6).

[9] D. L. Cronin, J. R. Wilkinson, L. J. Todd (J. Magn. Resonance 17 [1975] 353/61).

[10] D. Knol, N. J. Koole, M. J. A. de Bie (Org. Magn. Resonance 8 [1976] 213/8).

[11] G. Albertin, E. Bordignon, A. Orio, G. Pelizzi, P. Tarasconi (Inorg. Chem. 20 [1981] 2862/8).

[12] A. L. Spek (Cryst. Struct. Commun. 11 [1982] 413/6).

[13] S. N. Ganguly, M. Viscontini (Helv. Chim. Acta 67 [1984] 166/9).

[14] S. Kitagawa, M. Munakata (Inorg. Chem. 23 [1984] 4388/90).

[15] A. Bell, R. A. Walton, D. A. Edwards, M. A. Poulter (Inorg. Chim. Acta 104 [1985] 171/8).

[16] L. D. Detter, R. G. Cooks, R. A. Walton (Inorg. Chim. Acta 115 [1986] 55/63).

1.1.4.4.2 Compounds of the Type [(RNC)$_4$Cu(H$_2$O)$_2$]X$_2$

The compounds in Table 34 are prepared from Cu(ClO$_4$)$_2$ · 6H$_2$O and aqueous Cu(BF$_4$)$_2$, preferably in ethanol/ether at −15 to 0 °C, and must be separated immediately. Compounds No. 1 and 2 decompose on standing at room temperature, but they can be stored at −40 °C for several months. Compounds No. 3 and 4 liquefy to a brown syrup within a few hours at room temperature. All are insoluble in nonpolar organic solvents and decompose rapidly in polar solvents. Upon aging, they are transformed into Cu^I derivatives, as indicated from the change in the IR and ESR spectra. An X-ray powder pattern of compound No. 1, which is isomorphous with No. 2, revealed a tetragonal or pseudotetragonal unit cell with a = 15.9, c = 23.6 Å; d_m = 1.30 g · cm^{-3}, Z = 8. The ESR spectra indicate a complex structure for Nos. 3 and 4. This is also supported by their lower stability in the solid state. There is only one ν(NC) frequency in the IR spectrum for each compound. This suggests an octahedral structure with 4 equivalent RNC in the x-y plane and 2H$_2$O along the z axis. Compounds No. 1 and 2 have differing g values and ν(OH), which suggests some interaction between the Cu^{II} orbitals and the anions through the H$_2$O ligand.

Table 34
Compounds of the Type [(RNC)$_4$Cu(H$_2$O)$_2$]X$_2$.
For abbreviations and dimensions see p. X.

No.	R	X	yield; remarks
1	t-C$_4$H$_9$	ClO$_4$	93%; pink–purple, m.p. 79 to 80° (dec.) IR (Nujol): 2233.2 (NC), 3480 + 3560 (OH) ESR: g = 2.074
2	t-C$_4$H$_9$	BF$_4$	87%; pink–purple, m.p. 77 to 78° (dec.) IR (Nujol): 2234.8 (NC), 3520 + 3610 (OH) ESR: g = 2.040
3	t-C$_4$H$_9$CH$_2$C(CH$_3$)$_2$	ClO$_4$	80%; brown–purple, m.p. 52 to 53° (dec.) IR (Nujol): 2244.7 (NC), ca. 3400 (OH) ESR: g ≈ 2.04
4	t-C$_4$H$_9$CH$_2$C(CH$_3$)$_2$	BF$_4$	39%; brown–purple, m.p. 57 to 59° (dec.) IR (Nujol): 2237.2 (NC), 3500 + 3580 (OH) ESR: g ≈ 2.04

Reference:

R. W. Stephany, W. Drenth (Rec. Trav. Chim. **89** [1970] 305/12).

1.1.5   Compounds with Carbenes and Other Ligands Bonded by One C Atom

Carbene and Vinylidene Complexes

Complexes of the types (CH$_3$)$_3$PCH$_2$CuR and I with phosphane-stabilized carbenes are treated in "Organocopper Compounds" 2, 1983, p. 3 (type (CH$_3$)$_3$PCH$_2$CuR), or will be treated in "Organocopper Compounds" 4 (type I, M = Ag or Cu).

$$\begin{array}{c} H_3C \diagdown_{(+)} \diagup CH_2 - \overset{(-)}{Cu} - CH_2 \diagdown_{(+)} \diagup CH_3 \\ \quad\;\; P \qquad\qquad\qquad\qquad P \\ H_3C \diagup \;\; \diagdown CH_2 - \overset{(-)}{M} - CH_2 \diagup \;\; \diagdown CH_3 \end{array}$$

I

Other stable copper carbene complexes are not known. However, Cu^I and/or Cu^{II} based carbenoids are generally assumed as crucial intermediates in the copper–catalyzed decomposition and carbene transfer reactions involving diazo compounds, e.g., of the types R$_2$CN$_2$ [1, 4, 21, 29], RO$_2$CCHN$_2$ [2, 19, 22, 52], RO$_2$CCOCHN$_2$ [52], RO$_2$CC(N$_2$)COR' [52], (RO$_2$C)$_2$CN$_2$ [10, 52], or XC$_6$H$_4$COCHN$_2$ [3, 5 to 7, 14, 19]. In most cases, bis(acetylacetonato)copper has been used as the Cu component giving intermediate carbene complexes of the R'R''CCu(CH$_3$COCHCOCH$_3$)$_2$ type. However, other soluble Cu complexes [2, 4, 10] including undefined isocyanide complexes [9, 11] and salts such as CuCl or $CuSO_4$ [19] have been used as well. The reactions have been critically reviewed by Wulfman et al. [30 to 33]. For details see the references cites there and [1 to 7, 10, 14, 19, 21, 29].

Similar transient carbene complexes are assumed in the copper-catalyzed biosynthesis of cyclopropane rings by methylene transfer from the methyl group of S-adenosylmethionine to an unactivated alkene such as oleic esters [20]. They may also play a role in the catalysis of cyclopropyl → vinylcarbene isomerization by $(C_6H_5O)_3PCuCl$ or CuCl [41].

The isomerization of benzvalene II, benzobenzvalene III, and their deuterated analogues in the presence of Cu^0 indicates intermediate formation of carbenoids such as IV and V [26].

| II | III | IV | V |

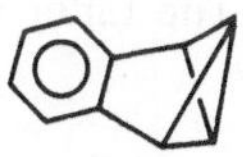

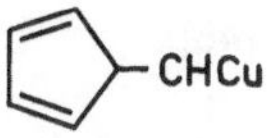

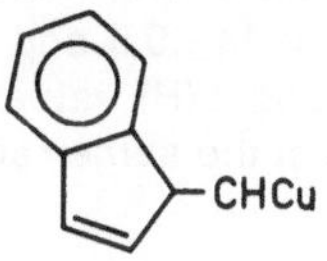

The energy of the vinylidene complex $[CH_2=CCu]^+$ has been calculated by the MO LCAO SCF method and compared to the isomeric ethyne π-complex $[CH\equiv CHCu]^+$. The latter is more stable. The $[CH_2=CCu]^+ \rightarrow [CH\equiv CHCu]^+$ isomerization barrier is 89.9 kJ · mol^{-1} [47].

Carbon Dioxide Complexes

The ion $[CuCO_2]^+$ forms under electron impact from Cu^I or Cu^{II} carboxylates in the gas phase. Mass spectrometric data in combination with calculations by the virial statistic method indicate the existence of a Cu-C bond. Based on these calculations, the structural stability decreases according to Scheme VI [45].

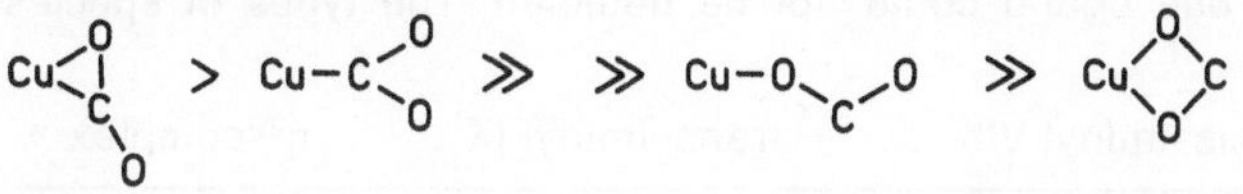

VI

For the ion $[CuCO_2(PH_3)_2]^+$, however, side-on or C coordination of CO_2 is excluded from ab initio MO calculations and energy decomposition analyses. End-on coordination should be favorable because of the strong electrostatic interaction between Cu and the negatively charged O of CO_2 [40].

A side-on coordination of bent CO_2 to Cu had been proposed [35] for complexes formed from CO_2 and Cu^{II} chelates with ephedrine and similar ligands [15, 24, 35]. Later investigations, however, indicated that unstable carbamato structures are formed in these reactions [23, 37].

The CO_2 fixation of RCu in the presence of phosphines D has been explained by intermediate formation of CO_2 complexes of the type $RCO_2Cu(CO_2)D_n$ (no further data given) [12, 13, 16 to 18, 25, 27, 28].

Alkenyl and Allyl Radicals

$C_6H_5\dot{C}H=CHCu$ radicals form by cocondensation of Cu atoms, $C_6H_5C\equiv CH$, and cyclohexane at 77 K in a rotating cryostat. There is no evidence for the formation of an alkyne complex. In contrast, from similar reactions of Cu and C_2H_2 only the complexes C_2H_2Cu and $(C_2H_2)_2Cu$ have been obtained. The ESR spectrum of $C_6H_5{}^\alpha CH={}^\beta CH^{63}Cu$ gives the expected quartet of doublets with a(Cu) = 133.7 G, a(H) = 45.1 G, and g = 2.0019. This is very similar to a(H) = 41.5 G of the β-hydrogen coupling constant for α-styryl and indicates that Cu adds to the unsubstituted end of the alkyne. Therefore a planar structure is assumed with the unpaired electron located in the p_y orbital on the α-carbon. By comparison with the value found for free Cu atoms in an inert hydrocarbon matrix, a 6 to 7% 4s spin density has been estimated for the Cu atom in the title product [42].

$CH_2\dot{C}(Cu)CH_2$ radicals (see VII). Bombardment of $CH_2=C=CH_2$ onto isolated Cu atoms trapped on the surface of an inert matrix in a rotating cryostat at 77 K gives Cu-substituted allyl radicals, but not the substituted vinyls $CH_2=\dot{C}CH_2Cu$. The main ESR absorptions are in the accordance with the proposed allyl structure: g = 2.0020, a(^{63}Cu) = 13±0.5 G and two a(H) values of 14±0.5 and 15±0.5 G. Less intense residual lines fit reasonably well with the same g and a(H) values, but a(^{63}Cu) = 8±0.5 G. The latter are interpreted by another trapping site of the same radical [48].

| VII | VIII | IX | X |

Iminyl Radicals ·N=CRCu and η²-Complexes RC≡NCu

·N=CRCu (R = H, CH_3, t-C_4H_9, C_6H_5) and η²-RC≡NCu (R = H, t-C_4H_9, C_6H_5). Cu atoms are condensed in an adamantane matrix at 77 K and bombarded with RCN (R = H, D, CH_3, $CH_2=CH$, t-C_4H_9, C_6H_5). The ESR spectra show the existence of three types of Cu-containing species. They have been assigned, by comparison with the analogous species obtained with Ag and Au, to the iminyl radicals ·N=CRCu (cis and trans isomers VIII and IX presumed, but not assigned) and to the η²-complexes X. End-on coordinated radicals or complexes with more than one ligand could not be detected. The types of species formed depend on R:

R	cis-iminyl VIII	trans-iminyl IX	η²-complex X	Ref.
H, D	+	+	+	[43, 44]
CH_3	+	+	−	[49]
$CH_2=CH$	−	−	−	[49]
t-C_4H_9	+	+	+	[49]
C_6H_5	possibly (poorly resolved)		+	[49]

ESR parameters of the species formed (a values in G, a(N) and a(C) not resolved):

	g factor	a(Cu)	ϱ(Cu)	Ref.
·N=CHCu	2.001	479	0.22	[43, 44, 49]
(one of the three species	2.001	280	0.13	[43, 44, 49]
not identified [44])	1.998	372	0.17	[43, 44, 49]
HC≡NCu	2.01 to 2.02	1386 to 1505	0.68	[44, 49]
·N=C(CH_3)Cu	2.0044	350.2	0.16	[49]
	2.0025	293.2	0.14	[49]
·N=C(C_4H_9-t)Cu	2.0022	419.5	0.2	[49]
	1.9984	339.4	0.16	[49]

The species ·N=CHCu and HC≡NCu decompose slowly at 77 and rapidly at 100 K [43, 44].

References on pp. 246/7

Intermediates Formed from Cu²⁺ and Radicals

The reaction of $\cdot CH_2CO_2^-$ radicals with Cu^{2+} (aq) has been studied by pulse radiolysis of NO–saturated aqueous $CH_3CO_2^-/CuSO_4$ solutions at pH = 6.0. The effect of Cu^{2+} on the kinetics of disappearance of the $\cdot CH_2CO_2^-$ radicals indicates the intermediate formation of $[O_2CCH_2Cu]^+$ and $O_2CCH_2CuO_2CCH_3$, which decompose in first–order reactions according to $O_2CCH_2CuO_2CCH_3 \rightarrow [O_2CCH_2Cu]^+ + CO_2 + \cdot CH_3$ (k = 90 s⁻¹) and $[O_2CCH_2Cu]^+$ $\xrightarrow{H_2O}$ Cu^+ (aq) $+ HOCH_2CO_2^- + H_3O^+$ (k = 2.8 s⁻¹) [34]. The formation of similar intermediates is indicated in the reactions of Cu^{2+} (aq) with $\cdot CH_2CO_2^-$ or $\cdot CH_2C(CH_3)_2OH$ in the presence of triglycine $NH_2(CH_2CONH)_2CH_2CO_2H$ [36] and, according to preliminary results, with $CH_3\dot{C}HCO_2^-$, $\cdot CHCl_2$, or $\cdot CCl_3$, but not with $R_2\dot{C}OH$ or $\cdot CO_2^-$ radicals [34].

Azaalkene η²-Complexes

$[(CH_3)_2N=CH_2Cu(CO)Cl]Br$ is formed from $[(CH_3)_2N=CH_2]^+[Cu_2Cl_2Br]^-$ and CO in THF at −40 °C. IR spectrum (THF, −60 °C): 2080 (CO), 1610 and 1640 (NC) cm⁻¹. At −15 °C the pale yellow solid releases CO quantitatively to give $[(CH_3)_2N=CH_2CuCl]Br$. In both compounds the Cu atom is assumed to be π–bonded to the N=C double bond. An equimolar amount of CO is evolved above −15 °C from (CO)CuCl in the presence of $[(CH_3)_2N=CH_2CuCl]Br$; the dinuclear $(CH_3)_2N=CH_2Cu_2(CO)Cl_2Br$ is formed [8].

Bimetallic Clusters

$(C_6H_5)_3P(CF_3SO_3)Cu(\mu\text{-}CC_6H_4CH_3\text{-}4)W(CO)_2C_5H_5$ (see XI) has been prepared by successive 1:1:1 addition of $CF_3SO_3Cu \cdot 0.5C_6H_6$ and $4\text{-}CH_3C_6H_4C\equiv W(CO)_2C_5H_5$ in CH_2Cl_2 to $P(C_6H_5)_3$ in CH_2Cl_2 with a 73% yield in the form of orange microcrystals, m.p. 113 to 115 °C. ¹H NMR (CDCl₃): 2.21 (s, CH₃), 5.77 (s, C₅H₅), 7.00 to 7.43 (m, C₆H₄+C₆H₅) ppm. ¹³C{¹H} NMR (CDCl₃): 22.0 (CH₃), 93.0 (C₅H₅), 120.2 (q, CF₃, J(FC) = 319 Hz), 128.9 to 149.8 (C₆H₄+C₆H₅), 215.5 (CO), 292.3 (μ–C) ppm. IR (CH₂Cl₂): 1941 and 2016 (both CO) cm⁻¹. From the NMR and IR spectra structure XI has been proposed [51].

XI

$[(C_6H_5)_3P(CH_3CN)Cu(\mu\text{-}CC_6H_4CH_3\text{-}4)W(CO)_2C_5H_5]PF_6$ has been prepared by successive 1:1:1 addition of $P(C_6H_5)_3$ and $4\text{-}CH_3C_6H_4C\equiv W(CO)_2C_5H_5$ in CH_2Cl_2 to a suspension of $[Cu(NCCH_3)_4]PF_6$ in CH_2Cl_2 with a 83% yield. In the presence of two equivalents of $P(C_6H_5)_3$, or of one equivalent of $(C_6H_5)_2PCH_2P(C_6H_5)_2$, no reaction occurred with the W carbyne. Not surprisingly, the W carbyne is regenerated by treatment of the title compound with $P(C_6H_5)_3$. Evidently, only $[CH_3CNCuP(C_6H_5)_3]^+$ is able to give a stable adduct with the carbyne [51].

Orange microcrystals, m.p. 100 to 102 °C. ¹H NMR (CDCl₃): 2.11 (s, CH₃CN), 2.28 (s, 4–CH₃), 5.71 (s, C₅H₅), 7.17 to 7.49 (m, C₆H₄+C₆H₅) ppm. ¹³C{¹H} NMR (CDCl₃): 2.2 (CH₃ of

References on pp. 246/7

CH$_3$CN), 22.1 (4–CH$_3$), 92.7 (C$_5$H$_5$), 119.8 (CN), 129.1 to 149.8 (C$_6$H$_4$+C$_6$H$_5$), 213.6 (CO), 294.9
(μ–C) ppm. ^{31}P{^{1}H} NMR (CDCl$_3$): −144.2 (sept, PF$_6$, J(PF)=713 Hz), +0.2 (P(C$_6$H$_5$)$_3$) ppm.
IR (CH$_2$Cl$_2$): 1946 and 2014 (both CO) cm^{-1}. Based upon the NMR and IR spectra, a cationic
structure analogous to XI (see p. 243) with CH$_3$CN instead of CF$_3$SO$_3^-$ has been proposed.
The absence of a ^{31}P–^{13}C coupling on the bridging carbyne resonance suggests that CH$_3$CN
is transoid and P(C$_6$H$_5$)$_3$ is cisoid to the μ–C atom [51].

Treatment with Cl$^-$, I$^-$, or CN$^-$ in CH$_2$Cl$_2$ led to the release of 4–CH$_3$C$_6$H$_4$C≡W(CO)$_2$C$_5$H$_5$.
Similar decompositions occurred with SR$^-$, BH$_4^-$, or H$^-$, or with Na naphthalenide in the
presence of CO [51].

(C$_6$H$_5$)$_3$PCu(μ$_3$–CC$_6$H$_4$CH$_3$–4)(W(CO)$_2$C$_5$H$_5$)$_2$ · CH$_2$Cl$_2$. Treatment of [(C$_6$H$_5$)$_3$PCuH]$_6$ with 4–
CH$_3$C$_6$H$_4$C≡W(CO)$_2$C$_5$H$_5$ (1:5) in C$_6$H$_5$CH$_3$ at room temperature affords a mixture of the title
compound (55% from CH$_2$Cl$_2$) and 27% C$_5$H$_5$W(η^3–CH$_2$C$_6$H$_4$CH$_3$–4)(CO)$_2$ [51].

Purple crystals. ^{1}H NMR (CD$_2$Cl$_2$): 2.12 (s, 4–CH$_3$), 5.23 (s, C$_5$H$_5$), 6.80 (m, C$_6$H$_4$), 7.29
(m, C$_6$H$_5$) ppm. ^{13}C{^{1}H} NMR (CD$_2$Cl$_2$): 22.1 (4–CH$_3$), 90.7 (C$_5$H$_5$), 123.2 to 135.8 (C$_6$H$_4$+C$_6$H$_5$),
168.4 (C–1 of C$_6$H$_4$), 222.1 (CO, J(WC)=171 Hz), 235.9 (CO, J(WC)=181 Hz), 339.6 (d, μ$_3$–C,
J(PC)=7 Hz, J(WC)=94 Hz) ppm. ^{31}P{^{1}H} NMR (CD$_2$Cl$_2$): 23.3 ppm. The ^{13}C{^{1}H} and ^{1}H NMR
spectra show one C$_5$H$_5$ environment and two CO sites present in solution. In view of the
X-ray diffraction results (see below) this implies that the title compound is fluxional in
solution, undergoing a rearrangement which brings about time-averaged C$_s$ symmetry on
the NMR timescale and racemizing the chiral structure in the solid state [51].

Monoclinic, space group P2$_1$/c − C$_{2h}^5$ (No. 14), a=10.034(4), b=21.514(8), c=17.964(6) Å,
β=91.73(3)°; Z=4, d$_c$=1.93, d$_m$=1.94 g · cm^{-3}. The structure is given in **Fig. 18** (CH$_2$Cl$_2$,

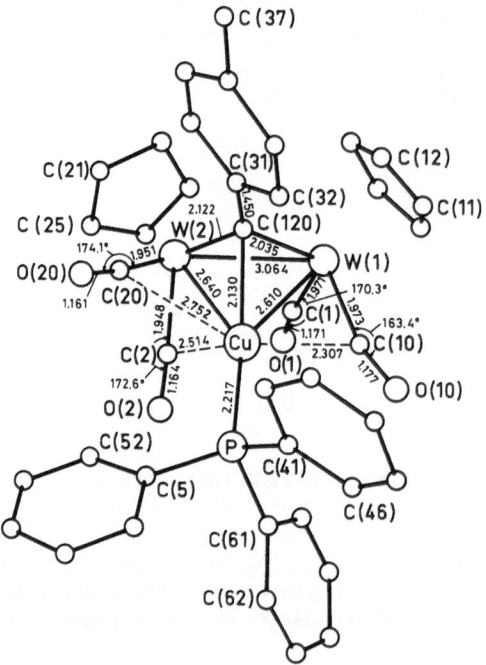

Fig. 18. Molecular structure of (C$_6$H$_5$)$_3$PCu(μ$_3$–CC$_6$H$_4$CH$_3$–4)(W(CO)$_2$C$_5$H$_5$)$_2$ · CH$_2$Cl$_2$ with selected bond lengths (in Å) and angles (CH$_2$Cl$_2$ omitted).

References on pp. 246/7

isolated by normal van der Waals distances, is omitted). The compound contains one Cu and two W atoms arranged nearly in an isosceles triangle, which is asymmetrically capped by the μ_3-CC$_6$H$_4$CH$_3$-4 ligand. The Cu is also weakly coordinated by two of the CO groups in a semi-bridging manner, and very weakly attached to a third. The coordination of the CO ligands to W is nonlinear; the degree of nonlinearity correlates with the Cu–C distances [51].

[Cu{(μ-CC$_6$H$_4$CH$_3$-4)W(CO)$_2$C$_5$H$_5$}$_2$]PF$_6$ (see XII) has been prepared from [Cu(NCCH$_3$)$_4$]PF$_6$ and 4-CH$_3$C$_6$H$_4$C≡W(CO)$_2$C$_5$H$_5$ in CH$_2$Cl$_2$. After addition of light petroleum ether, orange crystals are formed in a 82% yield. IR (CH$_2$Cl$_2$): 1942 and 2021 (both CO) cm^{-1} [51].

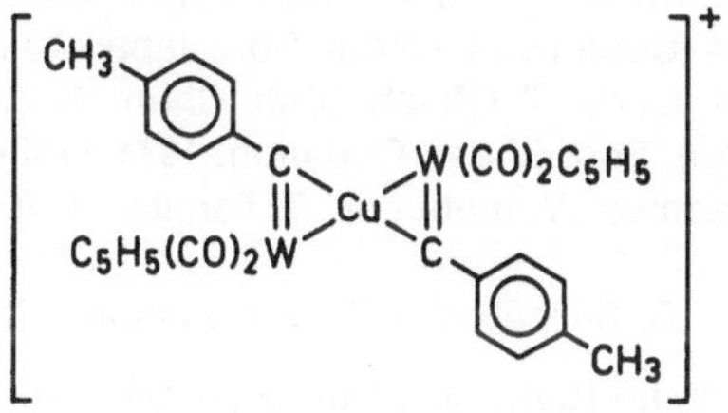

XII

[N{=P(C$_6$H$_5$)$_3$}$_2$]$^+$[(CO)$_{12}$Fe$_4$CuC(NCCH$_3$)]$^-$ [39] and [N(C$_2$H$_5$)$_4$]$^+$[(CO)$_{14}$Fe$_5$CuC(NCCH$_3$)] [39, 50] with the anion skeletons (CO omitted) XIII and XIV have been described in "Organoiron Compounds" C7, 1986, pp. 261 and 360.

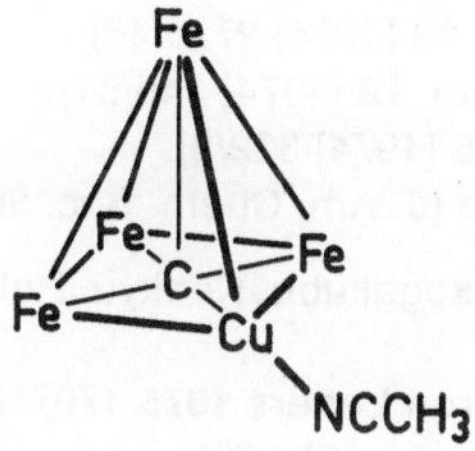

XIII

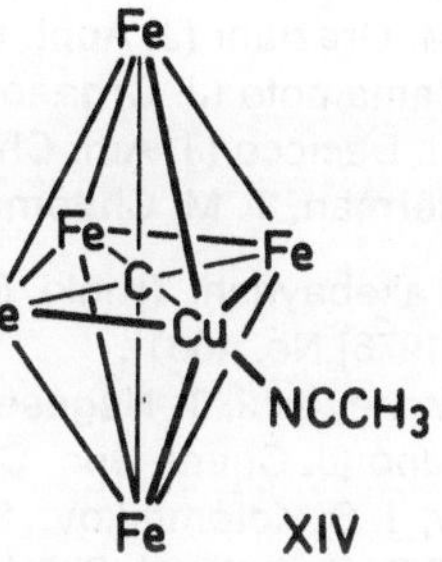

XIV

CH$_2$=CCu(Cl)Rh(C$_5$H$_5$)P(C$_3$H$_7$-i)$_3$ (see XV) has been prepared from the vinylidene complex CH$_2$=C=Rh(C$_5$H$_5$)P(C$_3$H$_7$-i)$_3$ and CuCl at room temperature with a yield of 71%. The orange-red crystals are air-stable for a short time. ^{1}H NMR (CDCl$_3$, 25 °C): 1.26 (dd, J(PH) = 14.6 Hz, J(HH) = 7.0 Hz) and 1.31 (dd, J(PH) = 13.9 Hz, J(HH) = 7.0 Hz; both for CH$_3$), 2.21 (m, PCH), 4.10 (dd, CH$_2$=, J(PH) = 2.3 Hz, J(RhH) = 2.3 Hz), 5.32 (dd, C$_5$H$_5$, J(PH) = 1.1 Hz, J(RhH) = 0.6 Hz) ppm [46].

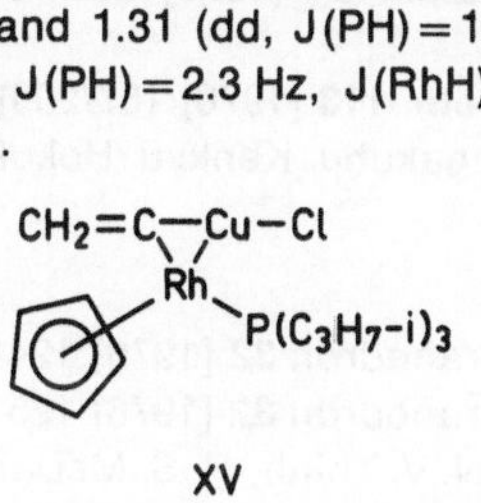

XV

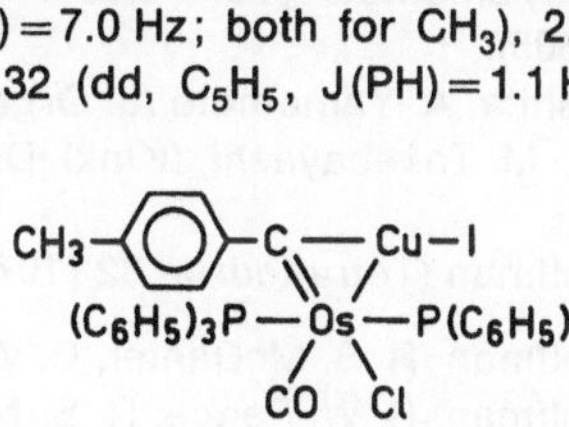

XVI

4-CH$_3$C$_6$H$_4$CCu(I)Os(CO)(P(C$_6$H$_5$)$_3$)$_2$Cl (see XVI) has been prepared from the carbyne complex 4-CH$_3$C$_6$H$_4$C≡Os(CO)(P(C$_6$H$_5$)$_3$)$_2$Cl and CuI in C$_6$H$_6$ at room temperature as a green solid. IR (Nujol): 255 (OsCl), 1912 (CO) cm^{-1}. The proposed structure is based on the X-ray

References on pp. 246/7

structure determination of the analogous AgCl complex which has a very similar IR spectrum [38].

References:

[1] H. Nozaki, S. Moriuti, M. Yamabe, R. Noyori (Tetrahedron Letters **1966** 59/63).
[2] H. Nozaki, S. Moriuti, H. Takaya, R. Noyori (Tetrahedron Letters **1966** 5239/44).
[3] M. Takebayashi, T. Ibata, H. Kohara, B. H. Kim (Bull. Chem. Soc. Japan **40** [1967] 2392/7).
[4] R. Noyori, H. Takaya, Y. Nakanisi, H. Nozaki (Can. J. Chem. **47** [1969] 1242/5).
[5] M. Takebayashi, T. Ibata, H. Kohara, K. Ueda (Bull. Chem. Soc. Japan **42** [1969] 2938/44).
[6] M. Takebayashi, T. Ibata, K. Ueda (Bull. Chem. Soc. Japan **43** [1970] 1500/5).
[7] M. Takebayashi, T. Ibata, K. Ueda, T. Ohashi (Bull. Chem. Soc. Japan **43** [1970] 3964).
[8] R. Mason, G. Rucci (J. Chem. Soc. Chem. Commun. **1971** 1132).
[9] T. Saegusa, Y. Ito, K. Yonezawa, Y. Inubushi, S. Tomita (J. Am. Chem. Soc. **93** [1971] 4049/51).
[10] I. Zugrăvescu, E. Rucinschi, G. Surpăteanu (Rev. Roumaine Chim. **16** [1971] 1099/105).

[11] T. Saegusa, K. Yonezawa, Y. Ito (Syn. Commun. **2** [1972] 431/9).
[12] A. Miyashita, A. Yamamoto (Kenkyu Hokoku Asahi Garasu Kogyo Gijutsu Shoreikai **23** [1973] 423/7; C.A. **82** [1975] No. 67594).
[13] A. Miyashita, A. Yamamoto (J. Organometal. Chem. **49** [1973] C57/C58).
[14] M. Takebayashi, T. Kashiwada, M. Hamaguchi, T. Ibata (Chem. Letters **1973** 809/12).
[15] J. Vičková, J. Bartoň (J. Chem. Soc. Chem. Commun. **1973** 306/7).
[16] A. Yamamoto (6th Intern. Conf. Organometal. Chem., Amherst, Mass., 1973, Abstr. P4).
[17] E. M. Cernia, M. Graziani (J. Appl. Polym. Sci. **18** [1974] 2725/46).
[18] T. Ikariya, A. Yamamoto (J. Organometal. Chem. **72** [1974] 145/51).
[19] L. T. Scott, G. J. Decicco (J. Am. Chem. Soc. **96** [1974] 322/3).
[20] T. Cohen, G. Herman, T. M. Chapman, D. Kuhn (J. Am. Chem. Soc. **96** [1974] 5627/8).

[21] Y. Kawai, M. Takebayashi (Kinki Daigaku Rikogakubu Kenkyu Hokoku No. 10 [1975] 31/4; C.A. **84** [1976] No. 4081).
[22] T. Aratani, Y. Yoneyoshi, T. Nagase (Tetrahedron Letters **1975** 1707/10).
[23] M. T. Beck, F. Joó (J. Chem. Soc. Chem. Commun. **1975** 230).
[24] M. M. Savel'ev, I. S. Kolomnikov, S. V. Vitt, M. E. Vol'pin (Izv. Akad. Nauk SSSR Ser. Khim. **1975** 486/7; Bull. Acad. Sci. USSR Div. Chem. Sci. **1975** 422).
[25] A. Yamamoto, T. Ito, T. Yamamoto, A. Miyashita, S. Komiya, T. Ikariya (Kenkyu Hokoku Asahi Garasu Kogyo Gijutsu Shoreikai **27** [1975/76] 55/66; C.A. **86** [1977] No. 5600).
[26] U. Burger, F. Mazenod (Tetrahedron Letters **1976** 2885/8).
[27] T. Ito, A. Yamamoto (Yuki Gosei Kagaku Kyokaishi **34** [1976] 308/18; C.A. **85** [1976] No. 136305).
[28] A. Miyashita, A. Yamamoto (J. Organometal. Chem. **113** [1976] 187/200).
[29] H. Sano, M. Takebayashi (Kinki Daigaku Rikoagakubu Kenkyu Hokoku No. 11 [1976] 49/52).
[30] D. S. Wulfman (Tetrahedron **32** [1976] 1231/40).

[31] D. S. Wulfman, R. S. McDaniel, B. W. Peace (Tetrahedron **32** [1976] 1241/9).
[32] D. S. Wulfman, B. W. Peace, R. S. McDaniel (Tetrahedron **32** [1976] 1251/5).
[33] D. S. Wulfman, B. G. McGibboney, E. K. Steffen, N. V. Thinh, R. S. McDaniel, B. W. Peace (Tetrahedron **32** [1976] 1257/65).
[34] M. Freiberg, D. Meyerstein (J. Chem. Soc. Chem. Commun. **1977** 127/8).
[35] L. M. Kachapina, Yu. M. Shul'ga, S. B. Echmaev, I. S. Kolomnikov, S. M. Vinogradova, Yu. G. Borod'ko (Koord. Khim. **3** [1977] 435/9; Soviet J. Coord. Chem. **3** [1977] 327/31).

[36] W. A. Mulac, D. Meyerstein (J. Chem. Soc. Chem. Commun. **1979** 893/5).
[37] E. G. Boguslavskii, A. A. Shklyaev, N. G. Malesimov, Z. M. Alaudinova, I. S. Kolomnikov, V. F. Anufrenko (Izv. Akad. Nauk SSSR Ser. Khim. **1980** 1231/4; Bull. Acad. Sci. USSR Div. Chem. Sci. **1980** 857/9).
[38] G. R. Clark, C. M. Cochrane, W. R. Roper, L. J. Wright (J. Organometal. Chem. **199** [1980] C35/C38).
[39] M. Tachikawa, R. L. Geerts, E. L. Muetterties (J. Organometal. Chem. **213** [1981] 11/24).
[40] S. Sakaki, K. Kitaura, K. Morokuma (Inorg. Chem. **21** [1982] 760/5).

[41] Yu. V. Tomilov, V. G. Bordakov, N. M. Tsvetkova, A. Ya. Shteinshneider, I. E. Dolgii, O. M. Nefedov (Izv. Akad. Nauk SSSR Ser. Khim. **1983** 336/43; Bull. Acad. Sci. USSR Div. Chem. Sci. **1983** 300/6).
[42] J. H. B. Chenier, J. A. Howard, B. Mile, R. Sutcliffe (J. Am. Chem. Soc. **105** [1983] 788/91).
[43] J. A. Howard, R. Sutcliffe, B. Mile (J. Phys. Chem. **88** [1984] 171/4).
[44] J. A. Howard, R. Sutcliffe, B. Mile (J. Phys. Chem. **88** [1984] 5155/7).
[45] Yu. S. Nekrasov, Yu. A. Borisov, S. Yu. Silvestrova, T. V. Lysyak, Ya. Ya. Kharitonov, I. S. Kolomnikov (J. Organometal. Chem. **269** [1984] 323/6).
[46] H. Werner, J. Wolf, G. Müller, C. Krüger (Angew. Chem. **96** [1984] 421/2; Angew. Chem. Intern. Ed. Engl. **23** [1984] 431/2).
[47] N. M. Vitkovskaya, V. G. Bernshtein, F. K. Shmidt (Kinetika Kataliz **25** [1984] 1000/3; Kinet. Catal. [USSR] **25** [1984] 855/8).
[48] J. H. B. Chenier, J. A. Howard, B. Mile (J. Am. Chem. Soc. **107** [1985] 4190/1).
[49] J. A. Howard, R. Sutcliffe, H. Dahmane, B. Mile (Organometallics **4** [1985] 697/701).
[50] M. Tachikawa, A. C. Sievert, E. L. Muetterties, M. R. Thompson, C. S. Day, V. W. Day (J. Am. Chem. Soc. **102** [1980] 1725/7).

[51] M. Müller-Gliemann, S. V. Hoskins, A. G. Orpen, A. L. Ratermann, F. G. A. Stone (Polyhedron **5** [1986] 791/8).
[52] M. E. Alonso, M. I. Hernandez, M. Gomez, P. Jano, S. Pekerar (Tetrahedron **41** [1985] 2347/54).

Table of Conversion Factors

Following the notation in Landolt-Börnstein [7], values which have been fixed by convention are indicated by a bold-face last digit. The conversion factor between calorie and Joule that is given here is based on the thermochemical calorie, cal_{thch}, and is defined as 4.1840 J/cal. However, for the conversion of the "Internationale Tafelkalorie", cal_{IT}, into Joule, the factor 4.1868 J/cal is to be used [1, p. 147]. For the conversion factor for the British thermal unit, the Steam Table Btu, Btu_{ST}, is used [1, p. 95].

Force	N	dyn	kp
1 N (Newton)	1	10^5	0.1019716
1 dyn	10^{-5}	1	1.019716×10^{-6}
1 kp	9.80665	9.80665×10^5	1

Pressure	Pa	bar	kp/m^2	at	atm	Torr	lb/in^2
1 Pa (Pascal) = 1 N/m^2	1	10^{-5}	1.019716×10^{-1}	1.019716×10^{-5}	0.986923×10^{-5}	0.750062×10^{-2}	145.0378×10^{-6}
1 bar = 10^6 dyn/cm^2	10^5	1	10.19716×10^3	1.019716	0.986923	750.062	14.50378
1 kp/m^2 = 1 mm H$_2$O	9.80665	0.980665×10^{-4}	1	10^{-4}	0.967841×10^{-4}	0.735559×10^{-1}	1.422335×10^{-3}
1 at = 1 kp/cm^2	0.980665×10^5	0.980665	10^4	1	0.967841	735.559	14.22335
1 atm = 760 Torr	1.01325×10^5	1.01325	1.033227×10^4	1.033227	1	760	14.69595
1 Torr = 1 mm Hg	133.3224	1.333224×10^{-3}	13.59510	1.359510×10^{-3}	1.315789×10^{-3}	1	19.33678×10^{-3}
1 lb/in^2 = 1 psi	6.89476×10^3	68.9476×10^{-3}	703.069	70.3069×10^{-3}	68.0460×10^{-3}	51.7149	1

Work, Energy, Heat	J	kWh	kcal	Btu	MeV
1 J (Joule) = 1 Ws = 1 Nm = 10^7 erg	1	2.778×10^{-7}	2.39006×10^{-4}	9.4781×10^{-4}	6.242×10^{12}
1 kWh	3.6×10^6	1	860.4	3412.14	2.247×10^{19}
1 kcal	4184.0	1.1622×10^{-3}	1	3.96566	2.6117×10^{16}
1 Btu (British thermal unit)	1055.06	2.93071×10^{-4}	0.25164	1	6.5858×10^{15}
1 MeV	1.602×10^{-13}	4.450×10^{-20}	3.8289×10^{-17}	1.51840×10^{-16}	1

1 eV/mol ≙ 23.0578 kcal/mol = 96.473 kJ/mol

Power	kW	PS	kp m/s	kcal/s
1 kW = 10^{10} erg/s	1	1.35962	101.972	0.239006
1 PS	0.73550	1	75	0.17579
1 kp m/s	9.80665×10^{-3}	0.01333	1	2.34384×10^{-3}
1 kcal/s	4.1840	5.6886	426.650	1

References:

[1] A. Sacklowski, Die neuen SI-Einheiten, Goldmann, München 1979. (Conversion tables in an appendix.)
[2] International Union of Pure and Applied Chemistry, Manual of Symbols and Terminology for Physicochemical Quantities and Units, Pergamon, London 1979; Pure Appl. Chem. **51** [1979] 1/41.
[3] The International System of Units (SI), National Bureau of Standards Spec. Publ. No. 330 [1972].
[4] H. Ebert, Physikalisches Taschenbuch, 5th Ed., Vieweg, Wiesbaden 1976.
[5] Kraftwerk Union Information, Technical and Economic Data on Power Engineering, Mülheim/Ruhr 1978.
[6] E. Padelt, H. Laporte, Einheiten und Größenarten der Naturwissenschaften, 3rd Ed., VEB Fachbuchverlag, Leipzig 1976.
[7] Landolt-Börnstein, 6th Ed., Vol. II, Pt. 1, 1971, pp. 1/14.
[8] ISO Standards Handbook 2, Units of Measurement, 2nd Ed., Geneva 1982.

Key to the Gmelin System
of Elements and Compounds

System Number	Symbol	Element
1		Noble Gases
2	H	Hydrogen
3	O	Oxygen
4	N	Nitrogen
5	F	Fluorine
6	**Cl**	**Chlorine**
7	Br	Bromine
8	I	Iodine
8a	At	Astatine
9	S	Sulfur
10	Se	Selenium
11	Te	Tellurium
12	Po	Polonium
13	B	Boron
14	C	Carbon
15	Si	Silicon
16	P	Phosphorus
17	As	Arsenic
18	Sb	Antimony
19	Bi	Bismuth
20	Li	Lithium
21	Na	Sodium
22	K	Potassium
23	NH_4	Ammonium
24	Rb	Rubidium
25	Cs	Caesium
25a	Fr	Francium
26	Be	Beryllium
27	Mg	Magnesium
28	Ca	Calcium
29	Sr	Strontium
30	Ba	Barium
31	Ra	Radium
32	**Zn**	**Zinc**
33	Cd	Cadmium
34	Hg	Mercury
35	Al	Aluminium
36	Ga	Gallium

System Number	Symbol	Element
37	In	Indium
38	Tl	Thallium
39	Sc, Y La–Lu	Rare Earth Elements
40	Ac	Actinium
41	Ti	Titanium
42	Zr	Zirconium
43	Hf	Hafnium
44	Th	Thorium
45	Ge	Germanium
46	Sn	Tin
47	Pb	Lead
48	V	Vanadium
49	Nb	Niobium
50	Ta	Tantalum
51	Pa	Protactinium
52	**Cr**	**Chromium**
53	Mo	Molybdenum
54	W	Tungsten
55	U	Uranium
56	Mn	Manganese
57	Ni	Nickel
58	Co	Cobalt
59	Fe	Iron
60	Cu	Copper
61	Ag	Silver
62	Au	Gold
63	Ru	Ruthenium
64	Rh	Rhodium
65	Pd	Palladium
66	Os	Osmium
67	Ir	Iridium
68	Pt	Platinum
69	Tc	Technetium[1]
70	Re	Rhenium
71	Np,Pu . . .	Transuranium Elements

HCl · CrCl₂ · ZnCrO₄ · ZnCl₂

Material presented under each Gmelin System Number includes all information concerning the element(s) listed for that number plus the compounds with elements of lower System Number.

For example, zinc (System Number 32) as well as all zinc compounds with elements numbered from 1 to 31 are classified under number 32.

[1] A Gmelin volume titled "Masurium" was published with this System Number in 1941.

A Periodic Table of the Elements with the Gmelin System Numbers is given on the Inside Front Cover